Language in Cognition

LANGUAGE FROM WITHIN

In this ambitious two-volume work, Pieter Seuren seeks a theoretical unity that can bridge the chasms of modern linguistics as he sees them, bringing together the logical, the psychological, and the pragmatic; the empirical and the theoretical; the formalist and the empiricist; and situating it all in the context of two and a half millennia of language study.

PUBLISHED

Volume I: Language in Cognition

IN PREPARATION

Volume II: The Logic of Language

Language in Cognition

PIETER A. M. SEUREN

OXFORD
UNIVERSITY PRESS

This book has been printed digitally and produced in a standard specification in order to ensure its continuing availability

OXFORD
UNIVERSITY PRESS

Great Clarendon Street, Oxford OX2 6DP

Oxford University Press is a department of the University of Oxford.
It furthers the University's objective of excellence in research, scholarship,
and education by publishing worldwide in

Oxford New York

Auckland Cape Town Dar es Salaam Hong Kong Karachi
Kuala Lumpur Madrid Melbourne Mexico City Nairobi
New Delhi Shanghai Taipei Toronto

With offices in

Argentina Austria Brazil Chile Czech Republic France Greece
Guatemala Hungary Italy Japan South Korea Poland Portugal
Singapore Switzerland Thailand Turkey Ukraine Vietnam

Published in the United States
by Oxford University Press Inc., New York

Reprinted 2011

ISBN 978-0-19-955947-3

Printed and bound in Great Britain by CPI Antony Rowe,
Chippenham and Eastbourne

to Pim Levelt
for his unfailing support, advice and friendship

Contents

Preface

The present book is the fruit of many years of, largely independent, thinking, probing, investigating, and trying to gain a grasp of those issues in the study of human language that have proved to be most resistant to scientific understanding. It brings together a wide variety of issues that are only rarely integrated into a comprehensive theory, yet are all directly relevant to an overall insight into the nature of language and speech and into the disciplines concerned with them. It does so, moreover, in a historical perspective, because the study of language has a long past but not one that has been cumulative in the sense that results, once attained, became accepted doctrine. On the contrary, valuable and interesting insights have often been lost and more blinkered views have often carried the transient day, our day included.

As the reader will quickly discover, the book is, on the whole, tentative and exploratory, pointing to the need for further research. Many questions are left open and many proposed answers are presented as hypothetical or suggestive. But quite a few of these open or half-open questions have never been mooted even though, once formulated, they are seen to touch on real issues. Perhaps I should have waited even longer and let the whole complex of questions mature even more before proceeding to publication, but since I am no longer, to speak with Dante, 'nel mezzo del cammin di nostra vita' but rather, in the nature of things, approaching the end of it, I have decided to present what I have in the way of considered thoughts on the matter, in the hope that others, younger than me, will take over and contribute their perspectives and their expertise.

The two volumes before you are, though they have sprung from one single source of inspiration, very different in character. Volume I is not very technical or formal but rather of a discursive nature, exploring and re-assessing the methodological, ontological, and cognitive aspects of the study of language, in particular the study of meaning. Volume II, by contrast, takes the reader into the more technical aspects of logic and discourse incrementation, with only the historical sections offering some respite. I have done my utmost to make the text accessible to less technically trained readers, devoting considerable space to the basic notions and motivations, but I do realize that readers without any previous training in logic and set theory will have difficulty following the arguments and analyses. I regret this, of course, and all I can say is that it only takes an elementary familiarity with logical notions and notations to gain an adequate grasp of what is proposed in Volume II. And I promise that it is worth the effort.

The book emphasizes the fact that linguistic studies must be basically ecologistic—a requirement that often stands in the way of fully formalized theories, even though a formal presentation of the analyses and theories developed remains the ultimate aim. For that reason, phenomena that have so far resisted formal treatment are systematically highlighted as warning signals for those formalists who think that the study of language stops where their formalisms stop. Those who are willing to take some time off from the daily fray of their work will find it rewarding to be made to realize that vast sections of the terrain still resist exact analysis, not only because of the nature of the material dealt with, which is often too vast and too divergent to be captured in a formal theory, but also because the right perspective for a fruitful formalization has not so far been detected. In Volume I the ground is prepared for the development of such new perspectives, while Volume II presents a few efforts at the actual designing of fruitful novel formalizations in the light of them.

I wish to express my gratitude to those who were kind enough to read drafts of chapters and discuss them with me, letting me have the benefit of their criticisms and suggestions, and their encouragement. First and foremost, I must mention my friend of forty years' standing Pim Levelt, to whom I have dedicated this book. I owe him a great debt of gratitude, both personally and professionally, not least because he made it possible for me to work at the Max Planck Institute for Psycholinguistics at Nijmegen after my retirement from Nijmegen University. His continuous support helped me over many a bout of failing courage and confidence. I must also thank my colleagues at the Max Planck Institute, in particular Peter Hagoort and Jos van Berkum, who looked over my shoulder when I wrote about the new neuro-imaging techniques, which to me are still a baffling *terra incognita*. I am indebted to Charles Collier of the University of Florida, Levin College of Law, whose incisive and inspiring comments on the chapters on speech acts and meaning were of great help. Then I must mention my friend and colleague Dany Jaspers of the University of Brussels, whose wide knowledge, well-formulated comments, and infectious enthusiasm were a constant source of inspiration. Ferdinando Cavaliere made many useful suggestions regarding predicate logic and its history. As regards the function and the importance of the lexicon, I found a kindred soul in my colleague Rob Schreuder, head of the Interfaculty Research Unit for Language and Speech. I am also indebted to my former student Cor Burger who provided me with the most valuable input on the relation between Russell and Meinong, described in Chapter 2 of Volume I. The many others who have helped me carry on by giving their intellectual, moral, and personal support are too numerous to be mentioned individually. Yet my gratitude to them is none the less for that.

P. A. M. S.
Nijmegen, 2008

1

Introduction

1.1 The gist

In this book, a radically new perspective is presented on the study of linguistic meaning, the place of logic in language and the embedding of language in cognition. It does not offer a full formal theory of language, but it does present a new analysis of the main parameters involved, pointing the way towards a fresh exploration of the territory and indicating the areas that are open to full formalization, given available techniques, next to those that have so far resisted formalization. The central innovating element is a systematic recognition of the MIND as the pivotal factor in language and meaning, along with the introduction of the PERSON as the holder of rights, duties, and responsibilities—an aspect which has likewise been neglected in existing theories of language and is elaborated in Chapter 4.

Those who have the stamina to read through the present book will find that some of the central tenets of twentieth-century Anglo-Saxon philosophy are argued to be basically flawed, precisely because the mind is not given its due. Matters come to a head in Volume II, where standard logic, which has acquired unique prestige during the twentieth century and forms the basis of formal semantics while being taken for granted by the pragmaticists, is deprived of its inviolate status. It is shown that there is an infinite gamut of valid logics, depending on the axioms adopted and on the semantic definition of the logical constants employed—a fact that throws the question of the logic of language wide open and makes it an empirical question. This requires a thorough revision of the axioms and other tenets considered sacrosanct in logic—an exercise that brings unexpected rewards. Throughout the two volumes, the intellectual traditions of the past, largely forgotten by today's protagonists, are revived and shown to contain unexpected treasures that are crucially relevant to the issues at hand.

1.1.1 *Natural ontology*

As far as I can see, there are two aspects of cognition whose neglect has been most damaging to the study of meaning, an ONTOLOGICAL and a

LOGICO-MATHEMATICAL aspect. As regards the ontological aspect, the basic point is the trivial observation that when we speak we speak about something, or else we do not speak. This seems obvious enough, but it gives rise to the question of the nature of the things we speak about. Do we only speak about things that really exist in the actual world, or do we also speak about things that do not exist in the actual world but are merely thought-up, either as thought-up individuals or as abstractions or reifications? And if we take the view that we also speak about nonexistent things, what sort of things are things that do not exist?

It does not take much reflection to see that we speak about, and quantify over, things and situations that are not part of the actual world with the same ease and naturalness as we do about things and situations that are. We are, on the whole, acutely aware of the difference between what we consider to actually exist or be the case and what we take merely to have been conjured up in our minds. In fact, much of our practical and intellectual effort is directed at establishing whether something that has been thought up actually exists or actually is the case. But the fact is that we refer to and quantify over thought-up entities, and we report on thought-up situations, as naturally as we do about actually existing entities and real situations. One only has to go through an ordinary text to see that a large portion of the references made are references to abstractions, reifications, or thought-up objects, which may again be thought-up abstractions or reifications.

Globally speaking, our, on the whole uncontroversial, position is that humans distinguish between things or situations that are taken to be actually there and things or situations that are not because they are recognized as being no more than the products of our (very creative) powers of imagination and cognitive representation. But this fact, though obvious, is not recognized at all in the formal study of meaning as it is practised today. And, as will become clear in the course of this book, this has serious consequences for a theory of meaning and for a reconstruction of the natural logic of language and cognition. There is a great deal of talk about 'possible worlds' and, yes, there is an intensional logic meant to take care of unreal things and situations, but the mind is never mentioned in this context. What is presented as 'intensional logic' is, in fact, a generalization of extensional, actual-world-related, logic over ill-defined 'possible worlds' that are seen as mathematical abstractions but subject to the restrictions and principles of the world that exists. What is not presented, though it should be, is an account of how the mind deals with thought-up entities and situations. If such an account were presented, it would make clear that thought-up entities and situations behave very differently from those that are recognized as being actual. Moreover, it has been known for almost thirty years that this 'programme of extensionalization of intensions' irredeemably founders over substitution in intensional contexts, as is explained in Sections 2.1 and 6.1. So the first thing that is needed is a full recognition of the fact that we, humans, speak not only about actual but also about thought-up things and situations and that these are not subject to Russellian and Quinean ontological principles. Existing formal theories of meaning have so far been unable to provide an adequate account of this fact.

An enormous obstacle, in this respect, is presented by the ontological views that are dominant in the Anglo-Saxon world, according to which nothing IS unless it is *actual* and thus either has individual existence or is reducible to things that have (or had) actual existence. The main champions of this brand of ontology were the influential twentieth-century philosophers Bertrand Russell and Willard Van Orman Quine. Both are the butt of severe criticism in Chapter 2, where the Kantian position is defended that there is no way of proving the real status and real nature of what is 'out there' because all we know and can know about the world is filtered through the mind. Hence, any ontological theory that pretends to be able to describe in adequate and objective terms what the world is really like, regardless of the way the mind processes information about it, is doomed to fail. All we can do is describe the way humans form reliable 'theories' about what is and what is not 'out there'—that is, theories whose predictions match our experiences and thus provide some reasonable degree of safety and control amidst the dangers threatening us and the calamities harming us all the time.

This is, of course, nothing but the EPISTEMOLOGICAL ARGUMENT that has dominated philosophy since the mid-seventeenth century and has found its most celebrated expression in René Descartes' 'cogito' principle, John Locke's 'veil of perception', and Emanuel Kant's 'unknowability of the thing in and of itself'. Russell and Quine attempted, with great power of thought and consummate formal prowess, to find a way around the inescapable dilemma posed by the epistemological argument, turning to mathematical and conceptual necessity. And their programme, one must say, was not totally unsuccessful. But then, carried away by hubris, they applied their findings to the human way of processing external data and internal ideas cognitively and linguistically, drawing countless followers in their wake.

What this means for us is that we must reconstruct the system according to which the natural, unsophisticated human mind specifies not only what is taken actually to exist or be the case, but also what, in any human's view, is not the case, or could be the case, or could never be the case, out there. This research programme is indispensable, because we also speak about those things and produce true and false statements about them. One can only agree with Johnson-Laird, who writes:

> We seem to perceive the world directly, not a representation of it. Yet this phenomenology is illusory: what we perceive depends on both what is in the world and what is in our heads—on what evolution has 'wired' into our nervous systems and what we know as a result of experience. The limits of our models are the limits of our world. (Johnson-Laird 1989: 470–1)

Our concern, therefore, is not the way the actual world is thought to be by certain philosophers, no matter how forceful their hypotheses, but the way humans construct a theory of the world by conjuring up and testing possible and impossible candidates, all of which provide ready reference objects. This, then, is the first respect in which existing studies grossly fail to draw the full consequences of the fact that language is lodged in the mind and integrated into the whole of cognition.

1.1.2 *Natural cognition and natural logic*

The second, closely related, respect in which the factor 'mind' plays a decisive role in semantics is to do with the logico-mathematical aspects of meaning and logic. Ironically, the formal study of meaning is pursued by professionals in logic and mathematics, not by linguists or psychologists. But although mathematical logicians are, or should be, best placed to make a distinction between abstract mathematical systems and concrete information-processing machines, they concentrate on the development of abstract systems which do what their makers want them to do. At the same time, the abstract systems are, without much reflection and without taking into account what psychologists have to say about the matter, projected onto the mind as if the mind itself were just another abstract system. The mathematical logicians thereby perpetrate the same act of hubris as is perpetrated by the mathematical ontologists. Despite the powerful cognitive claims made by some schools of formal semanticists, their work has, in actual fact, very little to do with the mind and everything to do with the development of new, sophisticated subtheories within the overall structure of standard logic and mathematics. The mind is, in other words, merely a playground providing an excuse for mathematical and logical diversions.

In her article 'Semantics—mathematics or psychology?', Barbara Partee writes:

> There is no reason why a psychological theory can't be expressed in mathematical terms. In fact,...it should be, since we are trying to discover something about the

structure of a certain mental faculty, and mathematics is the best available tool for describing structure. So I am not suggesting that there is any incompatibility between mathematics in general and psychology in general. (Hall-Partee 1979: 1)

Quite so, but the problem is that the mind is not any old structure or machine describable in terms of standard mathematical and logical techniques, but a 'symbolic system' (Johnson-Laird 1989: 469). The mind has its own natural mathematics and its own natural logic, and it does its computations according to a system that differs in many ways from the sophisticated mathematical and logical systems developed by modern specialists. It is very unlikely that the basic operators of 'natural' logic—that is, the quantifiers and the logical connectives—are defined in the same way for natural cognition (or natural language) as they are for the sophisticated forms of standard modern logic. This is what the tribes of researchers who are engaged in specifying the mathematical and logical aspects of psychological structures and processes have neglected to take into account. They believe that the only viable systems of logic and mathematics are the standardly accepted systems, not realizing that conceptual space allows for infinite varieties of sound logic and sound mathematics (and even more varieties of unsound logic and mathematics). If we want to understand the logico-mathematical aspects of meaning in natural language, we had better aim at a reconstruction of those varieties that reflect *natural* mathematics and *natural* logic.

The idea that a mathematical description of certain aspects of cognition might amount to formulating specific restrictions on, or modifications of, standard mathematical theory seems not to have occurred to present-day semanticists. Yet it is a perfectly reasonable thought. Modern logic and mathematics may be the ultimate achievement of the most sophisticated investigation into the foundations of logic and mathematics history has ever seen. But that does not make them the instruments of the unsophisticated minds of the noble or not so noble savages in whose midst language must have come about.

Cognitive scientists have recently developed this point of view as regards less deeply buried cognitive faculties, such as the natural arithmetical powers of infants, schoolchildren, and illiterate Brazilian Indians (Dehaene 1997; Butterworth 1999; Pica *et al.* 2004), the neurolinguistic development of the reading ability (Dehaene 2005), and the perception and processing of geometrical properties (Dehaene *et al.* 2006). There seems to be no good reason why this same point of view should not be valid for the human logical and mathematical powers, in particular as regards the way sets—that is, plural objetcs—are handled in natural cognition.

There is, therefore, an EMPIRICAL aspect to this question, and it is clearly in the province of psychology. Yet the focus of interest in cognitive psychology has, so far, been more on cognitive mechanisms than on cognitive principles. Johnson-Laird (1977) made a similar observation on the state of affairs thirty years ago. Since then, the situation has not changed much, despite a great deal of excellent work done by psychologists such as Evans (1989, 2004), whose work concentrates on the logical and inductive errors humans are prone to making, or Johnson-Laird, who has done much valuable work regarding the ways the mind may be taken to model situations and interpret utterances describing them. But the underlying logic and its concomitant mathematical notions have remained in the background. It is the intention of the present book to help to fill this gap. Given that the principles formulated in this book are immediately open to experimental testing, one hopes that more, and more precisely directed, experimental research regarding the validity and the details of these principles will be undertaken. Pending such research, all we can do is posit certain hypothetical principles, offering them for experimental testing.

This is the topic of Chapter 3 of Volume II of the present work, where the focus of interest is on the well-known discrepancies between natural logical intuitions and either traditional or standard logic. An attempt is made there to reconstruct the principles of what is called 'basic natural logic', which are derived from a 'basic natural set theory' defined by certain specific restrictions with respect to standard set theory.

Inevitably, this natural-logic hypothesis enters into competition with the existing and widely accepted account of the discrepancies in question in terms of the Gricean maxims. We consider these maxims to be insufficient as an explanation, if only because some of the discrepancies are totally unrelated to the maxims. One example must suffice here to illustrate this point. Consider the stark contrast, discussed more amply in Section 3.4.3 in Volume II, between sentence (1.1a), which is immediately seen by native speakers to be equivalent with (1.1b), and sentence (1.2a), which, in virtue of De Morgan's laws, is as equivalent with (1.2b) as (1.1a) is with (1.1b), yet can be seen to be equivalent in that way only after much reflection:

(1.1) a. He doesn't like planes or trains.
b. He doesn't like planes and he doesn't like trains.

(1.2) a. He doesn't like planes and trains.
b. He doesn't like planes or he doesn't like trains.

This obvious fact remains unexplained in terms of the Gricean maxims, which are all about propriety in social dealings. Yet they should provide an explanation, because that is what they have been set up for. It is shown in Section 3.4.3 of Volume II that the facts demonstrated in (1.1) and (1.2) follow directly from our reconstruction of basic natural logic, which thus appears to be better supported by empirical evidence than the Gricean approach.

1.2 The two volumes

Volume I begins by setting up certain central principles of the natural ontology according to which humans develop a theory of the actual world with the help of nonactual, thought-up 'worlds' or, better, situations. It then presents a view of how language expresses thoughts developed in terms of this natural ontology, with an emphasis on the mechanism of reference not only to extensional, actually existing objects but also to intensional, thought-up objects. An important aspect, elaborated in Chapter 4, is the fact that human communication does not, as is usually thought, consist in the transfer of propositionally structured information, but, rather, in a socially binding form of position-taking with respect to such information.

We then go on to a discussion of the general semantic aspects of language in the context of the total ecological 'architecture' of language, mind, and world. This leads to a cursory tour of lexical meaning and of the ways lexical meanings are structured so as to make linguistic utterances fit into given contexts or discourses.

Concentrating again on propositional content, basic principles of logic are introduced in Volume II, although not in the traditional fashion. A new and unconventional view of logic is developed there, in which the logical constants are treated as lexical items, in fact as lexical predicates, with the special property that their meanings allow for the computation of entailments. It is argued that this reduction of logic to lexical meaning shows better than anything else the relevance of logic and logical analysis for the study of linguistic meaning. This point of view is reinforced in Chapter 10 of Volume II, where it is shown that presuppositions are a general semantic property of lexical predicates and where it is argued that a proper theory of presuppositions requires a trivalent presuppositional logic. We then concentrate, in Chapter 3 of Volume II, on a reconstruction of the natural logic which nature may be taken to have instilled into human cognition.

This reconstructed logic is then placed in a historical perspective, which shows that basic natural predicate logic is, in fact, largely but not entirely identical with the logic proposed and defended by the Edinburgh philosopher William Hamilton in the nineteenth century. Aristotelian predicate logic is dissected in Chapter 5 of Volume II and reconstructed on the basis of

Aristotle's own texts, whereby it is found that Aristotle was not guilty of the logical error of undue existential import but left his logic incomplete. It is also found that the twelfth-century French philosopher Abelard completed Aristotelian predicate calculus in Aristotle's spirit, avoiding undue existential import in a way that leads to a logically sound system that is more powerful than standard modern predicate logic. Chapter 4 of Volume II shows that traditional predicate logic, with its undue existential import, has maximal logical power, in stark contrast to standard modern predicate logic, which has hardly any logical power left. It also shows that the logically sound Abelardian system of predicate logic has much greater logical power than standard modern predicate calculus, while still staying within the bounds of a strictly extensional ontology—a fact which raises questions regarding the status of standard modern predicate calculus in mathematics and mathematical logic. In Chapter 6 of Volume II it is shown that traditional predicate logic is also much more functional from the point of view of transmitting information than its standard modern counterpart.

The fact that, as a matter of principle, linguistic utterances need ANCHORING in context before they can be KEYED to a given situation and the objects in it, is first discussed in Chapter 3, in the context of Aristotle's concept of proposition. Chapters 7 to 9 of Volume II are devoted to a further theoretical elaboration of the context-sensitivity of natural-language sentences and utterances. The notion of presupposition is central in this respect. Chapter 10 of Volume II is devoted to the logical aspects of the context-sensitivity of language. A presuppositional logic is developed for both the propositional operators and the universal and existential quantifiers. In this logic, a distinction is made between, on the one hand, a default, discourse-restricted area of metalogical relations, which is taken to have some degree of psychological reality, and a purely theoretical area which has no psychological reality but is presented merely to show the character and properties of the logic involved.

1.3 The method

1.3.1 *Cognitive realism*

Some clarity is needed with regard to the notion of psychological reality. In the philosophy of science, there is a well-known opposition between REALISM and INSTRUMENTALISM. The former wants its theories to be somehow translatable into terms of actually existing structures. The latter is content when the theory's predictions all come out right and feels no need to press for any form of realist interpretation—although there is, of course, reason to feel satisfied when reality turns out to be like the formal theory. In principle, instrumentalism rejects any nonobservable reality that could provide a causal explanation of the data. Indeed, the very notion of causality is suspect to instrumentalists. Influenced by scientific positivism, which dominated the physical sciences during the first half of the twentieth century and considered any search for nonobservables unscientific, formal approaches to empirical questions tended for a long time to disclaim any relation to whatever underlying causal reality might be posited. The formal analyses are meant merely to yield a mathematically precise and maximally simple or 'elegant' *characterization of observed data*, preferably in the shape of an abstract, algorithmically organized system that will also correctly predict data still to be observed.

Over the past decades, instrumentalism has faded from at least the human sciences and has largely given way to realism, which holds that explanatory theories, whether formalized or not, must be seen as reflecting or describing or 'modelling' an underlying, nonobservable reality causally underlying any actual or future data. The way a theory models an underlying causal reality may differ from case to case. In some cases, the modelling can be taken to reflect directly the postulated physical reality. Let us call this HARDWARE REALISM. In other cases, a less committing or more circumstantial interpretation is called for, especially when little is known about the hardware (as in the case of linguistics and cognitive science, where knowledge about the corresponding physical properties and functions of the brain is still extremely limited). This we call NON-HARDWARE REALISM.

A specification of non-hardware elements in a theory in terms of physical hardware is known as REDUCTION. STRONG REDUCTIONISM requires a specification of the elements of non-hardware theories in terms of physical correlates in such a way that the data are explained by the physical properties of the correlates. We agree with Fodor (1975), who shows that this requirement often destroys the explanatory power of a theory. His example is the concept of money in economic theory. Clearly, a hardware specification of money will produce a list of many different materials, along with a list of nonmaterial specifications for cases where money is just a numerical specification in a network of books and computers. Economic theory is hardly helped by such a hardware specification, as money is a category in its own right in economics.

A similar example is provided by the phenomenon of greeting. A greeting may consist of a nod, the raising of one's eyebrows, the raising of a hand, a handshake, a pat on the shoulder, a syllable (*Hi!*), a few kind words, and so on. Generally, in the human sciences, what counts is the PHENOMENAL VALUE of elements, which is often only remotely related to their physical make-up (compare the qualia discussion in Section 8.2.2). For that reason, we adopt the position of WEAK REDUCTIONISM, which requires merely a physical BASIS for the appropriate elements of a theory but does not require the explanatory shift from the non-hardware theory to the physical properties of the correlates stipulated by strong reductionism. Or, in the words of Nancy Cartwright:

> If the world is full of properties that are not reducible to those of physics, but which causally interact with those that are, then the most accurate physics will necessarily be a *ceteris paribus* physics that can tell the whole story only about closed systems. (Penrose 2000: 181)

In the day-to-day practice of linguistic analysis and description, it is not always easy to keep non-hardware realism and instrumentalism apart. The first concern of non-hardware realism is to model the object of inquiry, thereby automatically characterizing the data. This may lead to a quandary when an elegant and predictively correct characterization of data is available which may or may not be taken to model the causally underlying reality. Suppose it has been found, on solid empirical grounds, that the data on distributive and group readings in the plural forms of a given language are adequately describable, with all predictions coming out right, in terms of a set of abstract mathematical functions. Can a non-hardware-realist interpretation now be claimed? The answer depends on the extent to which human cognition can be credited with abstract mathematical powers, and cognitive science obviously does have a say in this. Now suppose it would be unrealistic to credit human cognition with the powers necessary for carrying out the computations in question. Does this mean that the account, explanatory though it may be, is unacceptable to a non-hardware realist? Here the answer is no. First because to get to any postulated underlying reality explaining the data it is necessary to have a precise idea of what the data amounts to: data characterization cannot be missed. Moreover, the account may well be saved for non-hardware realism if it is possible to devise an apparatus which does stand a better chance of being implemented, one way or another, in the physical brain, and which produces the same results as the mathematical functions. The mathematical properties of the output of the apparatus are then said to be merely EMERGENT or EPIPHENOMENAL.[1] In many cases, the mathematical properties of a system are discovered before the actual system causing them is laid bare.

Let me give an example presented earlier in Seuren (2004*a*: 65–7). There is a particular arithmetical function, known as the *Fibonacci sequence*, which is found all over biological nature. The medieval mathematician Fibonacci, who

[1] Cp. Johnson-Laird (1989: 483): 'One central point to make clear is that the logical properties of a term are emergent properties of its meaning rather than an explicit part of that meaning.'

worked in Pisa and was instrumental in introducing Arabic arithmetic into Europe, presented the following riddle in his book *Liber Abaci* of 1202:

A certain man put a pair of rabbits in a place surrounded on all sides by a wall. How many pairs of rabbits can be produced from that pair in a year if it is supposed that every month each pair begets a new pair which from the second month on becomes productive?

If one assumes that he also meant the first rabbit pair to be productive from the second month on, then adding up the total number of pairs of rabbits in the model at each month results in the Fibonacci sequence, illustrated in Figure 1.1a.

The Fibonacci sequence is formed by starting with the sequence 0, 1 and subsequently adding the sum of the two preceding numbers:

0, 1, 1, 2, 3, 5, 8, 13, 21, 34, 55, 89, 144, 233, ... and so on

An important corollary is the fact that the ratio between a given Fibonacci number F^n and its predecessor F^{n-1} is more or less constant, approaching the asymptotic ratio shown in (1.3) as *n* gets higher:

(1.3) $F^n : F^{n-1} = \pm 1.618033988 \ldots : 1.$

For example, 144 : 89 = 1.6179775...: 1, whereas 233 : 144 = 1.6180555...: 1. This ratio has been known throughout history as the GOLDEN RATIO and the irrational number 1.618033988...is known as the number PHI. The Fibonacci sequence is the closest known whole-number approximation to phi. The approximation proceeds in a pincer-like fashion, in that the values are successively below and above phi.[2]

It is a well-known but nevertheless remarkable, and still unexplained, fact that one often finds, in biological nature, that petals, leaves, branches, seeds, and so on, are arranged according to the Golden Ratio. The question now arises of whether the organisms in question have, as part of their physical composition, an inbuilt computer carrying out an arithmetical calculus producing the sequence for any F^n or whether nature accomplishes this arithmetical feat in a different way. Knowing what we do about physiology, we are bound to

[2] There is at least one other whole-number approximation to phi. This is obtained from the model shown in Figure 1.1b by setting up the sequence that results from listing the total number of individuals in the model at each successive cycle: 0, 1, 2, 4, 7, 12, 20, 33, 54, 88, 143, 232, 376, 609,.... (The same sequence is obtained for cyclic totals F^n through the function $F^n = F^{n-1} + F^{n-2} + 1$, for $n > 1$ and starting with 0, 1—that is, the Fibonacci function of Figure 1.1b but with one element added on each cycle; note also that all values ≥ 2 of this function are Fibonacci numbers minus 1.) Now the ratios of the values of this function at any cycle n to its predecessor n–1 again approximate phi but in a descending manner, not pincer-like, as the Fibonacci sequence does. For example, 232 : 143 = 1.6223776...: 1, 376 : 232 = 1.6206896...: 1, and 609 : 376 = 1.6196808...: 1 and so on.

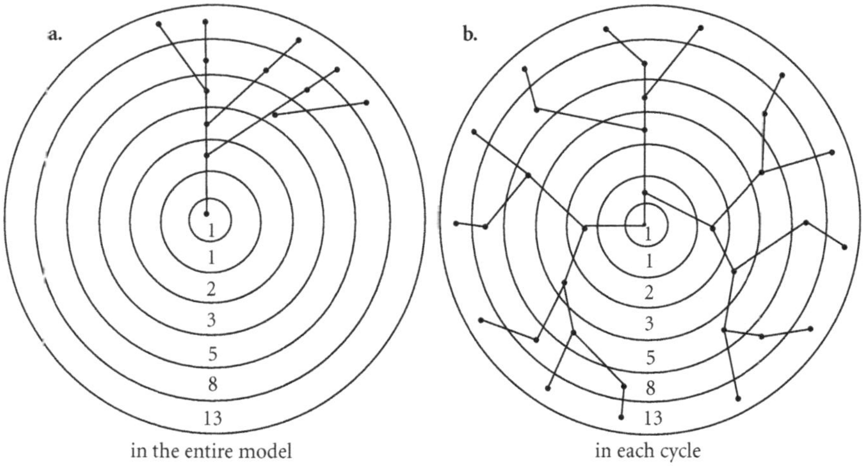

FIGURE 1.1 Two non-hardware-realist systems generating the Fibonacci sequence epiphenomenally: **a.** Fibonacci's own system; **b.** the new system

say that the latter must be the case. The organisms in question must be in possession of a different mechanism which produces the Fibonacci sequence epiphenomenally.

In fact, a simple hypothetical cyclic growth pattern does the job. What is needed is a growth pattern producing the Fibonacci numbers not as specifying the number of elements in the entire model at each cycle, as in Fibonacci's rabbit colony described above, but, rather, producing the Fibonacci numbers as specifying the number of elements in each cycle. This is achieved by a system of growth cycles which is built in such a way that each element reproduces itself twice, once on the next cycle and once on the cycle after next, as is shown in Figure 1.1b, where the Fibonacci numbers correspond to the number of elements in each successive cycle. Such a system produces the Fibonacci sequence epiphenomenally, and in a way that does not immediately exclude a hardware realist interpretation.

In Figure 1.1b, the apparatus starts with one element in cycle 1, which reproduces itself once in cycle 2, and again in cycle 3. So cycle 2 has again one element. But cycle 3 has two elements, as it receives the first reproduction of the element in cycle 2 in addition to the second reproduction of the initial element. And so forth. It is easily checked that the number of elements in each successive cycle corresponds to the proper Fibonacci number. This does not guarantee that the plants in question actually produce arrangements according to this sytem, but as long as specific biological 'hardware' research does not reveal the opposite, this possibility cannot be excluded either, whereas the calculus-machine hypothesis had to be excluded at first sight. As long as it is not known how nature actually does it, the most plausible hypothesis should be accepted.

This point of view is of relevance to science in general. It affects not only theoretical physics, which expresses itself mainly in mathematical terms, but also linguistics, in particular grammar and semantics. Abstract grammatical algorithms may be as good as they say, if they are psychologically implausible, it makes sense to search for an apparatus that produces the same results as the algorithmic rules but only epiphenomenally. The same applies, analogously, to semantics. Therefore, given the situation in the study of language and cognition, we opt for non-hardware realism, which is the best that can be achieved in the circumstances.

Besides the opposition between realism and instrumentalism, it makes sense, in the study of language and cognition, to draw a related opposition between FORMALISM and ECOLOGISM. In its purest guise, formalism regards a natural language as a computational system regrettably implanted into computationally imperfect human beings. By contrast, ecologists study language in its natural habitat. Fundamentalist or extreme ecologists shy away from formalisms and feel uneasy with any notion of system. For them, a natural language is, in principle, a collection of conventionalized cues enabling a speaker to guide the listener to a reconstruction of what the speaker wishes to communicate. Both views have convincing arguments supporting them, but they have also, in part, been developed without convincing arguments and sometimes even without any argument at all. Moreover, both parties seem to be guilty of overgeneralizing their conclusions. The arguments and analyses that do carry conviction are invariably restricted to certain aspects of language, while other, equally relevant, aspects are left out of account. As long as no absolute dominion is sought, the two 'philosophies' should turn out to be compatible and even complementary in the light of a more comprehensive analysis that gives both parties their due and rejects what cannot be upheld.

It is also useful to realize the degree of our ignorance: a complete and fully adequate scientific theory of the human language system and its various uses is still a remote ideal. This is an important methodological point, because the formal instruments that are at our disposal to make a theory 'scientific' are ultimately algorithmic in nature, and there is no knowing whether this may not, in the end, turn out to be too narrow a basis for formal analysis and description. It is not at all unrealistic to think that, in this respect, the future has a few surprises in store for us, in that the mind may turn out to be organized in ways that defy strict algorithmic limits and require different, more complex, formal techniques to be described adequately. In this respect (as in many others), the opinion of Roger Penrose must be taken seriously. For Penrose (see, for example, Penrose 2000), there are too many serious gaps in our understanding of the workings of the human mind for it to be plausible that it is merely a system of interrelated algorithms. Consciousness, in particular, is a basic property of the mind, yet it has so far firmly resisted all attempts at capturing it in algorithmic terms. Penrose ventures the idea that there may well be principles at work in the mind which are akin to those of quantum mechanics in physical nature. Being of a more practical disposition, and also because, unlike Penrose and more like Cartwright, I prefer weak reductionism (see above), I will not indulge in what must of necessity remain speculation in this regard. Yet I do feel that one must keep an open mind as to the possibility that new and more powerful formal insights and corresponding techniques will be developed which will do more justice to the facts of human cognition than has hitherto been the case.

Given this situation of formal uncertainty, I feel justified to defend the view that formalism and ecologism, far from being mutually exclusive, need each other in a complementary way if the purpose is to get closer to an adequate view of language and its use. But there is one preliminary condition. In order to achieve a proper integration of formalism and ecologism, some form of realism must be accepted by all concerned.

Instrumentalism and formalism are natural bedfellows, although they can also very well be separated. By contrast, instrumentalism and ecologism are intrinsically incompatible. Ecologism, being by definition about the realities of natural language, is necessarily realist: without realism there can be no ecologism. Therefore, for there to be a common ground between formalists and ecologists, it is necessary that the formalists embrace at least non-hardware realism. Henceforth, this is taken for granted. Instrumentalism is left out of account altogether. A necessary corollary of imposing realism on formalism in the study of language is that formalist approaches must subject themselves to the realities of this world—data, causality, and all. Realist formalists must accept that for a theory to be considered empirically adequate, it must somehow reflect the formal properties of the actually existing object of modelling, in this case the machinery taken to underlie the use of language. This view, which I call COGNITIVE REALISM, is the methodological basis of the present book.

Cognitive realism has one immediate important implication: there is no reason to insist on standard logical and mathematical means in the modelling of any underlying system taken to explain the data. There would be if all that is required were an instrumentalist characterization of the data, in which case the theory is no more than a metastatement about them, which is probably best presented in the neutral terms of standard logic and mathematics. But under a realist banner, much more is required of the theory. Now the theory has become a model describing and defining not the data but the object of enquiry which is not open to direct inspection but is taken to cause the data. Such a model must reflect the formal properties of the machinery underlying the use of language. A grammar, for example, had, therefore, better be specified not in terms of standard model theory, but in terms that seem best to approximate the actual structure of grammars and the way they function. There is no doubt that grammars can be described in standard model-theoretic terms, since the combined language of logic and mathematics is powerful enough to describe any structured object (see the quote from Hall-Partee 1979 in Section 1.1.2 above), but such a description would not be a proper modelling as it would fail to reflect the structure and functioning of the object described. A generative-transformational grammar, for example, could be described in set-theoretically based model-theoretic terms, but such a description would lose explanatory power as it would fail to capture the generalizations expressed in the transformational rules, which are not sensitive to set-theoretical structures or properties but to constituent-structure configurations and to instructions that come with lexical predicates.

More to the point, in case some subsystem, such as the semantic system of language, proves to have a logic that deviates from standard logic, the modelling had better be presented in terms of that logic. It is important to realize that there are infinite ranges of formally well-defined varieties of logic—that is, consistent systems for the formal derivation of entailments—all incorporating standard logic one way or another, so that it is an empirical question which logic within the infinite space of formal logics is the one chosen by nature for the purpose of human language. This not only lends a novel interest to the study of so-called 'deviant' varieties of logic, it also shows that one is entitled to require of cognitive realism that it be entirely open-minded with regard to such varieties, there being no *a priori* certainty that the logic of language is identical with standard logic.

This requirement is insufficiently appreciated in today's world of learning, where unorthodox or deviant varieties of logic are still looked upon with suspicion, and sometimes even with derision. But if the analyses presented in this book are correct, one must conclude that today's world of learning has this at least all wrong. In the present book it is argued that the logic of language deviates from the standard system in ways that show a high degree of functionality with regard to the practical use of language in everyday life.

A further consequence of painting formalism in realist colours is the necessity to acknowledge the limited power of the formal techniques and the underlying insights available today. Not only is there the fact, pointed at above, that the formal means at our disposal are likely to turn out to have insufficient power to serve the purpose of a formal description of the workings of the mind, it is also true that formalists like to think that only formal theories or descriptions are scientific, forgetting that full formalization is, in fact, the final step in the scientific procedure. Full formalization is often not feasible because the object of enquiry is too complex or too opaque for the time being, or because the formal means are themselves too weak to do the job. Human cognition is a prime example. No fully formal description of human cognition, whether instrumentalist or realist, is available to date, mainly because the object is too complex and too many questions about it are unanswered. But this does not mean that a scientific study of human cognition is impossible, only that cognitive science finds itself in a stage of development that is, in many ways, not yet ripe for full formalization.

In the empirical sciences, formalization is nothing but the provisional rounding off of a usually long process of making oneself familiar with the field, trying out the extent to which it is empirically accessible, formulating causal questions, exploring possible answers while testing them out in the light of available evidence. Cutting out the preparatory stages and jumping at formalization straight away is apt to create a skewed view of the object of inquiry, while practitioners and outsiders are made to believe that such a view is correct as it carries the authority of 'science'. This is what has happened too often in the study of language during the twentieth century. As a result, most current formal theories of language turn a blind eye to the cognitive embedding of language, which has so far largely resisted formalization. The fact that the mind plays a role is recognized only to the extent that the formal machinery is seen as being handled and used by humans, with all their weaknesses, illogical quirks and habits of convenience. Such theories may achieve a degree of formal precision but fail to achieve what they set out to do—present a faithful picture of language and the way it is part of human nature.

Over a century ago, the Cambridge philosopher Alfred Sidgwick issued a warning regarding formalization, pointing his finger at logic, where the first signs of 'symbolic' formalization began to manifest themselves:

The chief habit of thought antagonistic to a regard for special context is, beyond dispute, I suppose, that due to the attempt to make Logic Formal, or (worse) symbolic. Whatever value these developments of Logic undoubtedly have is bought at a cost which deserves to be reckoned rather than ignored. But . . . there is hardly a suspicion in the minds of formal logicians that they have any price to pay. (Sidgwick 1895: 282)

One may well feel that Sidgwick was a little too apprehensive about formalization. Yet there is a large element of truth in what he said, even with regard to logic, where formalization is constitutive, and hence indispensable. Sidgwick warned that the urge to formalize created a Procrustean bed for logic, where the logical aspects of context-dependency turned out to escape the formalizing powers of logicians, with the result that the object of inquiry, which to him was the 'natural' logic of language, was mutilated into the formalized logical system that became available—a critique not too far removed from the position taken in the present book.

This point is of major importance, since it appears that until the present day it has only been possible to provide formal models of the language system to the extent that it involves so-called MODULES, or special-purpose, self-contained, algorithmically organized machines, insulated from general cognition—exchangeable spare parts of the mind/brain, so to speak. Top-down or generative grammar, in particular, appears to be such a special algorithmic module, as is argued in Section 7.2. Not so, however, for semantics, which is better regarded as a system linking up the grammar module with general cognition. An adequate semantic theory appears to be viable only when it is accepted that many elements in the semantic machinery contain open parameters whose values are to be retrieved from situational, contextual, and world knowledge shared by speaker and listener (Clark 1992). Yet one must regrettably admit that the ways in which this is achieved still largely escape us.

Again, as has been stressed before, it is important to realize that the formal techniques that are at the scientist's disposal became available as a result of the great formal breakthrough in the foundations of logic and mathematics that took place during the late nineteenth and the early twentieth centuries and is associated with names like Boole, Frege, Russell, and Gödel. If one takes history seriously, one must admit that there may well be further breakthroughs, bringing more insights and techniques that will enable scientists to tackle new categories of data in formally well-defined ways. One thinks, for example, of formal methods to determine inductive conclusions—a problem that has so far proved refractory. This would not only be an important step forward in the understanding of human perception and creative intelligence, it might also be applied to resolve ambiguities in language, and thus be of great use in the development of automatic parsing programmes. Students of formal parsing techniques have so far systematically found themselves up against unassailable barriers—a fact that may well be due to the inductive, and hence nonalgorithmic, or partially algorithmic, nature of utterance comprehension. One should realize that the formal top-down modelling of grammars, which has proved remarkably successful, was impossible until the generalized notion of the algorithm became available during the 1920s. As we are reminded in Penrose (1997, in Penrose 2000), there is no guarantee that further developments in formal thinking will not be applicable to the analysis and description of cognitive processes that have so far escaped formalization.

1.3.2 *Grammatical and semantic formalism*

1.3.2.1 *Grammatical formalism* In the study of language, formalism has two branches: GRAMMATICAL and SEMANTIC. I shall deal with the grammatical branch only cursorily because the present book is not specifically about the technicalities of syntax, morphology, or phonology, but about the foundations and the semantics of language. Grammatically, formalism sees a language as either a (top-down) GENERATIVE computational system which defines how to combine symbols into well-formed strings or, alternatively, as a (bottom-up) PARSING system which defines how incoming strings of symbols are to be parsed into structures that allow for a (formal) semantic interpretation. The generative tradition is about half a century old and is represented by a number of theories, the most well-known among them, no doubt, being Generative Grammar and its offshoot known as Generative Semantics. This tradition has its roots in the work by Leonard Bloomfield and Zellig Harris, and also, to some extent, in the early work of Noam Chomsky. In Bloomfield's and especially in Harris's hands, this tradition was clearly instrumentalist, in that linguistic theory was expected to provide the means for a characterization of the data in terms of a maximally succinct system.

Chomsky hesitantly introduced a realist perspective, in a bid to return to the ecologistic roots of the discipline. He proposed that the instrumentalist systems of his predecessors, which were meant to describe data and not an underlying machinery, should somehow correspond to such an underlying machinery. But Chomsky has always remained highly ambiguous on this score (see Botha 1989: 159–64), probably because he failed to consider the possibility of non-hardware realism (Seuren 2004*a*: 61–71). Chomskyan Generative Grammar did not develop any semantics. Its offshoot Generative Semantics was well on its way towards doing so when it was nipped in the bud (see Seuren 1998: 493–512 and the references cited there for this curious episode in the recent history of linguistics).[3] A similar theory was propounded around

[3] See, for example, Harris 1993; Huck and Goldsmith 1995; Seuren (1998: 493–512, 2006, 2007).

the same time by the Californian linguist Wallace Chafe (1970), who developed a, perhaps somewhat idiosyncratic, system for the meaning-to-surface generation of sentences, with a heavy emphasis on underlying semantic content. This work points to the importance of discourse for an adequate theory of grammar and meaning (Chafe 1970: 347) and is thus a harbinger of the studies in discourse that were to come soon after. It does not take a stance, however, on the question of the possible psychological reality of the system proposed.

The parsing tradition, in so far as it has been of a formal nature, has always been entirely instrumentalist. It is represented mainly by Categorial Grammar, which derives from work done by the Polish logician Ajdukiewicz during the 1930s but branched into a variety of related formal grammatical theories such as Montague Grammar, Tree Adjoining Grammar, Lexical Functional Grammar, and Generalized Phrase Structure Grammar, all of an instrumentalist persuasion. Pure Categorial Grammar is a parsing system that checks the sentences of well behaved formal languages for grammaticality and incorporates a model-theoretic formal semantics. Unfortunately, natural languages are not so well behaved, owing to the many ways in which natural language meaning and utterance interpretation depend on nonlinguistic knowledge. In recent years, this family of grammatical parsing theories seems to have been relatively stagnant. The parsing tradition is now largely owned by the computational linguists, who have a better eye for the quirks of natural language but tend to take only a limited interest in questions of theoretical foundation, preferring to focus instead on statistical methods.

1.3.2.2 *Semantic formalism* What interests us more directly, in the present context, is semantic formalism, which differs radically from its grammatical counterpart. (The uninitiated reader may well feel lost at the following exposé, but that should be no reason for despair, as the gist of the matter will become clear anyway.) Classical formal semantics treats a natural language as part of a model-theoretic system containing an algorithmically defined language whose well-formed formulae ('sentences') are semantically defined by TRUTH CONDITIONS requiring an ontological state of affairs or 'world' to produce truth. Predicates are defined by their SATISFACTION CONDITIONS, which require either individual elements or sets of (sets of...) elements to produce a true formula ('sentence'). Logic is incorporated in that objects are generalized over by means of variables and quantifiers that are defined in the standard way, and in that Boolean operations are carried out over truth (Boolean 1) and falsity (Boolean 0) according to the standard definitions of the propositional operators.

To know what a sentence means is, in the eyes of model-theoretic semanticists, to know under what conditions it is true. For this reason, formal semantics is often called 'truth-conditional'. Its roots lie in the powerful and deservedly prestigious studies in the foundations of logic and mathematics carried out a century ago. In these studies, terms like *language*, *meaning*, and *interpretation* occur in a technical, partly metaphorical, sense, but some later practitioners, especially Donald Davidson and Richard Montague, thought it interesting to see if the metaphorical element could be eliminated by postulating that a natural language is, in fact, a formal system of the kind defined in logical, model-theoretic semantics. Formal semantics resulted from an exploration of that idea. Since it is part of formal semantics that the 'language' of the formal system be defined algorithmically, formal theories of natural-language semantics often incorporate a fledgling algorithmically organized 'grammar'. These grammars, however, bear little relation to the grammatical complexities of natural languages. In the early days, attempts were made to establish a link with the world of professional linguists. But these attempts met with little success. As it is, formal semantics lacks any serious grammatical theory, just as formal linguistics lacks any serious semantics. Formal semanticists and linguists are still possible worlds apart.

One naturally wonders why this should be so, and the answer could fill a tome or two. The story would describe the entirely different intellectual traditions and concerns of the two disciplines. Modern linguistics, tracing its origins to comparative philology which, in its turn, was spawned by the late eighteenth-century romanticist curiosity of both geographically exotic and prehistorically ancient languages (see Seuren 1998: 79–104), never took leave entirely of its ecologistic roots, despite the introduction of all kinds of formal methods. New formal elements were incorporated, such as the notion of an algorithmic theory to reconstruct the native speaker's competence in forming sentences. But in the new context of cognitive science, which replaced behaviourism during the 1960s, such formal elements were not *a priori* incompatible with (non-hardware) realism, since (a) the mind was now credited with at least some modularly insulated computational powers, and (b) even without that assumption, there was no knowing whether there would not be a psychologically plausible apparatus producing the results epiphenomenally. In any case, the natural habitat of language was always remembered to some extent—unless, of course, the aim was to develop computer programs that could be usefully applied.

By contrast, logic, the parent of formal semantics, had just gone through a period of rigorous mathematization and, above all, depsychologization. The positions were no longer held by traditional philosophers but by men and women trained in the logic and mathematics that goes under the banner of Bertrand Russell, perhaps with a modicum of philosophy—and that only of the Quinean 'desert landscape' type popular in the Anglo-Saxon world. And mathematics had no truck with the mind, only with the metaphysical necessities of sets and cardinalities.[4] When leading figures like Donald Davidson and Richard Montague turned to natural language, what they were seeking was not an adequate insight into the nature and functioning of human language—which would have meant a radical break in their intellectual concern—but, in principle, a formalization of the metaphysics of truth in natural language (although there was also the excitement of having a new playground for building formalisms, as natural language provided interesting and challenging handicaps).

Moreover, a notion of 'meaning' had been developed in logic that was restricted to set-theoretically defined truth conditions of well-formed formulae in well-defined extensionalist models. This rather unworldy notion of meaning was transferred by the logicians to natural language without much ado and presented as the object of what they called 'formal semantics'. Discrepancies and recalcitrant facts were considered to be of lesser concern and relegated to the new discipline of pragmatics, which holds that speakers' sense of practicality and social propriety will override logical necessities. The linguists were, on the whole, taken by surprise and felt out of their depth. Their reaction was either to embrace the new ideology, repressing any feelings of discomfort, or turn away from it on the grounds that this could not be the right way to go about studying language.

Meanwhile, formal semantics went its own way, which was gradually leading towards a greater openness with regard to realism and the role of (shared) knowledge. The most obvious manifestation of this trend is the assumption, in some recent versions of formal semantics, of a WORKING MEMORY accumulating information provided by previous discourse, with the purpose of accounting for the fact that, for semantic reasons, most natural language sentences are usable only in certain classes of contexts, variously called DISCOURSE REPRESENTATIONS (Kamp 1981; Kamp and Reyle 1993), DISCOURSE DOMAINS (Seuren 1972*b*, 1975, 1985), or MENTAL SPACES (Fauconnier 1985).[5] Discourse-dependent semantics is mainly supported by facts to do with presuppositions, anaphora, and topic–comment modulation.[6]

[4] Apart from the Dutchman Luitzen Brouwer's mind-based Intuitionism, generally dismissed as a respectable but odd idiosyncrasy (see Van Dalen 1999; Dummett 2000 for relevant information).

[5] The insight that utterance interpretation is heavily dependent on contextual factors goes back to outspokenly ecologist authors such as Wegener (1885), Stout (1896), Gardiner (1932), Firth (1957). See Seuren (1998: 125–7, 169–77) for a detailed discussion.

[6] Including anaphoric identification across subdomains, such as the fact that one may *believe* there to be a burglar in the attic and *hope* that *he*—the real or imagined burglar believed to be in the attic—is not armed. See Sections 7.2.2 and 10.5 in Volume II for further comment.

The recognition of the importance of discourse in sentence meaning and utterance interpretation is, of course, an important step towards an adequate recognition of the role of cognition in language. A realist element, therefore, inevitably pervades these discourse-sensitive versions of formal semantics, as opposed to classical, Montagovian formal semantics, which is an exclusively instrumentalist discipline to which the question of the actual existence of an apparatus answering to the instrumentalist description is of no concern. This new, hesitantly realist, reorientation has, in its turn, revealed a number of quandaries that are now forcing discourse-oriented formal semanticists to look across the divide to more ecologically oriented approaches. Since this aspect is of particular relevance in the present context, I will expatiate a little.

The information contained in each new utterance is said to be INCREMENTED to the existing representation (domain, space), which thus changes with each new increment. On the whole, discourse-oriented formal semanticists take it for granted that the incrementation of new utterances (sentences) to a given discourse representation is fully determined by (a) preceding linguistic input and (b) the utterance (sentence) itself. For them, the meaning of a sentence is a *function*, in the mathematical sense of the term, from given to new discourse representations—that is, from old to new truth conditions for the discourse as a whole.[7]

Although this sounds neat and elegant, it neglects the essential fact that external cognitive input is required for incrementation to take place. The building up of discourse representations (domains, spaces) requires open access to available shared situational, contextual, and world knowledge. This means that truth-conditions are often co-dependent on nonlinguistic knowledge to be retrieved from outside the semantic system. Sentence meaning can be considered to be a discourse incrementation *function* only in an abstract sense in which the whole of any amount of shared knowledge is taken to be part of the function, which makes the function not only nonoperational but also noncompositional in principle.

It is traditionally believed, in formal semantics, that natural language semantics must be taken to be COMPOSITIONAL. By this is meant that the meaning of a complex structure is computable from the meanings of its axiomatically given elements. These have usually been taken to be the complex structure's own constituents, but other input is allowed as well, as long as all input is derived from linguistic elements in well-defined linguistic structures,

[7] This notion of sentence meaning was first presented in Seuren (1975: 228–39) (English translation in Seuren 2001: 3–11).

perhaps discourse structures. Since formal semantics does not usually make a distinction between type-level sentence meaning and token-level utterance interpretation, the compositionality thesis implies that the interpretation of an utterance should also be a function of given linguistic input only, which may include elements in the given discourse. But it is now widely accepted, at least among ecologists, (a) that a distinction must be made between type-level sentence meaning and token-level utterance interpretation, and (b) that at least utterance interpretation is NOT compositional in the sense defined. And the insight is gaining ground that (c) not even type-level sentence meaning is fully compositional, since the type-level meaning definitions of many lexical predicates contain OPEN PARAMETERS, sometimes called FREE VARIABLES, that link the system up with other parts of cognition in that the values of the parameters or variables must be retrieved from available shared knowledge.

Many ecologists react to the noncompositionality thesis by proclaiming that no computation is involved at all in the determination of the meanings or interpretations of complex structures. Any 'digital' analysis is rejected and an appeal is made to a necessarily imprecise 'analogous' series of largely introspective associations. I consider this to be an overreaction and I also think it is misguided, first because a loss of precision is always to be avoided and secondly because a more than superficial inspection of the data shows that the complexity in form–meaning relationships far exceeds the powers of introspection. To a very large extent, the regularities observed and the generalizations that can be formulated have no counterpart in conscious experience, which makes it imperative to resort to some form of computation for an explanation. The question is not whether forms of computation are involved but, rather, how much is contributed by modularly restricted computation and how much by external input. Much of the debate between pragmatists and those formal semanticists who are prepared to doubt the compositionality thesis revolves around this question.

The facts that show the untenability of the strict compositionality thesis are, on the whole, well known. It is, for example, well known (cp. Bendix 1966: 57; Clark 1992: 311) that the possessive genitive in an expression like *John's book* may express a large variety of truth-conditionally different relations between John and the book in question. It may be the book owned by John, written by John, held by John, borrowed by John, reviewed by John, just mentioned by John, and so on. The important fact is that the range of relations expressible through the genitive's in English is not linguistically definable—that is, as part of the description of the language—but is cognitively open in that, for each token occurrence, an appropriate value must be identified on grounds of preceding context and shared knowledge of the world and of the situation at hand. Or, in the words of Herb Clark:

> The possibilities are in theory unlimited in number and cannot be enumerated, and what it [i.e. *John's dog*] is taken to mean on any occasion relies heavily on the coordination of the speaker and the addressee. Possessives, in short, are contextual expressions. (Clark 1992: 311)

This goes for possessive predicates in general, including the English verb *have*, the prepositions *of*, *with*, *without*, and in some cases also *to*, genitive and dative case morphemes in flectional languages, possessive pronouns, and so on. For example, whereas a sentence like *This hotel room has a bathroom* requires a one-to-one relation between the room and the bathroom for truth to arise, a grammatically analogous sentence like *This student has a supervisor* is content with a many-to-one relation for truth. The reason is not to be sought in the linguistic system (whoever does not know these facts is not helped by remedial courses in English) but in world knowledge about hotels and universities (Seuren 1985: 21). It is probably correct to say that every possessive predicate in every language contains an open parameter referring the listener, within certain limits, to what may be considered a well-known relation of appurtenance for the categories of object denoted in each token occurrence (Janssen 1976, 2007).[8]

Similar examples are easily found. Adjectival predicates like *short*, *many*, *old*, *long*, *sufficient* (but not, for example, *criminal*, *sad*, *healthy*) often derive part of their satisfaction conditions from what may be called a class- or situation-dependent prototypical 'norm': a ship's model can be huge as an object in my study but be dwarfed in an exhibition hall (see Section 9.3 for a novel solution to this problem). The predicate *flat* has different satisfaction conditions according to whether it is applied to a tyre, a road surface, a landscape, or a face—quite apart from metaphorical and other 'nonliteral' uses (see Section 9.3). These are instances of the widespread phenomenon of the COGNITIVE DEPENDENCY OF LEXICAL MEANINGS, discussed in Chapter 9 and in Section 7.4 in Volume II. It is easy to see that such phenomena are nasty

[8] The same, incidentally, applies to nominal compounds: what *tennis* is to *court* in *tennis court* differs radically from what *tennis* is to *elbow* in *tennis elbow*. However, since compounds are often conventionalized to become lexical items in their own right, the intended relation is, in such cases, not to be retrieved from a contextual or encyclopedic knowledge base but is fixed by lexical meaning description. When they are not conventionalized but put together ad hoc, Clark's observation applies (Clark 1992: 311): '... compound nouns are contextual expressions since their meanings aren't denumerable and what they mean on any occasion depends on the close coordination of the speaker and addressee'. In sum, compounds are noncompositional.

gremlins in the systems proposed by the formalists, while they are grist to the ecologists' mill.

1.3.3 *The various forms of ecologism*

1.3.3.1 *Introspection versus underlying system: cognitivism* Just as the formalists are requested to embrace realism, the ecologists, on the other side of the divide, are called upon to take their realism seriously and face the fact that their analyses, no matter how sharply to the point and how satisfying on an intuitive level, will fail to provide an explanation of the observed data as long as they are based on INTROSPECTION—that is, as long as they are couched in the phenomenological phraseology of human experience and are not shown to follow from an underlying system or machinery, which is the view taken in the present book. Speech, or the use of language, is not like making a choice between jam and marmalade, or deciding to visit an aunt, but rather like using a specific tool to achieve a specific end. To understand language is to understand the tool. To understand speech is to understand how the tool can help to achieve the end. One may perhaps explain a speaker's decision to say *I hate you* by appealing to the practice and principles of general psychology, which are largely open to introspection. But general psychology does not tell one how these words express the underlying thought and owing to what system such an utterance is understood the way it is. The answer to such questions does not consist in introspectively relating linguistic to other experiences, as extreme ecologists do, but in reconstructing how the machinery works, the machinery itself being largely beyond any threshold of awareness.

Extreme ecologism is like Freudian psychiatry. It works, at least some of the time, as long as self-recognition is what one strives for. But self-recognition, though no doubt therapeutically wholesome, is not the aim of science. The insight sought by science is more detached. Cognitive-realist science is about noncircular, systematic, and mutually interacting causal relations, within the constraints imposed by physical nature as well as by the *a priori* principles of analytical necessity. It looks, in principle, for ways to formulate and test a relation of compatibility with, and a translation into the terms of, what has been found in other sciences, including the physical sciences, even when such a 'translation' takes away the explanatory force of a theory or a concept, as with Fodor's money, or with the qualia discussed in Section 8.2.2. Without this condition of physical reduction, the theory would not be weakly reductionist but phenomenological.

The question goes back to early twentieth-century European linguistics, when introspection was rife as a method of investigation. In his *Sprachtheorie* of 1934, the German psychologist-linguist Karl Bühler, arguing against the introspective method, expressed this point in the following words:

> One should not make one's results ultimately and forever dependent on the condition of a highly refined ability to describe one's own thought experiences. On the contrary, one should make an effort to make them accessible also to less subtly trained eyes. Even more importantly, what counts is *objective* verification.
>
> (Man darf den Befund nicht für alle Zeit an die Bedingung einer hochgezüchteten Feinheit des Beschreibens eigener Denkerlebnisse knüpfen, sondern muß danach streben, ihn auch weniger subtilen Augen zugänglich zu machen und noch mehr: es gilt ihn *objektiv* zu verifizieren.) (Bühler 1934: 254)

Nowadays we say that intuitive notions may have the status of data and that the analysis is meant to make the data explicit through a theory generating them. It seems that in this respect modern ecologism, in its various manifestations, by and large fails the standards of scientific methodology.

A prime example of reliance on introspection and rejection of the notion of system is the relatively recent school of COGNITIVISM (e.g. Langacker 1987, 1991), closely followed by CONSTRUCTION GRAMMAR (Goldberg 1995, 2006; Croft 1991, 2001). I am singling out these schools for special critique because of the popularity boom they have enjoyed lately in linguistic circles and beyond. Cognitivism has its origins in linguistics but has expanded into semantics and pragmatics. Academically speaking, this trend is primarily driven by considerations of psychological plausibility and aims at a full reduction of human language to the general principles of human cognition, while any kind of formal analysis or description is either rejected out of hand or treated with distrust. Sociologically speaking, cognitivism appeals to those academics who abhor anything expressed in symbols that do not belong to the ordinary alphabet. 'Sharing experiences' is the word, even if this means giving up basic criteria of scientific method, such as notional precision or falsifiability.

From a purely intellectual point of view (see also Seuren 2004*b*: 593–4), grammar and semantics are taken to be of such a nature that they can be fully accounted for in terms of general cognition, without any modular, special-purpose machinery or circuitry. Grammatical structures are taken to suggest or reflect semantic relations directly (often iconically), without any intervening grammatical or semantic computation. In keeping with its extreme antiformalism, cognitivism rejects any form of 'deep' structure that specifies the sentence's meaning in a precise and analytical way and is to be transformed through a formal, algorithmical grammar into a surface structure reflecting meaning only indirectly, in virtue of speakers' grammatical competence. In short, Cognitivism and Construction Grammar reject grammatical derivations.

Semantics is restricted to the psychology of concepts, without any appeal to logic or the notion of truth. To the extent that semantic structures are recognized, they are treated as not being formally precise but suggestive and associative and to be interpretable in terms of conscious experience rather than hidden structural principles of cognitive processing. Form–meaning relations are not formulated in any precise or falsifiable way but in intuitive terms (which then again have imprecise relations to their meanings). Moreover, there is no innate language faculty. Whatever is universal in language follows from the necessities inherent in the overall genetic make-up of the human species and the physical structure of the world. Language acquisition is taken to be based on statistical frequency of linguistic input—the theory of 'usage-based grammar' (Tomasello 2003).

It will be clear that such a frame of mind is more appropriate for lexical than for grammatical studies. And, indeed, cognitivism has booked notable success in lexical semantics and, more generally, in those areas of language study where such largely unformalizable factors as viewpoint, functionality, prototypicality, and the like play a dominant role. Where I begin to have reservations is when cognitivists try to amalgamate lexicon and grammar by turning grammar into a form of lexicon. Competence in a language is said to consist in the memorization of idioms and frequently occurring phrases and language acquisition is said to be subject to frequency measures of linguistic input supported by general cognitive categorization skills. Goldberg does this by dint of merely terminological stipulation—although the terminology is not consistently sustained—(Goldberg 2006: 5, 18; italics and boldface original):

> All levels of grammatical analysis involve constructions: learned pairings of form with semantic or discourse functions, including morphemes or words, idioms, partially lexically filled and fully general phrasal patterns....
>
> What makes a theory that allows constructions to exist a 'construction-based theory' is the idea that the network of constructions captures our grammatical knowledge of language *in toto*, i.e. **it's constructions all the way down.**

But such a stipulation, of course, hardly suffices for what is meant to be an overall theory of grammar and language.

The danger for cognitivism is that it will repeat not only the views of 'direct-experience' syntax that were current during the end of the nineteenth and the beginning of the twentieth centuries, but also the frequency-based syntactic notions developed in the context of behaviourism (Skinner 1957). I have nothing against reviving old theories. On the contrary. But I do object to reviving theories that have been shown, by arguments, facts, or subsequent findings, to be wrong without addressing the arguments, facts, and subsequent findings in question. European 'direct-experience' grammar was overtaken by developments in structuralism and generative syntax that showed the untenability of the introspective method. Behaviourist 'grammar' bit the dust because of arguments against behaviourism as a whole and against its application to grammar in particular.[9] Yet despite this *déjà-vu*, large numbers of linguists and psychologists have joined the cognitivist ranks, perhaps because of a deep-seated emotional resistance to anything formulaic or formal.

All this taken together makes cognitivism highly problematic. Semantic content and semantico-grammatical relations are presented in vague terms of direct experience, often with the help of ad hoc schemata that have no basis in theory.[10] 'Constructions', allegedly associating form and meaning, are postulated but seldom, if ever, explicitly formulated. Counterevidence, such as the blocking of reflexivization in topic–comment structures like (1.15) below, is not eagerly jumped upon, as it should be, but ignored as long as possible. The methodology is one of looking for confirming evidence and hardly ever for crucial counterevidence. We all know that confirming evidence can go on being found for false theories till doomsday comes, but science consists precisely in spotting and, if necessary, actively searching for, possible crucial counterevidence. And this is something that is conspicuously absent in the writings of the cognitivist grammarians.

Operator scope is a fruitful source for crucial counterevidence, but it is left unmentioned by most cognitivist authors, with the result that scope-related facts, such as those mentioned in (1.4)–(1.7) below, or the fact that, for example, **I understood one word not* is, though comprehensible, ungrammatical in English, while its literal equivalents in German and Dutch are both comprehensible and grammatical, are either left undiscussed or discussed inadequately.

Langacker (1991: 107–41) is an exception in that he devotes a lengthy section to relative scopes of quantifiers and negation. In treating the facts concerned, Langacker avoids any logical parlance or formalisms, probably because these

[9] The first basic criticism of behaviourism is found in Karl Lashley's historic paper read at the 1948 Hixon Symposium (Lashley 1951; see also Gardner 1985: 10–14, 191–3). The arguments proposed by Lashley were largely taken over (without proper attribution) in Chomsky's review of Skinner's *Verbal Behavior* (Chomsky 1959), which pertained more to behaviourism as a general method and philosophy of psychology than to Skinner's version of it (MacCorquodale 1970).

[10] A semblance of technicality is created by terms like *conceptual* versus *perceptual space*, *conceptualization*, *frame of reference*, *profile*, or *procedure*, but these terms are not defined in any precise or testable fashion. All that is said is that they 'appeal to only obvious and independently established cognitive abilities' (Langacker 2003: 25). It is clear that the critique uttered by Karl Bühler in the quote given earlier fully applies to cognitivist grammar in all its varieties.

are considered incompatible with his ideology. He couches his analysis entirely in terms of his no doubt 'highly refined ability to describe [his] own thought experiences', rejected as a valid criterion by Karl Bühler in the quote given earlier. While his observations are sophisticated (but unoriginal because derived from the formal-semantics and the philosophical literature), his solutions are in many ways defective. Thus, his discussion of the subaltern entailment from ALL F is G to SOME F is G is restricted to the comment that, in the sentence *Some even numbers are divisible by two*, 'naive intuition urges the replacement of *some* by *all*' (1991: 109). And if one attempts to extrapolate from his accounts of some cases to other, similar, cases, one finds that, for example, **All cats are numerous* should be a well-formed sentence understood as 'the set of all cats is numerous' (1991: 117), yet one has no way of knowing if such an extrapolation is acceptable to the author because further surprises may always be in store. The ad-hocness of the descriptive schemata and the concomitant lack of precise predictions are, in fact, the main problems with Langacker's notion of quantification theory. There is, moreover, the total neglect of the monumental work by formal philosophers like Tarski and Carnap regarding the relation between meaning and truth. One would expect at least a reply to such authors' considerations and arguments.

Goldberg also touches on phenomena of quantifier scope, pointing to the fact that mature speakers know that a sentence such as (1.4a) is scope-ambiguous but the corresponding (1.4b) is not (Goldberg 2006: 31–2, 155–61) (I have added (1.4c) for good measure):

(1.4) a. I gave one text to every student.
b. I gave every student one text.
c. To every student I gave one text.

Goldberg proposes that 'because scope is strongly correlated with topicality, the information-structure properties of constructions predict their predominant assignment of scope' (Goldberg 2006: 161). Yet she only considers scope phenomena of the existential and universal quantifiers, paying no attention to the powerful scope effects of negation and other scope-bearing operators. And even for the two classical quantifiers, her analysis leaves much to be desired.

While one observes in both Langacker and Goldberg a desire to develop a 'natural' predicate logic based on naturally given logical powers, one sees at the same time that the method applied to attain such a goal is inappropriate, if only because the enterprise is not backed up by the necessary familiarity with logical notions and techniques.

Apart from that, however, it seems relevant to point to the fact that the difference between (1.4a) on the one hand and (1.4b, 1.4c) on the other appears to correlate with the hierarchical status of high (adverbial, prepositional, clausal, or nominal) sentence constituents in constituent trees: a high S-final peripheral constituent (such as *to every student* in (1.4a)) is allowed to overrule the basic left-to-right scope ordering constraint, which explains the scope ambiguity of (1.4a) (see Seuren 1972*c*, 2004*a*: 186–9, 207 and especially 1996: 124–6 for ample discussion; see also examples (8.12)–(8.15) in Section 8.6.2). The same phenomenon is encountered in sentence pairs like (1.5a, 1.5b), (1.6a, 1.6b) or (1.7a, 1.7b), where the high peripheral adverbials *for six hours*, *at six*, and *for six weeks* lead to scope ambiguity in S-final but not in S-initial position:[11]

(1.5) a. He didn't sleep for six hours.
b. For six hours he didn't sleep.

(1.6) a. The secretary had left at six.
b. At six, the secretary had left.

(1.7) a. I lent him my bicycle for six weeks.
b. For six weeks I lent him my bicycle.

The fact that high peripheral S-constituents lead to scope ambiguity is, in all likelihood, not an isolated or arbitrary fact of natural language, but should rather be seen as part of the general phenomenon that high peripheral constituents tend to swivel easily from a left-peripheral to a right-peripheral position and vice versa, which again demonstrates the cognitive and syntactic fact that they do not belong to the *nucleus* (Seuren 1969) or *matrix structure* (Seuren 1996) of the sentence. As such, this explanation should be grist to Goldberg's mill, were it not for the fact that it requires tree structures and derivations, which are distinctly out of favour.

Examples such as (1.6a, 1.6b) and (1.7a, 1.7b) constitute further evidence in that one of the operators at issue is hidden in underlying forms. In (1.6a), there is an underlying scope-sensitive tense operator standing either in or over the

[11] *Because*-clauses follow the same pattern:

(i) She didn't leave because she was afraid.
(ii) Because she was afraid she didn't leave.

As pointed out by Goldberg (2006: 144), (i) is ambiguous (*pace* intonation), but she fails to consider (ii), where the right high-peripheral scope-bearing constituent *because she was afraid* has swivelled to the left, eliminating the small-scope reading of the *because*-clause under the negation. Goldberg attributes the ambiguity of (i) to the subjective notion of 'backgrounding', ignoring, among other things, the rather serious question of the difference in truth conditions between the two readings.

scope of *at six*, the latter position resulting in the reading where *at six* takes only the tenseless structure [the secretary leave] as its scope. In (1.7a, 1.7b), the space needed for the scope difference is provided by the internal analysis of the lexical predicate *lend*, analysed as containing (but not as being identical to) the structure 'allow-to-have' (see Section 8.6). Here, (1.7b) unambiguously represents the wide-scope reading for *for six weeks*, while in (1.7a) both this reading and the one in which *allow* takes scope over *for six weeks* ('I allowed him [for six weeks he have my bicycle]') are admitted. It is not hard to conclude that Goldberg's treatment of operator-scope phenomena is fragmentary and defective and, therefore, lacks convincing power.

Yet, on the whole, cognitivism has been able to bring to the fore those aspects of language that have so far escaped formalization and have, as a consequence, been neglected in the more formally oriented branches of linguistics. Yet it has failed to forge a link with them. On the contrary, cognitivist writings are imbued with a staunch refusal to acknowledge the value of any kind of formal treatment and a corresponding proprietary attitude with regard to the subjectivist approach advocated in them.

By and large, theoretical linguists agree, given the impossibility of having introspective access to the rules and principles of grammar, that whatever grammar system is postulated is likely to be only partially related to, or governed by, general principles of cognition. Formal grammarians assume, correctly in my view, that, quite apart from any possible underlying cognitive motivation, the rules of top-down grammar operate 'blindly', as a purely formal, algorithmically organized module, beyond any form of conscious control or external cognitive input (see Section 7.1.4 for a defence of this view and a discussion of possible counterevidence). Both language comprehension and language production are highly sensitive to deviations from the norm, which, as it has turned out, can be formulated only in fairly abstract and formal terms, and not in terms of incidental, nonsystematic cues, games, or improvisation.

Since it is very common for sentences to be produced in some garbled form and yet be interpreted without fault, the very notions of 'garbling' and 'correct interpretation' demand an explanation. A language is defined not by garbled utterances but by judgements of correctness of expressions for given meanings. Both good and not so good speakers of English will be able to make proper sense of a garbled utterance like *Me not water in. Me no can swim!* But only good speakers will produce an utterance like *Please don't push me into the water. I can't swim!* (see Seuren 2004*b*: 616). Examples like these can, of course, be produced at will. They show that a language is defined by good *speakers*, not by good *listeners*.

But even if a language were badly defined and would admit of all sorts of improvisations, as seems to be the case with the improvised contact languages called Pidgins, it would still be impossible to account for the speakers' ability to produce and comprehend utterances in such an ill-defined language without the assumption of a mental machinery converting thoughts into sounds and vice versa—activities that far transcend the realm of guesswork on the basis of cues.

In this respect, the genesis of Creole languages is highly instructive. As one knows, Pidgins necessarily become nativized and turn into Creole languages when children acquire them as a native language. When this happens—we know from documented sources that this may happen in one generation—the degree of improvisation diminishes and strict grammatical and phonological rules take over in ways that sometimes do and sometimes do not betray an underlying cognitive motivation. All these Creole languages, no matter what their input languages have been or where they arose, have the basic constituent order S(ubject)–V(erb)–O(bject). They all have a system of preverbal particles (mostly derived from original verbs) of T(ense)–M(odality)–A(spect)—their famous TMA-system. They usually lack any flectional morphology and if they have one it is rudimentary (derivational morphology is often richer). Plurality and intensive or repetitive aspect on verbal meanings are occasionally expressed by means of reduplication—a clear case of underlying cognitive motivation (iconicity). Grammatical exceptions or specific sentence structures (such as comparatives or, for example, *The more he drinks the more he talks* or *Like father like son*) are absent in the early stages of Creoles and only begin to turn up after a considerable lapse of time. And more typical features of Creole languages can be enumerated.

What one witnesses here is the emergence of grammar in communities, not in individuals growing up in a well-established linguistic environment. And a conservative estimate of what one may conclude from this is that the emergence of grammars in communities is subject to, but far from determined by, cognitive and other possible functional pressure.[12] Attempts to argue that Creole grammars, and by extension the grammars of all languages, are *fully determined* by such pressures have so far been unsuccessful. The prevailing consensus, in this respect, seems perfectly sensible: grammars are

[12] Jackendoff's claim (Jackendoff 1993: 134) that 'the grammar of creole *is* Universal Grammar' fails to reckon with the fact that (a) what is called 'universal grammar' does not define a language or group of languages but consists of a set of constraints on grammars, and (b) some machinery is required anyhow to transform the abstract SA-structures with their multiple embeddings into more linearly arranged 'flat' surface structures appropriate for acoustic output (Seuren and Wekker 1986).

self-contained, cognitively codified modules (some call them 'autonomous') whose make-up has probably been influenced by external functional factors but which have long since taken over from general cognition and have gone their own way.

In general, there are plenty of grounds to hold, as I do in the present book, that language use in any kind of human society is invariably supported by a powerful machinery lodged in the mind and consisting of interrelated special-purpose subsystems. Science is in many ways like police work. If speech were a crime, the police would be well advised to assume the existence of a powerful and well-organized criminal organization coordinating instances of speech and providing speakers with technical support. If the police were content with regarding instances of speech as isolated occurrences involving lone individuals who take pleasure in doing evil while using makeshift means, they would be guilty of a serious dereliction of duty.

1.3.3.2 *Pragmatics: its strength and its weaknesses* While the opposition between formalism and ecologism has been there from the moment language began to be studied (Seuren 1998: 23–7), the modern form of this opposition differs greatly from its historical antecedents. On the semantic side, modern ecologism started as a reaction to the artificiality felt to adhere to the formalist method employed in model-theoretic, logic-based semantics. The most notable representative of early modern ecologism is no doubt Herbert Paul Grice (1913–1988), whose work gave rise to the discipline of PRAGMATICS during the 1970s. The Gricean form of 'classical' pragmatics was mainly intended to make up for the counterintuitive results that arise when standard logic is applied to natural language. Classical pragmatics does not cast doubt on standard logic, nor does it reject formal methods of analysis and description, but it claims that the formal semantic machinery by itself is insufficient to explain all the phenomena observed in the use of language. Grice introduced his famous MAXIMS, which are in effect principles of good practice, deemed to be naturally taken into account by humans when they operate with the formal systems at work in language, especially standard logic. These principles are based on, and formulated in terms of, common sense and common decency, honesty, or sincerity. Their application is taken to produce the well-known Gricean IMPLICATURES—noncogent inferences drawn by the listener on the basis of the implicit assumption that the speaker observes the maxims. Formalization is, on the whole, not striven for (the few attempts in this direction that were made by some of Grice's followers have been signally unsuccessful). Questions of grammar are entirely beyond the horizon of classical and other forms of pragmatics.

The paradigm of classical pragmatics has proved remarkably resistant. Apart from the fact that, especially in the beginning, it provided a home for the many practitioners who turned away in despair from the technical complexities and the obvious inadequacies of semantic formalisms, its main strength seems to lie in the picture it presents of humans handling an abstract and intrinsically rigorous system in the rough reality of forever varying everyday life. The abstract system can perhaps be formalized but 'life', in its infinite variety, cannot, or so it is believed.

A central and extremely important element in classical pragmatics is the realization that linguistically determined meaning is insufficient to explain the ways in which actual token utterances are interpreted: *meaning underdetermines interpretation*. This underdetermination manifests itself *inter alia* in the fact that any semantic system underlying the use of language must contain open parameters of the kind mentioned in Section 1.3.2.2. That there is a trade-off between type-level, linguistically defined meaning and token-level information, whether situational, contextual, or encyclopedic, seems certain. Again, the question is not whether there is such a trade-off but rather how much is contributed to actual, token-utterance interpretation by either element.

Why, then, do some semanticists, including the present author, object to the Gricean maxims? The reason lies in their superficiality and in the uncritical attitude they manifest regarding standard logic. Both faults adhere to Gricean pragmatics as birthmarks that need to be operated away. Since this issue affects the core of Gricean pragmatics in all its manifestations, I must again beg leave to expatiate a little. The gist of this excursion is the proposal, elaborated in Chapter 3 in Volume II, to replace the Gricean maxims with general principles of human cognition, which is taken to be subject to restrictions of naturalness regarding the way it deals with objects in general and with plural objects (sets) in particular.[13]

This view requires an attitude of resistance with regard to standard logic, which was developed regardless of naturalness restrictions. We consider standard modern logic NONNATURAL or CONSTRUCTED and therefore *prima facie* an unlikely candidate for the post of logic of language. It is the result of highly developed forms of abstract, formal thinking and has lost many of its links with ordinary cognitive and linguistic functioning. Indeed, it totally lacks any empirical dimension. Since pragmatics accepts standard modern logic without much questioning, I feel that the charge of gullibility with regard to standard

[13] Although this idea has been with me for decades, it has been slow in maturing and what I can present in this book is, though hopefully promising, still highly tentative.

logic is justified. What is needed is less awe and a greater openness to the ways a logic can achieve informational consistency and linguistic functionality. But in order to attain that state of mind, it is also necessary to delve into the foundations of logic—an activity largely shunned by pragmatists (and others). In this sense, the charge of superficiality can be levelled against current pragmatics. The study of what are still called 'deviant' logics is not only useful but also necessary for a proper understanding of natural language.

It is in this way that I intend to account for the implicatures that form the object of many pragmatic studies. I do not wish to deny the reality of the implicatures and I fully acknowledge that something has to be done to account for the intuitions they are based on. But I also feel that any explanatory principles are to be sought at a deeper level than the Gricean maxims, as part of the cognitive and interpretative machinery that generates them. Owing to their superficiality, the relation of the Gricean maxims with the results they are meant to produce is, in the last resort, opaque. It is quite possible, perhaps even probable, that common sense, decency, honesty, and sincerity are deeply ingrained driving forces of human behaviour. Since we recognize them as being honourable and indicative of good citizenship, they have a special appeal and one easily feels happy to see them embedded in human nature as principles guiding linguistic interpretation. The problem is, however, that there is an explanatory gap between them and the phenomena they are meant to explain. The maxims simply fail to do the job they were hired for.

This gap appears even at a mundane level. Consider, for example, the implicatures 'not all' from the use of *some* and 'not and' from the use of *or*. If these are explained by what Grice (1975: 45) calls the maxim of quantity—'Make your contribution as informative as is required'—then why is there no implicature 'John has not been killed' from the use of *John is dead*? No one will deny that *John has been killed* entails, and is hence more informative than, *John is dead*, just as *all* induces an entailment to *some* and *and* to *or*, which makes the former more informative than the latter.

The answer normally given by pragmatists is that there is no scale, in the sense of Horn (1972, 1989), ranking *be dead* lower than *have been killed*, while there is one for *some* and *all*, the former of each pair being ranked lower than the latter. And why is there no scale for *be dead* and *have been killed*? The answer to that is that the semantics of these predicates does not involve any appeal to the notion of quantity. All right, but then, what notion of quantity is involved in *and* and *or*? And why should my 'contribution' be required to be maximally informative regarding quantity and not regarding other aspects of the world, such as manner of death? I find it hard to avoid the conclusion that the maxim in question, even when reinforced by a theory of scalarity, fails to explain the implicatures at hand. By contrast, the distinction between natural and constructed cognition, as developed so far, revolves primarily around the mathematical notion of set and thus applies by definition not only to the quantitative aspects of the objects spoken about but also to the logic derived from the underlying set theory.

More concretely, why should it be a sign of cooperativity to infer the implicature 'not all' from the use of *some*? Imagine a CEO who invariably takes disastrous decisions being addressed by one of his underlings who, politely and very cooperatively, puts it to him that *some* of his decisions may have unfortunate results, thereby clearly not implicating 'not all'. Or take my computer, which tells me that *Some items in this location have the same name as items you're moving* even when this applies to *all* the items in this location. How informative is my computer required to be? Or consider the case of me being about to go through the passport control of some country and seeing a notice saying *Travellers with Belgian or French passports must fill out form B.* Does my wish to be a good citizen now make me understand that travellers with dual Belgian and French nationality do not have to fill out form B? Of course not. Such an inference would be an indication of bad citizenship. If defaults are overridden in such cases, one would like to know more precisely what such overriding amounts to and what the defaults are based on. The Gricean maxims fail to provide an answer. I claim that an investigation such as is presented in Chapter 3 in Volume II of the formal principles according to which humans deal with sets and, accordingly, with the logical constants of natural language offers a more promising perspective, allowing a dependency on speakers' perception of the interpretative strictness expected in certain situations.

A clear case showing the explanatory deficiency of the Gricean maxims for the conflict between standard logic and natural intuitions is provided by the sentences in (1.1) and (1.2) presented above in Section 1.1.2 and repeated here for convenience:

(1.1) a. He doesn't like planes or trains.
b. He doesn't like planes and he doesn't like trains.

(1.2) a. He doesn't like planes and trains.
b. He doesn't like planes or he doesn't like trains.

In terms of standard logic, (1.1a) and (1.1b) are equivalent for the same reason as (1.2a) and (1.2b) are: both look like instances of De Morgan's laws. Yet the equivalence of (1.1a) and (1.1b) is immediately recognized by speakers, while that between (1.2a) and (1.2b) requires careful reflection. I am not aware of any psychological experiment to test this difference, but I am confident that any such experiment will bear out the intuition. The point is that the Gricean maxims, which are all about social propriety and communicative functionality, have no bearing at all on these examples, whereas they should have if they are to explain the discrepancies between standard logic and natural intuitions. By contrast, the system of basic-natural propositional logic developed in Section 3.4.3 of Volume II does account for this difference.

One invariably finds, not only in pragmatics but also elsewhere, a curious resistance, probably grounded in diffidence, to any proposal involving non-standard varieties of logic. The notion of a sound logical system more tailored to the functional needs of human language and its use than standard logic is either not considered at all or, when it is, it is dealt with in an imprecise way. Pragmatists generally stay away from a detailed investigation of the formal aspects of semantics and its underlying logic, that being considered the business of the formal semanticists. In similar manner, formal semanticists do not wish to be involved in matters of practical use, that being considered the (lesser) concern of the pragmatists. Both pragmatics and formal semantics thus miss the opportunity to push forward the frontiers of the formalizable.

The aversion to formalism has reached a peak in a movement called RADICAL PRAGMATICS, which claims a lineage to Wittgenstein as he was after 1930. Radical pragmatics prefers to see all utterance interpretation as being determined by context-bound properties of token utterances. If it is admitted that there is a system behind the phenomena, it is with a grudge and any such system is accorded a minimal and subordinate role. Questions about how precisely token interpretations are arrived at by humans are seldom taken up and never answered. It is as if one lived in a world in which motorists believe that their vehicles run the way they do not because of the engine and other technical appliances, but because the drivers know where they want to go.

1.3.3.3 *Functionalism in grammar and the status of data* In the study of grammar, ecologism is most clearly represented by a movement traditionally known as FUNCTIONALISM, which aims to explain most or all of human language by an appeal to pressure exerted by the functional needs of language. Unfortunately, however, functionality as a concept in linguistic theorizing is as elusive as it is indispensable. It is elusive because, after centuries of linguistic study, it is still not known what purpose or purposes are served by language. And even if this were known, we do not know how these purposes are best served, as they may well impose contrary requirements. Speakers, for example, are helped by coding rules, but listeners are helped by semantic transparency—that is, the contrary of coding rules. Then, our notion of the procedures involved in the actual production and interpretation of linguistic utterances is at best fragmentary, which makes talk about the functionality of such procedures premature, if not outright speculative. Yet the viewpoint and criterion of functionality cannot be missed because it can hardly be denied that language has a *raison d'être*, no matter how hard it proves to zoom in on it. The fact that there is a *raison d'être* makes it legitimate and even mandatory to look at language from a functional point of view.

Yet despite the lack of clarity surrounding the notion of functionality, functionalism is riding high in present-day linguistics, probably as a result of a deeply ingrained attitude of resistance to formal theories of cognitive functioning and a corresponding wish to see language as a product of general, unformalizable, mental life. Functionalists themselves openly recognize and discuss their opposition to formalism (Darnell *et al.* 1999; Clark 2005; Laury and Ono 2005). It seems, however, that this opposition is based more on prejudice and sociological pressure than on strictly academic argument. Functionalism is based on the premiss that the prime *raison d'être* of language is its function in fulfilling communicative needs. Taken by itself and depending on what is meant by 'communicative needs', this premiss is harmless enough, perhaps even incontrovertible, and perfectly compatible with realist-formalist approaches. Yet, in practice, the functionalists and the (realist) formalists are at war. In the words of Frederick Newmeyer:

> Each school accuses the other of an a priori exclusion of certain types of data, certain methodologies, and certain modes of explanation that it sees as pivotal. For example, formalists are said to reject a priori the possibility of explanations that appeal to facts outside the linguistic system proper.... Functionalists, on the other hand, are said to lack the agnosticism of the true scientist in that they demand that there be an external motivation for any grammatical phenomenon, whether one is justified or not.... My view is that there is some truth to the charges from each camp, particularly if we look at what is considered 'mainstream' work in each area. (Newmeyer 1999: 470–1)

Newmeyer's position, with which I concur, is that one should make an effort to be sensible and break through the petty restrictions of established practice:

> I argue that there is no fundamental incompatibility between the central tenets of the formalist and the functionalist approaches. In a nutshell, there can be an autonomous structural system at the center of language and syntactic structure can be motivated functionally. Hence each of the two approaches to language can make fundamental contributions to our understanding of grammar. (Newmeyer 1999: 469)

It is, unfortunately, true that all or most formalist theories in existence pay too little attention to the functional requirements imposed on language and that many or most functionalist theories on the market baulk at the thought of a formal system. But this is a question of attitudes and prejudices on both sides, and certainly no rational ground for a wholesale proprietary rejection of either formalist or functionalist theories or explanations. Formally structured systems that are lodged in biological organisms are almost by definition subject to, but not determined by, functional constraints and requirements. It is commonly accepted that they owe their evolutionary development largely to the functional pressure for survival and optimization of existence. Since language is a prime example of a formally structured system lodged in biological organisms, there should be no opposition between functionalism and formalism, even though in practice it does exist, often manifesting itself in virulent ways.

Like so many other functions or faculties in the human brain, the language faculty is best seen as incorporating a large amount of evolutionary 'tinkering' or EXAPTATION in the more technical jargon: structures that served a given function in earlier species are used, or duplicated (Marcus 2004: 112–14, 2006; Fisher and Marcus 2006: 10), for different functions in subsequent species. This alone speaks against an exclusively functionalist rationale for human language as it exists today.

But there are also language-internal factors working against an optimization of functionality. As argued in Seuren (2004*a*: 147–9) and in Seuren and Hamans (2008), sociolinguistic factors often lead to antifunctionalist features in certain language varieties, owing to the fact that groups of speakers sometimes find it convenient to create a language variety that is more difficult to acquire and use and hence less efficient or functional than the variety it deviates from. This happens, for example, when a higher social class wants to prevent intrusion from lower classes. The higher class will then introduce lexical and grammatical irregularities whose acquisition requires a type of schooling that is inaccessible to the lower classes.

Sometimes also, speakers want to be able to communicate without ordinary citizens or the police being able to understand what they say, as in the well-known case of London-criminals' 'rhyming slang'. Conservatism and group pride likewise often contribute to the introduction or maintenance of nonfunctional or even dysfunctional features, in social life as well as in language. In Great Britain, for example, a monetary system of pounds, each pound consisting of twenty shillings, each shilling consisting of twelve pence (the so-called LSD system), remained in use until as late as 1971. The question is: why and how did this obviously nonfunctional system come into being in the first place? The question is related to that of the development-through-culture of linguistic numeral systems, which are mainly based on the number 10, although the bases 5, 6, 12, and 20 are also found (Bauer 2004). For each of these counting systems there may have been a functional explanation in certain given contexts, such as the counting of specific objects (sheep, wheat sheaves, bales), but such functionally motivated origins soon lost their function and the number names crystallized into systems that have proved hard either to eradicate or to improve upon. Magic—hardly a source of functionality—may have played a role as well, as numbers have often been endowed with a positive or negative magical force (one thinks of the number 13). In addition, one frequently sees the mixing of numeral systems, as is evident in the British LSD system, or in the standard-French numeral system, which is essentially decimal but partially also vigesimal. Such mixtures are almost by definition nonfunctional, yet they are seen to crop up with great regularity (Bauer 2004).

As regards language in general, the conclusion must be that, next to purely functional motivations regarding the expression of socially committing intents, many other forces are at work in the development of natural languages, owing to the complexities of the communities using them. The most rational procedure, in this field of conflicting tendencies, would seem to be:

(a) find out how linguistic systems are built and how they function;
(b) analyse the aims that language can be taken to serve; and finally
(c) see how well language serves those aims.

This procedure is, however, often not followed by functionalists. Rather than investigate functionality *a posteriori*, many impose *a priori* functional requirements on the theory of language, interdicting from the outset any analysis, description, or explanation that fails to include, or follow from, a possible functional motivation. And even when functionalists do accept the more rational procedure, they too often neglect the first step, sometimes even denying that there is a system. The second step is almost universally neglected by all parties, including the formalists, who usually take it for granted, without further analysis, that the purpose of language is the transfer of factual information. The functionalists, by contrast, have a knack of saying that language serves to 'express experience'—the term *experience* being taken in the widest possible sense, presumably including the 'experience' of forming propositional thoughts.[14] The first two steps of the procedure having been either neglected

[14] In fact, as is argued in Chapter 4, the central function of language is defined by speech act theory. When producing an utterance, the speaker either takes on or imposes a commitment vis-à-vis the audience with regard to the propositional thought expressed in the utterance. This is the basis not just of all language use but of all conscious use of established signs. The combined structure of

or dealt with superficially, the final phase is then entered without proper preparation.

Functionalism has manifested itself recently in a flurry of publications, emanating largely from California and advocating the notion of 'usage-based grammar'. They are, on the whole, united with the cognitivist movement discussed in Section 1.3.3.1. The underlying idea is that a grammar does not form an algorithmically organized, rule-driven system producing a surface output from a propositional input (or rather, from an intent that incorporates a proposition; see Chapter 4), but a collection of 'constructions' representing form–meaning associations, whereby the meanings need not be propositional in nature.

The authors in question typically appeal to a perhaps natural yet unjustified popular instinct to the effect that 'abstract' systems fail to do justice to language, whereas they study language 'as it really is', thereby assuming a spurious air of authenticity—like boy scouts around the camp fire, who feel that this is 'real life'. Language, they say, is found in actual usage, with its quirks and statistical frequencies, rather than in formal algorithms. One easily grants, of course, that usage involves a great deal more than just formal grammar. Normally, one learns to cope with usage situations while acquiring the grammar of one's language—though the two processes are likely to be of a very different nature. It is, therefore, perfectly legitimate to study usage as it is, as many do, but then one studies the USE made of language, not language itself. Actual traffic is different from the system according to which it has been regulated. Traffic may be light or heavy, but such predicates do not apply to the underlying system, just as linguistic usage may be either highly original or repetitive, again predicates that do not apply to the underlying system of grammar. One may learn the traffic system by taking part in actual traffic under proper guidance, but the result of that learning process is not only an ability to cope with a set of traffic situations but also, essentially, a set of rules. Or, as Newmeyer put it, in his article 'Grammar is grammar and usage is usage':

> I think that some arguments in favor of usage-based models can be dismissed right away. Most importantly, one that can be laid to rest is based on the fact that since properties of grammars are functionally motivated, grammar and use are necessarily inextricable. Such a view seems to assume that once a system is characterized as discrete and algebraic, a functional explanation of that system (or its properties) becomes impossible. But that is simply not true. Indeed, it seems to be only linguists who have

the commitment element and the proposition with regard to which the commitment is taken on or imposed I call the underlying INTENT of an utterance.

this curious idea. In every other domain that I am aware of, formal and functional accounts are taken as complementary, rather than contradictory. (Newmeyer 2003: 686–7)

I will not go over all the arguments, both for and against, in the discussion that followed Newmeyer (2003) in the March 2005 issue of *Language* (Clark 2005; Laury and Ono 2005; Meyer and Tao 2005), but I will discuss what I consider to be the main question arising from that discussion, namely that of the status of data: what are the data relevant for the study of the underlying language system (competence data), and what are performance data or data of use?

In principle, the answer is not too difficult. We consider competence data to consist in JUDGEMENTS of, or judgemental reactions to, phonological, morphological, and syntactic correctness of expression for any given meaning. If such judgements or judgemental reactions are shared by speakers in a speech community, we have data about the language (or dialect) of the speech community in question. By contrast, we consider performance data to consist in facts of actual linguistic behaviour. This means that observations taken from real-life linguistic interaction (such as (1.11) below) can count as competence data only if backed up by the acceptance or rejection by native speakers as correct or incorrect in the language or language variety at issue.

It is clear that reliable tapping of correctness judgements or judgemental reactions is a delicate and difficult enterprise, which must be undertaken with the greatest possible methodological care—a requirement that is, regrettably, not given the attention it deserves in the various schools of theoretical linguistics.

Meyer and Tao (2005: 226) discuss the following example:

(1.8) Conservatives argue that the Bible regards homosexuality to be a sin.

This sentence is taken from *The New York Times* and thus constitutes performance data, since it has actually occurred as a written utterance. It is remarkable because the English verb *regard* is generally taken not to allow for Subject-to-Object Raising. Yet in (1.8) *regard* has induced precisely this rule of syntax. Does (1.8) also constitute competence data? That is, is the underlying sentence type a correct expression for what the token utterance of (1.8) is obviously meant to express? The answer depends on whether there is a group of speakers who can reliably be taken to accept (1.8) as a correct expression for the intended meaning. If there is not, (1.8) must be considered to contain a performance error. But let us assume there is such a group. Then what this means is merely that for those speakers of English the verb *regard* is indexed in their mental lexicon as allowing for the rule of Subject-to-Object Raising. Just as there are many speakers of American English who accept sentences like *He is probable to arrive tomorrow*, which are unacceptable to British speakers of English, even though the intended meaning is clear. All this means is that the adjective *probable* is marked, in the mental lexicon of the speakers in question, as allowing for the rule of Subject-to-Subject Raising, just as *likely* in all other forms of standard English—a mere instance of internal variation in the vast population of speakers of English.

Most grammar systems have gaps in that they fail to specify the correct expression in certain cases, often to do with plural or gender agreement. Meyer and Tao (2005: 227) present the case of disjunctions with a singular and a plural disjunct, as in:

(1.9) Either Harry or his parents [is/are] coming.

Some speakers accept *is*, some *are*, some neither and some don't know. If these judgements properly reflect the competence of the speakers in question, then there are three groups of speakers: those whose grammar yields the singular *is*, those whose grammar yields the plural *are* and those whose grammar fails to specify a correct form. It is very useful that such cases should be brought to the attention of 'abstract' grammarians like Newmeyer or myself, because we tend to gloss over them too easily. But in no way do they constitute counterevidence to the claim of grammar as a modular, algorithmic system.

Then there is the question of a possible statistical preference of one form over another when the two forms are mere grammatical variants without any distinction in meaning or conditions of use. Such variant pairs do occur, although they are less frequent than is often thought, as there often are subtle differences. A good example of a really fully equivalent pair is found in Dutch subordinate clauses, where a perfect tense auxiliary may either precede or follow the past participle, without any detectable difference—as far as we know—in meaning, function, or condition of use, whether context-determined or of a sociolinguistic nature. The two variants in (1.10) are, to all intents and purposes, exactly equivalent:

(1.10) Ik dacht dat Arthur [gelogen had / had gelogen].
I thought that Arthur [lied had / had lied]
I thought that Arthur had lied.

It may well be that a corpus shows a statistical preponderance of one order over another, but this is no guarantee that a different corpus will show the same. What this means for an algorithmic theory of grammar is merely that the grammar allows for a free choice of either variant. Yet sociolinguistic factors may lead to a preference for one variant over another. When this happens, one variant may disappear in some varieties of the language in question, while the other is marked for one or more different varieties, or one variant may simply become categorical throughout. But there is no reason at all to assume that the statistics of the phenomenon in question is part of the grammatical system incorporated by the speakers in question. The statistics follows from the grammar system and may influence its change over time, but it is not a part of it.

On the whole, I agree with Newmeyer in principle, though not in every detail.[15] In particular, I agree with Newmeyer when he defends 'the classical position ... that speakers mentally represent full grammatical structure, however fragmentary their utterances might be' (Newmeyer 2003: 682). In Newmeyer (2005: 234) an apparent counterexample to this principle is discussed, namely the problem of: 'why, in response to the question *What did you talk about?*, one can say *That war is imminent*, even though the sentence

(1.11) *We talked about that war is imminent.

is ungrammatical.' So as to dispense with this obstacle, Newmeyer appeals to the general observation that 'the anaphoric dependencies that are mentally represented can be quite complex' (2005: 234), following this up with some examples from Modern Greek whose relevance is at least doubtful. A much simpler answer would consist in the observation that a sentence like (1.12) appears to be flawless:

(1.12) What we talked about is that war is imminent.

This fits in with the view that, in discourse, answers fill in the missing value in preceding questions and that the canonical semantic form of questions has a topic–comment structure that can be expressed as a (pseudo)cleft sentence (see Section 3.2 and Chapter 11 in Volume II). What seems to block matrix incorporation in (1.11) is the absence of a nominal argument for the preposition *about*. In English, this little problem is often solved by the insertion of *the fact* (even when what is at issue is not a fact), as in:

(1.13) We talked about *the fact* that war is imminent.

This somewhat clumsy solution is, in fact resorted to by Newmeyer himself in the quote just given, where he says 'Most importantly, one that can be laid

15 For example, I object when Newmeyer sees the phenomenon of usage-based grammar as 'the resurgence of many of the same ideas that were the hallmark of generative semantics' (Newmeyer 2003: 683). In fact, no generative semanticist has ever denied the algorithmic character of grammar. On the contrary, that was just about their most central tenet. Yet they did raise the valid and important question, somewhat underemphasized by Newmeyer, of what are performance and what are competence data.

to rest is based on *the fact* that since properties of grammars are functionally motivated, grammar and use are necessarily inextricable'—even though he subsequently denies the factual character of the *that*-clause. Other languages, such as German, solve this problem by inserting a cataphoric pronominal argument:

(1.14) Wir sprachen *dar*über, daß Krieg bevorsteht.
we spoke *it*-about that war is imminent
We discussed the imminence of war.

Apart from such details, however, it is important to note that the position taken by Newmeyer (and myself) does not imply that language-acquisition processes cannot be influenced by usage phenomena such as frequency or experience-based association or the speech of role models. On the contrary, those are contributing factors determining the growth of grammars in infants—and often also in second-language acquisition—and thus of language change. As De Saussure already pointed out:[16]

> Without doubt, these two objects are closely linked and presuppose each other: the language system is necessary for speech to be intelligible and to produce all the effects it has; but the latter is necessary for the language system to establish itself; historically, speech always precedes the system.... Finally, it is speech that makes any language system change.... There is thus an interdependence of the language system and the use made of it; the former is at the same time the instrument for and the product of the latter. But all this does not stop these two from being absolutely distinct. (De Saussure 1916: 37–8; translation mine)

The important point is that the end product of the language acquisition process in the brain of a language learner is a modular structure and not a motley collection of form–meaning associations. There probably is such a motley collection in the speakers' brains, since it is easy to demonstrate that speakers tend to fall back on set phrases, especially when the phrases in question have recently occurred, but collections of set phrases do not constitute grammars. This is shown by the general fact that speakers have no difficulty understanding phrases or constructions that one may safely assume not or only rarely to have occurred in their linguistic experience with the frequency required for the status of 'set phrase'.

A clear example is (1.15), which was observed in real life (see Seuren 2004*a*: 184; 2004*b*: 608–9). A father has just admonished his young son that he should

[16] De Saussure's *parole*, translated here as *speech*, stands for what is called *usage* in the present context.

not cry because, he said, 'Well-educated boys don't cry'. The boy then snaps back:

(1.15) *I* didn't educate me, *YOU* did!

Any competent speaker of English immediately understands this sentence as being equivalent to the pseudocleft construction 'the one who educated me isn't ME but YOU' and will grasp the difference with (1.16), which is equivalent to 'the one who educated himself isn't ME but YOU':

(1.16) *I* didn't educate myself, *YOU* did!

This case is instructive in many different ways. It falsifies Chomskyan binding theory, which has no provision for topic–comment structure. It likewise falsifies Chomsky's claim (1972: 88–99) that heavy accent can be assigned to any constituent, even to single morphemes and phonemes and is, therefore, a surface-generated phenomenon. It shows, moreover, that the corresponding (topic–comment) pseudocleft structure must be taken to be closely related to the structure underlying the surface sentence (1.15) with its contrastive accents (a point that is further elaborated in Chapter 11 in Volume II).

In the present context, (1.15) is relevant for at least two reasons. First, it is entirely unrealistic to assume that all those speakers of English, including young children, who understand (1.15) immediately and correctly do so because of their frequent past exposure to cases where comment or focus accent on the subject blocks reflexivization.[17] Secondly, the algorithmic mechanism leading from an underlying cleft construction to the surface structure *I didn't educate me* in (1.15) is quickly formulated in terms of MATRIX INCORPORATION of the comment into the topic, which is contained in the matrix structure. All it takes is to assume an underlying cleft construction not [be I [the x [x educated me]]], with the comment [be I] as the main predicate and the topic [the x [x educated me]] as the nominal subject term containing the matrix-S [x educated me]. Here, *me* is not reflexivized because it is not

[17] It is true that sentences like *I know me* (with or without contrastive accent on *I*) do occur, which shows that the reflexivization rule is not absolute. (In fact, in Section 4.5.1—and also in Seuren 1989—I argue that reflexivization is a deep cognitive process and that reflexive pronouns are not referential.) Such sentences constitute crucial evidence against the claim that their grammaticality and proper interpretation even by young speakers is explained by input frequency. Sentences like *I know me* or (1.15) above, can hardly be taken to be sufficiently prominent in the linguistic input properly processed by young children across the board to give rise to their systematic ability to get both the semantics and the syntax right. Whoever objects to this argument carries the burden of proof and has to show that they do occur with sufficient frequency in the input corpora of all children learning English. It is worth noting, in this connection, that a sentence like *I know me*, without contrastive accent on *I*, still requires the reflexive *mezelf* when translated into Dutch, whereas it requires the nonreflexive *mij* when there is a contrastive accent on the subject pronoun, as in the Dutch rendering of (1.15). For Dutch, therefore, the claim that cases of unexpected nonreflexivity, as in (1.15), are rare, very rare, carries even more weight.

coreferential with the subject term of its clause, that term being the variable x. The comment is then incorporated into the matrix-S, yielding *Í didn't educate me*. (For more detailed comment see Sections 4.5.1 and 7.2.2 and Chapter 11 in Volume II.) The algorithmic treatment thus explains the data in a nontrivial sense, whereas the usage-based account has nothing to offer.

The same argument applies to the intricacies of verb-clustering as found in Dutch and German (see Seuren and Kempen 2003). These constructions are open to finely tuned and, on the whole, uniform grammaticality judgements, even when the frequency of occurrence of the cases at hand does not differ significantly from zero.

Examples of this nature abound: one only has to look out for them. And one must do so, if it is one's objective to do science: science lives by *crucial* confirming and disconfirming evidence. Apart from the cases just discussed, a rich supply of such crucial evidence is found in Newmeyer (2003) and Seuren (2004*b*).

Given that the usage-based philosophy of grammar has recently gained great popularity in the linguistic community, one feels relieved to find that functionalism does not form one single homogeneous school but manifests itself in a large variety of currents whose adherents differ considerably from each other not only as regards their attitude towards formal analysis and theory but also as regards the weight attached to criteria of functionality. I have often found functionalist analyses and explanations to be of great value in the perspective of my own methodological position, which combines ecologism with a moderately formalist form of cognitive realism. To take just one example (though many more could be cited), John Hawkins's work (e.g. 1983, 1988, 1994), shows the functional advantages of many constituent-order phenomena in the production and comprehension (parsing) of utterances, and there is every reason to suppose that these advantages have causal effects on language structure. Such work is of great value, as it aims to detect functionality after the fact, rather than imposing functionality as a prerequisite.

1.4 The 'innate hypothesis' and the genesis of language

1.4.1 *The 'innate hypothesis'*

The reader may wonder why, in a book of this nature, no attention is paid to the widely debated issue of innatism in language. In this section I give my reasons for this apparent omission. But first, what is this 'innate hypothesis' (IH) defended by Chomsky in a variety of versions over the past forty or fifty years? It is not the hypothesis that humans require certain innate structures to be able to acquire a natural language, because that is not a hypothesis but necessary by definition, if only because one needs a congenitally acquired brain to learn anything. It is, rather, a *specific* hypothesis, in the sense in which, for example, animal instincts and natural drives are *specifically* innate, about the innate structures that humans require in order to become competent in a language.

Chomsky and his followers have stipulated that, for example, grammars are innately structure-dependent and not dependent on, say, linear counting of elements in sentences. They have, in general, maintained that whatever looked, at any time, like a grammatical universal of any kind must be considered innate in the true biological sense of the word—that is, as a specific prewired feature of what is, or was, supposed to be a 'language organ'. Lately, however, with the introduction of the so-called 'Minimalist Program' (Chomsky 1995), this strong version of IH has been watered down to a point where it has become vacuous. Now, one understands, one should, in the spirit of Ockham's razor, look for ways to derive the universal properties of grammars from constraints that are imposed by cognition in general, by the peripheral machinery of speaking and hearing and by the ecological setting of language in an environment of humans who use language for certain purposes.

Ironically Chomsky (1995) repeatedly stresses that his Minimalist Program is likely to be successful, in that it will lead to the conclusion that language as a whole is derivable from the constraints of cognition, environment, and peripheral machinery, without any residue that would fall outside that explanation. He does not say, however, that if that were to turn out to be so, there would be no room left for any specific IH as a driving force in first-language acquisition, as all universals would be reducible to independent constraining factors. Therefore, Chomsky's IH can stay alive only in the hope that the Minimalist Program is unsuccessful. One cannot at the same time argue in favour of Chomsky's IH and expect success for the Minimalist Program.

Be that as it may, I eschew the innateness debate because to the extent that it has substance it is in the hands of specialists in molecular genetics, brain-scanning techniques and related fields, which are way beyond my own expertise. This does not mean that I consider all talk outside these specialized subjects about a possible specific IH senseless. On the contrary, the possibility of a specific, prewired 'faculty of language' helps to create a perspective for discussions about certain questions, although I do feel that this perspective has been overexploited in recent years, so much so that debates and even research programmes have become speculative to a degree that risks exceeding admissible standards.

In fact, I think that the arguments in favour of the general thesis that the human brain is prewired for the acquisition and creation of specifically human languages beyond any already present external constraints, perhaps as a result of evolutionary exaptation, carry a great deal of conviction. The fact that, the moment children grow up in a Pidgin-speaking environment, Pidgin languages develop into real, Creole, languages that show a marked uniformity across the world, as said in Section 1.3.3.1, forms a robust argument. Equally robust, it seems, is the argument derived from the discovery, some ten years ago, that a deficiency in a single-gene, the so-called FOXP2-gene, may cause severe language and speech disorders and even impair normal language acquisition (Marcus and Fisher 2003).

Yet it is, in a general sense, way too early to attempt to formulate any falsifiable claim regarding *specific* universal properties of human languages that do not have to be learned because they are already there in the prewired brain. The best one can do is follow the inductive course: try to establish what the universal properties are of human languages and their grammars, then see to what extent these properties are derivable from general cognition, from peripheral physical or physiological conditions or from the uses to which language is put in daily practice, and then, finally, see what, if anything, remains in the way of irreducible but real universal properties of languages and their grammars (see Scholz and Pullum 2002). And since we have hardly made a start with these research programmes, the question of any *specific* IH can only be a red herring. But let me give my reasons in detail.

First, we must know WHAT is to be learned in WHAT ecological setting before we can make sweeping statements about HOW it is learned. We are only beginning to find out what language amounts to and what functions it has. The socially committing aspect of speech, for example, discussed in Chapter 4, has been totally neglected in the IH debate. It is generally forgotten that every utterance is the expression of a causally underlying speech-act intent, consisting of a speech-act modality and a mental proposition with regard to which the speech act commits the speaker. The nature and structure of the underlying mental proposition, in particular the fact that a proposition consists in the attribution of a property to one or more entities, is never discussed. Yet this fact deserves at least consideration as a possible source of the structure-dependency of grammatical rules.

Another, related, misconception in the discussion around the IH regards the nature of syntax, which, under the influence of the Chomskyan school of linguistics, is seen as an autonomous set of rules and principles determining the ways morphemes can be put together into larger morphological and syntactic structures. What is forgotten is that such a notion of syntax is not only very recent but also basically wrong, as the brain, when preparing for an act of speaking, does not grab a handful of morphemes and then sees if they can be made to form a grammatically correct structure. It is very clear—and no amount of twentieth-century theorizing has been able to undermine that age-old insight—that utterances result causally from a decision on the part of the speaker to express a given propositional thought under a social-commitment operator in the form of speech sounds, drawing lexical items from the available lexicon and putting them together according to a rule system transforming the propositional input into a well-formed surface structure, which is then realized as sound through the phonology of the language in question. This being so, it makes little sense to look for (an equivalent of) autonomous syntactic rules in the structures and processes of the brain.

This lack of knowledge about what actually makes a language, regrettably unrecognized by the participants in the IH debate who all claim total territorial rights as regards the theory of language, makes it premature to discuss the question of the total package of equipment infants need to acquire the language of their environment.

Second, the arguments in favour of or against any specific IH are, for the most part, mere rhetoric. What one hears is 'Language is so complex that Universal Grammar must be innate' or: 'Children learn their native language in such a short time that the learning process must be supported by an innate predisposition for learning specifically human languages.' Or, on the other side of the fence: 'Constructions must be learned, since they are acquired so late and in such a piecemeal fashion' (Goldberg 2006: 73), or: 'What is truly remarkable is the degree to which human languages differ from one another, given that all languages need to express the same types of messages' (Goldberg 2006: 16, quoting Tomasello 2003). This latter impressionistic statement may well be set against the words of the French grammarian Nicolas Beauzée :

> Constantly following this method I found everywhere the same views, the same general principles, the same universality in the laws common to Language. I saw that the differences of specific languages, the idioms, are nothing but different aspects of general principles or different applications of common fundamental laws, that these differences are limited, based on reason, reducible to fixed points, that therefore all the peoples of the earth, despite the differences of their tongues, speak, in a sense, the same Language, without anomalies or exceptions, and that, finally, one can reduce to a relatively small number *the necessary elements of Language*, and the teaching of all languages, to a simple, brief, uniform and easy method. (Beauzée 1767,1: xvi–xviii translation mine)

Such arguments may have some heuristic value but they are merely impressionistic, of the same nature as the argument, much heard before the 1960s, that one cannot get to the moon because it is so far away that one simply cannot get there. In fact, such arguments are felt to be stronger to the extent that one's imaginative powers are more restricted (see Sampson 1997: 33 for a clear rebuttal of this type of rhetoric). Chomsky's formulations may suggest that he regards, or regarded, this type of argument as 'conceptually necessary' (a key rhetorical term in his 1995), but that just shows his restricted horizon.

Third, next to nothing is known about the learning machinery. The brain structures that enable learning either in a general or in a specific sense are largely unknown. Therefore, the leap to biology, standardly found in the writings of Chomsky and his school, is a leap into the unknown.

Fourth, the hypothesis is ill defined. There is no null hypothesis, given the lack of knowledge regarding the general cognitive endowment of humans. Or, in terms of the historical philosophical debate between rationalists and empiricists, the supposed *tabula rasa* of the empiricists is undefined and, in fact, undefinable, because it is impossible to specify how a nonspecialized, in particular a non-object-oriented, brain would process perceptual input, or how much nonspecialized computational capacity would be required to process any arbitrary symbolic language. Any brain structure resulting in any form of intelligence must be specialized one way or another. Or, to look at it from a closer angle, nobody knows, either in software or in hardware terms, what it would take a nonhuman primate's brain to develop into a brain capable of acquiring a human language. Given the absence of any kind of null hypothesis, there can be no well-defined specific hypothesis.

What happens in practice is that the opponents of specific innate hypotheses try to get rid of language universals, at least of those proposed by the Chomskyans, which pertain to grammars rather than to surface phenomena. The surface-structure universals sought after by the language typologists are treated with less suspicion, provided they look as if they can be attributed to external motivating factors. Meanwhile, as has been said, the antinativists try their best to magnify the differences among languages and to describe the full linguistic competence of mature speakers as fully determined by statistical input frequencies—the theory of usage-based grammars—whereby their notion of full linguistic competence is not elaborated and is, in fact, often reduced to a Mickey Mouse version of what full linguistic competence actually amounts to.[18] Data that show the abysmal inadequacy of their notion of full

[18] Compare, for example, Postal (2004: 6):

Dismissiveness with respect to the difficulty and importance of reaching adequate descriptions can only undermine the proper valuation of real work that actually advances descriptive adequacy. I am thinking here, for example, specifically of work on negative polarity, which I have tried to educate myself about over the last half-dozen years or so. Here it is uncontrovertible that really serious insights

linguistic competence are either not mentioned or are reasoned away with the help of specious arguments or a reversal to pre-Bloomfieldian 'intuitive' syntax. The ideological *a priori* thus favours partisanship rather than the dispassionate advancement of knowledge.

Fifth, those who argue in favour of a specific IH—the Chomskyans—and therefore bear the onus of proof, show a distinct lack of interest in the details of language acquisition by children, who normally start by displaying their prowess in naming objects by means of single-word utterances, soon following this up with using single words for making their desires known. Then come two-word utterances, mostly involving some form of propositional subject–predicate structure. At this stage, they understand much more than they can actually say, especially when spoken to in a way that directly involves them in active linguistic interaction, as opposed to 'adult speech', which largely passes them by. One can observe that children, at quite an early age, 'bootstrap' themselves to higher levels of competence.

It is reasonable to assume that they recognize an intended meaning pragmatically, then actively proceed, at a deep level of unawareness, to reconstruct the processes involved in the utterance heard by way of repetition and analysis-by-synthesis, thus unwittingly forming hypothetical rules about 'the way it's done' in the language of their environment. These hypotheses are then stored in the children's nascent grammar, often with overgeneralization, as with morphological paradigms, or with undergeneralization, as with incorrect trial-and-error uses of deictic and anaphoric expressions. Defenders of a Chomskyan IH should wonder when and how, in this continuous bootstrapping process, such alleged specific universal constraints as the A-over-A principle or the Subjacency Constraint or any form of Island Constraint come into play, but this question is hardly ever raised.

By contrast, those who argue against any form of IH—the cognitivists, constructionists, and the proponents of usage-based grammar—have shown an active interest in the actual details of language acquisition up to full competence (for example, Tomasello 2003). In this, they are helped by the fact that most of the players in this field are, by trade and training, language acquisitionists, not linguists. The onus of proof, however, is not borne by this side of the argument. All the opponents of any specific IH can do to invalidate any version of a specific IH is try to demonstrate the *absence* of any data that might disturb the statistical peace. Perhaps understandably, no great enthusiasm is displayed, in these circles, to ferret out data or details,

have been gained.... But enormous problems remain and no one who even samples the literature on this topic can, I believe, be long deluded into shallow assumptions about the *ease* of finding viable descriptions.

such as the sentences discussed above in (1.4)–(1.7) or (1.15), in early language acquisition or in mature linguistic competence which weaken the position of usage-based grammar as an overall theory and strengthen theories based on the assumption of a derivational mechanism.

Sixth, the Chomskyans' favourite 'argument of the poverty of the stimulus', also called by the pretentious misnomer 'Plato's problem',[19] can be sustained only to the extent that the linguistic input received and processed by young children can indeed be shown to be insufficient as an explanation of their ability to pick up specific, nonobvious semantic distinctions and to produce well-formed sentences accordingly (see, in particular, Pullum and Scholz 2002). If that has been shown, what follows is merely that there must be a rule-driven machinery for the formation of sentences on the basis of a semantic input. What enables early language learners to extrapolate from the linguistic input received and processed to a wider, more encompassing ability to produce and interpret novel utterances is then still an open question. Learners may do so because the rule system they are in the process of building unexpectedly generalizes to cases that are not part of the original input, or because nonlinguistic, more general cognitive factors may, in hitherto unknown or unsuspected or unclear ways, act upon the tender rule system still under construction. It may also be because the brain is equipped with a machinery that is innately constrained in specific, empirically contingent ways. But the poverty-of-stimulus argument is in itself insufficient to decide what causes this ability of language learners to extrapolate from the given input.

In sum, as long as empirical results are inconclusive, discussions about any specific form of innate hypothesis regarding the acquisition of a language by young children had better stay at a level of methodology and conceptual analysis. It seems to me that the discipline is better served with scepticism in regard of any specific innate hypothesis. It can then direct its efforts to the impartial study of language as it is and how it is acquired, instead of drawing premature conclusions.

1.4.2 *The genesis of language*

The same, in principle, applies to the debates that have been raging over the past fifteen or so years about the natural origins of human language in

[19] This is a misnomer because the problem discussed in Plato's dialogue *Meno* deals with *a priori*—that is, conceptually necessary—mathematical (and moral) knowledge, rightly taken by Plato to be retrievable by means of rigorous introspection (although Brouwer's intuitionism has shown that, as far as mathematics is concerned, such a procedure does not necessarily lead to the standard forms of mathematics that dominate the market today). In Chomsky's case, we have to do with contingent, not conceptually necessary, 'knowledge' presumed to be innate as a matter of empirical fact (see Pullum and Scholz 2002: 17, note 5).

the context of new findings regarding the genetically fixed brain structures involved in the use of language and their correlates in other, subhuman, species. The recent explosion of new results in evolution studies based on the discovery of DNA structure during the 1950s, together with the availability of new brain-scanning and imaging techniques, has placed the question of the paleontological and evolutionary origins of human language in a new light. This has naturally enticed many philosophically minded researchers in the field of DNA biology and biochemistry to indulge in speculations in this regard. Unfortunately, however, the term 'speculations' is apposite. Again, as in the case of the Chomskyan innate hypothesis, it is mandatory to know first WHAT human language amounts to before one can develop general and falsifiable theories about HOW it came into being. And here, one has to admit, the authors in question fall short of what is required. Their ideas about language are, on the whole, badly underdeveloped and to the extent that they have read up on linguistics and related subjects, they draw their information largely from the writings of Chomsky and his followers—not a very reliable source, to say the least. But even if they had used better sources, they would not have found the answers needed for their theories, because nobody, whether linguist or nonlinguist, has an adequate and sufficiently complete idea of what natural language actually amounts to.

In a number of recent publications, Botha (1997*a*, 1997*b*, 1998*a*, 1998*b*, 1999, 2003) has laudably and, I think, successfully attempted to sort out the notional and methodological aspects of the recent literature on the evolution of human language, including Chomsky's own semi-creationist fantasies. But Botha, wisely, does not attempt to formulate any empirical conclusions.

In the chapters that follow, I occasionally (as in Section 4.4) venture a thought, explicitly presented as speculative, regarding the evolution of human language, but this book is, as a whole, concerned with finding out what language amounts to, leaving questions of what was needed for it to evolve, or is needed for it to be acquired by infants or second-language learners, to later generations.

2

Virtual reality

2.1 The creative mind as part and creator of ontology

2.1.1 *Natural human ontology is basically intensional*

Although the dominant strands in twentieth-century Anglo Saxon philosophy have been beneficial to the study of language in many ways, they have also done untold harm. This chapter is about the harm, less about the benefits and the harm comes mainly from ontology, or the theory of being. For almost a century, the study of meaning has been shackled by ontological views developed in the context of dominant Anglo Saxon schools of philosophy. Although seemingly remote from the day-to-day practice in academic work, these views have deeply affected the human sciences and the study of linguistic meaning in particular. This chapter is devoted to an analysis of this state of affairs and of its history and of the conditions that must be fulfilled for improvements to be achieved.

One way to break the matter open is to consider the notion of truth in relation to virtual reality. If truth is seen as correspondence between what is thought or said on the one hand and what is the case on the other—a view I have no intention of criticizing at this point—then it is important to be clear on the question of what is meant by the phrase 'being the case'. To answer that question is the task of ontology. Few ontologists deny that in order to specify what is the case, it is necessary to assume that there are 'things', usually called 'objects', or 'entities'. And it is here that opinions parted about a century ago.

Influenced by positivism, whose ontology was restricted to observables and which was already a powerful force in the philosophical study of scientific methodology, but also under the influence of the mathematical, strictly extensionalist, turn in logic, some philosophers began to preach that for anything to qualify as an object it has to satisfy the following criteria:

(i) it is subject to the physical laws of causality;
(ii) it is fully defined, in the sense that it has values on all the parameters entities of that nature are supposed to have—even if not all the values of those parameters are known;

(iii) it has a well-defined identity, in that it is identical with itself and distinct from all other objects;
(iv) it cannot both have and lack a given property (or: have different values on the same parameter) at the same time.

A universe consisting exclusively of such objects is called a strictly extensionalist universe and the objects or entities in such a universe are said to be extensional in the sense that they either have actual, tangible, individual existence, or are reducible to such objects.[1] A strictly extensionalist ontology is a metaphysical theory saying that only objects satisfying the criteria (i)–(iv) have being: they are all there is. Who first proposed such an ontology is not clear, but the name most often heard in connection with it is that of Bertrand Russell. Accordingly, and following Parsons (1980: 1) and Zalta (1983: 5, 172 n.4), I will freely use the adjective *Russellian* in connection with such a philosophical view. The apparent clarity and matter-of-factness of this ontology gave it great popularity in those parts of the world where matter-of-factness was held in high esteem, the English-speaking part of North America being foremost among them. It was there that around the middle of the twentieth century the American philosopher Willard Van Orman Quine summarized, streamlined, and propagated Russellian ontology with great vigour, not to say passion. The result has been that dissenting voices, whether in philosophy or in the areas affected by it, were not or only weakly heard.

My argument is not primarily that this ontology is wrong—though I have grave reservations about it—but rather that it is irrelevant for the study of language. Since, in the eyes of the extensionalist philosophers, everything that 'is' or 'has being' in the universe satisfies the criteria (i)–(iv), speakers will have no choice, when they utter a proposition, but to speak of such objects only (or else they speak about nothing, which is impossible). And the truth or falsity of any proposition expressed must, therefore, be truth or falsity with respect to such objects: if they possess the properties assigned to them in a proposition p, then p is true; otherwise p is false. This makes it impossible to speak about nonexistent objects because there aren't any such things. Talk about 'properties' is all right, *prima facie* at least, since that is construable as talk about sets of extensional objects. But when it looks as if one talks about a nonexistent entity, as when one utters sentence (2.1), then,

[1] Extensionalism in the sense defined here should not be confused with materialism, although the two are not unrelated. Materialism, as a philosophical doctrine, is the doctrine that all there is is ultimately reducible to physical matter, through either strong or weak reductionism (see Section 1.3.1). Strict extensionalists must be materialists but those who, like myself, defend a multilayered natural ontology that includes various forms of nonextensional being may be materialists but need not be.

according to the strict extensionalists, what one actually does is talk about nothing. (Possible-world semanticists may well flare up in protest at this point, since, they will say, possible-world semantics fully allows one to talk about nonexistent objects, these being objects in nonactual possible worlds. I must ask them, however, to hold their fire until I argue, as I do below, that their defence fails.)

(2.1) The god Apollo was worshipped in the island of Delos.

This is the crux of the matter. For even if philosophers maintain that (2.1) is about nothing, this is not the opinion of speakers of natural languages. For them, the sentence just quoted is about the god Apollo, who happens not to exist (not unlike Russell's barber, in the paradox of that name commented on below, is about a barber who cannot exist). Moreover, sentence (2.1) is true for natural speakers, while sentence (2.2) is false:

(2.2) The god Poseidon was worshipped in the island of Delos.

Sentence (2.1) is true despite the fact that the referring expression *the god Apollo* has no referent in the physical world, now or in the past. The state of affairs that makes it true consists in the fact that some people in the past were in a mental state of reverence for a nonexisting entity, the god Apollo, and gave physical expression to that state of mind by erecting a statue representing Apollo, building a magnificent temple complex for him and congregating there from time to time while performing rites and singing hymns in his honour. They did all this for Apollo, not for Poseidon, though neither has ever existed.

In the hands of the philosophers, the concept of truth varies with the kind of ontology they embrace. But ordinary speakers of natural languages are, on the whole, unaffected by philosophical analyses. They have their own intuitive, perhaps even naïve, 'theory' of what is and what is not the case, what kinds of objects populate the world and hence what is and is not true when they make assertions in their language. The philosophers may think they know better, but as long as they are unable to brainwash every speaking individual in this world, from the Amazonian Indians to the members of the Concertgebouw Orchestra, there is an EMPIRICAL REALITY to be discovered by the scientist and constituted by how humans, as a species, construe sense data and construct a world theory. As a necessary corollary, there is the EMPIRICAL QUESTION of how humans construe the notions of truth and falsity. This is the question that occupies us here. What the philosophers have to say about this question may be interesting, it may even be of some help in the setting up of categories in the analysis, but it is basically beside the point. To understand language one has to find out what human world construal underlies the use of language, not what is the best philosophical theory—though the two may coincide.

Some philosophers did worry about natural language. Gottlob Frege, for example, Russell's senior by about a quarter century, was as much a philosopher of language as a philosopher of mathematics. In his view, a sentence such as (2.1) does not express a proposition and lacks a truth-value because the property of being worshipped in the island of Delos is attributed to nothing. Therefore, there is no ascription, hence no proposition, hence no truth-value. However, to uphold this view one has to maintain that sentences like (2.1) or (2.2) lack a truth-value, which is clearly not so. A student of ancient Greek religion will fail on answer (2.2) but pass on answer (2.1). Therefore, Frege's view on reference to nonexistent objects must be deemed inadequate.

Russell tried to save natural intuitions of truth and falsity as much as possible by dispensing with the—otherwise very natural and intuitive—notion of 'aboutness'. With that purpose in mind he presented his famous THEORY OF DESCRIPTIONS, which is, in fact, a programme aimed at the elimination of all definite referring terms, like *the god Apollo*, from the expressions of logic, including those that purport to render the meaning of natural language sentences. A sentence such as (2.3a) should then be read as 'there is a god Apollo, who lives in Kathmandu and who is identical with nothing else', rendered formally as (2.3b):

(2.3) a. The god Apollo lives in Kathmandu.

b. $\exists x[\text{God Apollo}(x) \wedge \text{Live in Kathmandu}(x) \wedge \forall y[\text{God Apollo}(y) \rightarrow x = y]]$

But this analysis is basically inadequate. For besides all kinds of linguistic, semantic, and pragmatic objections, it fails to account for the natural intuition that (2.1) is true. It may perhaps be thought to account for the natural intuition that (2.3a) is false, since no one will doubt that it is not true that there is a god Apollo who lives in Kathmandu and is unique of his kind, but it fails, as a matter of principle, to account for the truth of (2.1). In Russell's analysis, sentence (2.1) comes out as 'there is a god Apollo, who was worshipped in the island of Delos and is identical with nothing else'. But in the concomitant extensionalist ontology, where nothing can be said to have being unless it actually exists, the existential quantifier 'there is' induces an entailment of actual, tangible, existence for individual objects. Sentence (2.1), therefore, should entail the actual, tangible, existence of the god Apollo. And since Apollo never tangibly existed, (2.1) turns out false in Russell's analysis. Yet speakers of English, provided they are sufficiently informed, recognize sentence (2.1) as unambiguously true.

Russell states:

> All propositions in which Apollo occurs are to be interpreted by the above rules. If 'Apollo' has a primary occurrence [has large scope; PS], the proposition containing the occurrence is false; if the occurrence is secondary [has small scope; PS], the proposition may be true. (Russell 1905*a*: 491)

But the theory of descriptions shows in no way how *Apollo* in sentence (2.1) can possibly be assigned small scope in such a way that (2.1) turns out true. Nor is it likely that any form of internal lexical analysis (see Section 8.6) will achieve such a feat, as (2.1) is in no way ambiguous.[2]

Quine, in his (1960), carried Russell's programme further and named it the *programme of elimination of particulars* (or *singular terms*), meant to reduce all occasion sentences, whose interpretation requires contextual or situational backing, to eternal sentences, which, lacking as they do any definite referring terms, can be said to be true or false without any appeal to context or situation. Quine even went so far as to define actual existence in terms of modern predicate calculus. He writes (1953: 13): 'To be assumed as an entity is, purely and simply, to be reckoned as the value of a variable', summarizing this two pages later into his famous slogan 'To be is to be the value of a variable'. This is, of course, extremely odd, mainly because it makes 'being' (or 'existence') dependent on the existence of variables, which are thus given an ontological status beyond and above the objects in Quine's 'desert landscape'. Yet the ontological status of variables and other elements in logico-mathematical formulae is never elaborated by Quine. Apart from this difficulty, however, if, in terms of Russell's analysis, which Quine adopts, sentence (2.1) is taken to be true, which it clearly is, the god Apollo must be the value of a variable and must, therefore, have being. But since Quine emphatically withholds the property of being from Apollo, he is unable to account for the native speaker's intuition that (2.1) is true.

Richard Montague, in his once famous essay 'The proper treatment of quantification in ordinary English' (Montague 1973), thought of a remedy without having to revise the extensionalist foundations of logic. His

[2] This critique of Russell's theory of descriptions is in full agreement with Zalta (1988: 11):

> [O]ne might offer Russell's infamous theory of descriptions as the means of analyzing away the propositions in question. Unfortunately, this theory not only fails to do justice to the apparent logical form of the propositions in question, but more importantly, when applied generally, it fails to preserve the intuitive truth-value of a wide range of other propositions. For example, it turns the historical fact that Ponce de León searched for the fountain of youth into a falsehood. Results such as this suggest that the theory of descriptions is, at best, not general and, at worst, false.

programme of extensionalization of intensions saw intensional—that is, thought-up—entities like Apollo as real in a class of possible worlds not containing the actual world and allowed certain predicates, such as *be worshipped*, to yield truth with regard to such entities. Standard extensional logic could thus be extended in a parallel fashion to worlds other than our own. But we know that this programme foundered. As regards nonexistent objects, it foundered because, as a matter of principle, it cannot account for the fact that one can perfectly well speak about objects with contrary properties, such as a round square cupola, or Russell's well-known barber who is himself a male villager and who shaves all and only those men in the village who do not shave themselves. Since this barber cannot exist (he would have to shave and not-shave himself), he does not occur in any of the postulated possible worlds and thus should not be a possible object of reference. Yet he is one, notably in Russell's paradox.

In a more general sense, it founders because of the problem of PROPOSITIONAL ATTITUDE sentences. This problem stands out most clearly when one considers the embedding under an intensional operator of necessarily true sentences, which are true in all possible worlds, or necessarily false sentences, true in no possible world. In Montague's theory, all necessarily true sentences are substitutable *salva veritate* under intensional operators and so are all necessarily false sentences. But this means that if Thomas holds the necessarily false belief that there are finitely many prime numbers, he must also hold the necessarily false belief that all bachelors are married, even if he himself—the best judge of his own beliefs—insists that he believes the former but rejects the latter. This theory thus has unacceptable consequences for natural language. Dowty *et al.* recognized the fatal force of this obstacle, saying:

> We must acknowledge that the problem of propositional attitude sentences is a fundamental one for possible world semantics, and for all we know, could eventually turn out to be a reason for rejecting or drastically modifying the whole possible world framework. (Dowty *et al.* 1981: 175)

Montague deserves credit for having carried his programme through to its logical conclusion. Had he not done so, the basic inadequacy of the programme would not have stood out with the same clarity. Thanks to him, we now have a situation that is well-nigh unique in the human sciences: a research paradigm has been shown to be basically flawed and—unheard of in the humanities—that conclusion is universally accepted. Yet no alternative solution has come into sight (and possible-world semantics has dwindled accordingly).

The failure to see that artificial ontologies stand little chance of elucidating the nature and function of natural language has distorted the entire picture of what language is and how it functions in the minds and bodies of human speakers who interact linguistically. When one looks at the natural ontology of humans, one quickly sees that natural ontology differs in important ways from the 'desert landscape' ontology defended by our extensionalist philosophical friends.

A similar argument applies to logic. Given that standard modern logic has also been deeply affected by the strictly extensionalist ontology that prevailed during the twentieth century and has, in its turn, decisively shaped the theoretical study of meaning in language, one cannot but conclude that current notions of linguistic meaning profoundly distort linguistic reality. This is why, in Volume II, a beginning is made in investigating the LINGUISTIC REALITY OF THE LOGICAL CONSTANTS that define any logic, so as to come closer to a determination of what 'natural' logic—that is, the logic inherent in natural language—amounts to.

But to revert to ontology, what, one will wonder, is the great difference between the extensionalist version of philosophical ontology and 'natural' ontology that has allegedly caused the distortion of the overall picture of natural language? The answer to that question is both simple and complex. The simple answer is that 'other minds' are an inalienable element in natural ontology, but they have found no place in extensionalist ontologies. Karl Popper, perhaps the most nonconformist philosopher in the Anglo Saxon world, does recognize the separate status of 'other minds', and minds in general, making them form his World 2, as opposed to his World 1 of extensional objects. But Popper was not unique in this. The point is also made by the nonconformist British philosopher John N. Findlay, who speaks of 'realism' when referring to what is called strict extensionalism here, and of 'idealism' when we speak of an ontology containing virtual reality:

> Realism in English-speaking countries has in general contented itself with exposing the fallacies of idealistic arguments, with showing that it is perfectly possible for our minds to gain a foothold in a world of things to which reflection is alien; it has then proceeded to deal with the world of objects, among which minds are included as an eccentric species, without troubling itself much further about epistemology. (Findlay 1963: 2)

Strictly extensionalist ontologies regiment the actual and any other possible world into sets of objects of the kind defined by the criteria (i)–(iv). Apart from the question of whether this is tenable in the light of the many abstractions and reifications we work with in our daily lives, one monumental fact, essential for natural ontology, is overlooked in such ontologies. This is the fact that at least in the actual world—and, one would expect, therefore also in many other possible worlds, if such a notion makes sense—certain 'objects' are found, in particular higher vertebrate animals including humans, that are endowed with the faculty of thought.

The complex answer begins by recognizing that the fact that these animals (humans) think has enormous consequences for natural ontology. One consequence is that, by definition, thinking implies the building of theories about the material world (and about other things as well, such as mathematical necessities). Thinking minds, therefore, produce theories—Karl Popper's World 3.

But thinking also implies the ability to call up, and thus call into some form of shadowy being, virtual entities and virtual events and situations about which they think and about which they say and ask things and build theories. These virtual entities or objects are part of the natural ontology of thinking animals. Perhaps, Popper should have set up a separate 'World 4' for them, but he did not. But even if he didn't, the fact that there is this 'fourth world' means that when thinking animals have language, as humans do, they speak with an equal ease and naturalness about entities they consider nonexistent as they do about entities they consider existent, making true and false assertions about them.[3]

This undeniable fact stares one in the face. Professional philosophers, logicians, and semanticists have tried to deny it for a long time, but it keeps coming back. It simply is true that the nonexistent Muses are nine in number, that nonexistent deities are worshipped, or that children are sometimes frightened by nonexistent bogeymen. If statements to that effect were statements about nothing they could not be either true or false. This may be an unwelcome message for those who cherish clinically regimented ontologies, but it cannot be ignored if it is one's aim to gain an adequate insight into the nature of human language and cognition.

Why philosophers, logicians, and semanticists have insisted with such force and such passion on a strictly extensionalist universe is an interesting question. The answer must probably be sought partly in the fear, which arose about a century ago, that the things we hold forth about are of such diversity that an ontology accepting them all as entities or objects would be in danger of exploding. We expatiate about the military-industrial complex, about the beauty of nature, about the average Australian, about differences and similarities, about gods and devils, about mysteries, problems, and solutions,

[3] Humans also QUANTIFY with an equal ease and naturalness over entities they consider nonexistent as they do over entities they consider existent, as the following example shows (Bill Vallicella to Edward Buckner, http://maverickphilosopher.powerblogs.com/posts/1177106402.shtml): 'Not every logically impossible object is identical to the round square; ergo, there is at least one logically impossible object that is not identical to the round square'. The conclusion must be, as Vallicella and Buckner observe, that logic is metaphysically neutral.

about the triangle in general, and so on ad infinitum. If all such 'things' were to be accepted as objects, the universe would, these philosophers feared, become a tangle that could never be unravelled—Plato's famous beard. It was felt, therefore, that a proper theory of the world should stick to concrete entities and do away with fantasies and even with abstractions. Only on the assumption of a 'reality' cleansed of such 'disorderly elements', as Quine has it, would it be possible to devise formal systems capturing its workings and structure. Unfortunately, however, there are some 'disorderly elements', namely minds, that must be reckoned to belong to any extensionalist version of the actual universe, yet have so far resisted attempts at capturing them in terms of a formal system.

A second, related, factor was the flight from the mind and anything mental, which was beginning to gain momentum about a century ago. This flight—understandable enough in the light of what has been said—soon led to behaviourism, first in psychology, then throughout the human sciences. One reverted to basic observables, very much in the spirit of positivism which was then riding high in the sciences in general, but was soon found to be unworkable as its ontology was too barren to serve the sciences, just as it was too barren to suit most varieties of extensionalist ontology. As one knows, behaviourism had a devastating effect in the human sciences and was largely abandoned around 1960, having lost its positivist basis. Yet extensionalism, a weaker form of positivism but still ill at ease with mental phenomena, has lasted till the present day. This has created a 'professional' picture of human language in which, paradoxically, the mind figures prominently as the seat of language, yet is totally absent from its semantics (other than in the guise of sterile formalisms that stand in no relation to whatever else goes on in the mind). In the present book I contend that, as far as semantics is concerned, this picture, although backed by the prestige of what are taken to be the 'professionals', is fundamentally misguided and that to put it right one has to go to the ultimate roots of the affliction—that is, to the very foundations of language, logic, and ontology. The reward of this expedition is a surprisingly elegant and arresting perspective which is not only new and intriguing but also re-establishes historical links with the long and fascinating Western tradition in philosophy and language studies—links that were unwisely severed during the twentieth century.

2.1.2 *The social reality of languages and dialects*

Paradoxically, the kind of strictly regimented extensionalist ontology described above forbids talk about language in general or languages in particular—a fact which, to my knowledge, has never been brought up. The ontological status of language in general, and in particular of each separate language and its internal varieties, with their forms and meanings, has never been clarified. This is a serious point, since neither languages nor linguistic forms nor linguistic meanings are the kind of object that would pass the extensionalist philosophers' muster. In particular, the criteria (i) and (iii) cited at the outset of this chapter are not satisfied by languages, linguistic forms, and linguistic meanings, which are only partially subject to the physical laws of causality and often lack a clearly delimited identity setting them off against all other objects of the same category. Every linguist knows that languages are usually not clearly distinguishable from one another or from their dialects and that it is often impossible to tell different words apart. Phonemes, in particular, though relatively well defined as perceptual—that is, mental—units of speech sound interpretation, are notoriously elusive with regard to their physical identification. Little wonder, therefore, that the logically oriented branches of language study never had a grip on phonology, where the units of analysis—the phonemes—escape identification in terms of an extensionalist ontology. The same goes for lexical meanings, whose identity, if any, still escapes us, as is shown in Chapters 8 and 9: there is precious little literature on lexical meaning that has emanated from formal semantics and similar logically oriented endeavours in linguistics.

In practice, questions such as these are avoided in logical and formal-semantic quarters. All one hears is that a language is a formally well-defined system of strings of symbols, but the corresponding ontological status remains unaccounted for. The best one can make of it is to assume that, in the disciplines in question, any given language is treated as a mathematical object—that is, an object in an assumed realm of mathematical reality. But no statement to that effect is to be found in the relevant literature. One is ultimately left with the impression that language forms a separate level of reality, descended from the sky and establishing relations of truth and falsity with regard to the 'desert landscapes' of the formally defined 'worlds'.

In early generative grammar, languages were treated, as 'theories', in both a formal and a scientific sense, which would probably make them part of Popper's World 3. But languages are not 'theories' in any possible sense of the word. Languages are, if anything, sets of conventions, possibly involving bits of formal computational machinery, regarding the expression of thoughts (or, rather, the manifesting of various forms of socially binding position-taking vis-à-vis listeners with regard to any given proposition *p*—the 'intents' of Chapter 4). These conventions are, moreover, subject to norms and values relating to the propriety of the expressions used. But to what 'world' do such convention systems belong?

One way to approach this problem area is to present the following dilemma. It makes sense to distinguish between a natural language L as an *object* on the one hand, and *command of L* on the other (where L is seen as a set of conventions subject to norms and values, and where 'command of L' is to be understood as the ability to use L as a means of expressing and comprehending intents, as indicated above). Such a distinction makes sense because one can both be competent in a language and still make new discoveries about it. Thus, a competent speaker of English may discover new words or new meanings of words that were unknown to him before. He may discover, for example, that there is a socially and geographically defined group of speakers who identify themselves, and are identified by others, as speakers of English and who use the word *gormless*, unknown to many other speakers of English, in the meaning 'lacking intelligence' or 'stupid', although only in informal contexts and preferably in a context of British English.

On the other hand, however, there is the curious fact that a language cannot exist unless there is at least one person who has a command of it. Being in the command of at least one person is a necessary condition for any language to exist: no speaker, no language. This condition is of a cognitive nature, since 'having command of a language'—often called 'knowing a language'—is a cognitive state. The extensional objects defined by the conditions (i)–(iv) formulated at the outset of the present chapter are not considered subject to such a condition. Unless, of course, like Bishop Berkeley, one adheres to a fully idealist ontology where objects are thought to exist only in virtue of being the object of a cognitive state and where God is believed to save the permanence of objects by being omnipresent all the time. But such an idealist ontology is a poor candidate for the status of 'natural ontology'. We thus face the *prima facie* paradoxical situation that a distinction is made between an object L and cognitive possession (or 'knowledge') of L, while L cannot exist without there being at least one person having cognitive possession of (or 'knowing') it.

The answer seems to lie in the concept of SOCIAL REALITY. Social reality is typically characterized by the fact that it cannot exist without there being a community of individuals who have created it and uphold it: no community, no social reality. When there is only one person with a command of a language L, this person cannot discover anything new about L. But when there is a, possibly large, number of persons competent in and using L, then L exists as a social reality within that group of people. While one speaker suffices to make a language L, a community of speakers makes L part of the social reality of the group in question. Now individual speakers may make new discoveries about L, especially when the community is so large that individual members cannot be acquainted with what lives in each subgroup.

We thus say that being in the competence of and being usable by the members of a community is a necessary condition for any language to exist as part of the social reality of the community in question. And, as we know, virtually all natural languages are in the competence of and used by a, more or less large, group of speakers and thus are part of the social reality within the community involved. A community whose members are competent in and use what they consider to be 'the same language' forms a speech community.

But what, then, do we mean by the term *social reality*? The 'objects' of social reality, including natural languages, come about as the result of shared norms, values, and conventions in a socially definable community. Just as there is no language L if nobody is competent in L, there is no social reality if there is no community whose members have incorporated a common set of norms, values, and conventions into their mental machinery. Obligations, reputations, values, taboos, status, wealth, good manners, traffic rules, laws are typical objects of social reality, along with languages—although natural languages are unique among objects of social reality owing to their complexity and owing to the fact that they are largely inaccessible to introspection. (Popper might have set up a World 5 for this class of elements in natural ontology.)

This concept of social reality is obviously in need of a great deal of further conceptual analysis and refinement against the background of what is known about the complexities of sociological phenomena. It seems clear, for example, that processes occurring in social reality are not fully reducible to the individuals making up the community in question: in social reality, the total does not equal the sum of the parts. Social structures, states, and processes appear to have a degree of autonomy that has not so far been investigated. But such an investigation cannot be our aim in the present study. All we can do here is point out that when we speak about a language or a dialect, we speak about a bit of social reality in the community served by it.

It would seem that the ultimate, basic, defect of formal semantics and related logic-based approaches to the analysis of natural languages lies in their failure to acknowledge that languages are primarily objects of social reality and not mathematical objects. Formal semantics and its kin treat languages as objects belonging to mathematical reality, or at least as objects defined by mathematical properties. But this is far too strong a position. Although it is no doubt correct to say that languages display many surprising and intriguing mathematical properties, they are not mathematical objects in whatever sense of the term. Nor are they defined by their mathematical properties.

2.2 The reality of virtual being

It seems undeniable that in natural language, extensional and virtual objects are referred to and quantified over in exactly the same way. This is a fact which the extensionalist view has always been unable to account for. The extensionalist view immediately produces the quandary of true negative sentences, as Russell was quick to see a century ago. Sentence (2.4) is no doubt true:

(2.4) The god Apollo does not live in Kathmandu.

In 1905, Russell quipped about the alleged baldness of the present king of France:

> By the law of the excluded middle, either 'A is B' or 'A is not B' must be true. Hence either 'the present King of France is bald' or 'the present King of France is not bald' must be true. Yet if we enumerated the things that are bald, and then the things that are not bald, we should not find the present King of France in either list. Hegelians, who love a synthesis, will probably conclude that he wears a wig. (Russell 1905*a*: 485)

We may paraphrase this for the case at hand by saying that if one searches among the inhabitants of Kathmandu one will not find Apollo there. Yet if one looks among those who live elsewhere one will not find him either. Nor will a Hegelian synthesis that makes him be of no fixed abode help much, because Apollo will be equally absent from the vagrants of this world. The issue centres around the question of who or what is the recipient of the ascription of the property of living in Kathmandu in any true or false utterance of the sentence *The god Apollo lives in Kathmandu.*

This, however, is not the end of the extensionalist's woes, since the quandary of true negative sentences is parallelled by the quandary of true positive sentences, unattended to by Russell and almost the entire tradition of twentieth-century formal semantics. Yet these clinch the issue. Sentences like (2.1), *The god Apollo was worshipped in the island of Delos*, are a fatal thorn in the flesh of modern logic and the semantics based on it. They are fatal for Frege, because the obvious truth of (2.1) proves that there is an underlying proposition—propositions being the primary bearers of truth-values. They are fatal for Russell because his theory of descriptions forces him to accept actual existence for the god Apollo. They are fatal for Quine because to get truth for (2.1), Apollo has to be the value of some variable, which would give Apollo the 'being' Quine is anxious to deny him. They are not directly fatal for Montague, provided one grants him his extravagant possible world ontology. But they are indirectly fatal, as has been shown, because Montague's analysis founders on cases of equivalence, including cases of necessary truth and necessary falsity, where the theory predicts substitutivity *salva veritate* but language fails to oblige.

2.2.1 *Back to Meinong and Russell*[4]

To disentangle this web of puzzles and paradoxes, we must retrace the steps taken by the logicians, back to definite referring terms, back to intensional or virtual entities and back to the central role of the mind in the production and comprehension of utterances. In short, we need to go back to the beginning of the twentieth century. So let us zoom in on that period and focus on Cambridge in England, where we see Russell writing his *Principles of Mathematics* which appeared in 1903. He still adheres to the old mode of thought which leaves room for a certain amount of idealism allowing for virtual beings in virtual realities, but is on the brink of embracing the new mode of thought, which is much more positivist and says that being and tangible existence are of a piece: nothing IS that does not EXIST. His specific dilemma is whether to exclude or include the mind in determining the status of propositions that are true or false with respect to 'what is'. The new ideology is already beginning to break through:

> In the discussion of inference, it is common to permit the intrusion of a psychological element, and to consider our acquisition of new knowledge by its means. But it is plain that where we validly infer one proposition from another, we do so in virtue of a relation which holds between the two propositions whether we perceive it or not: the mind, in fact, is as purely receptive in inference as common sense supposes it to be in perception of sensible objects. The relation in virtue of which it is possible for us validly to infer is what I call material implication. We have already seen that it would be a vicious circle to define this relation as meaning that *if* one proposition is true, *then* another is true, for *if* and *then* already involve implication. The relation holds, in fact, when it does hold, without any reference to the truth or falsehood of the propositions involved. (Russell 1903: 33)

This passage is about the fact that entailment is a metaphysical property of propositions whose meanings have been defined in advance: once meaning is fixed, entailments follow as a matter of metaphysical necessity. Propositions are hypostatized as elements in their own right, 'descended from the sky'. Entailment, likewise, is seen as not being subject to a criterion of psychological reality, since there are entailments in standard logic that are obviously not psychologically real. Necessarily true sentences, for example, such as *All bachelors are unmarried*, are entailed by any sentence, and necessarily false sentences,

4 In writing this section I was helped considerably by my former student Cor Burger, who has done much work on the relation between Russell and Meinong.

such as *All bachelors are married*, entail any sentence, including themselves. This much has become part of standard logical teaching, and I have no wish to raise any objection in this regard—but for the observation, elaborated in Chapters 1 and 3 in Volume II, that some entailments clearly are subject to a criterion of psychological naturalness and thus constitute observable data.

Yet Russell, and with him the entire logical tradition of the twentieth century, devotes remarkably little thought to the fact that meanings must be fixed before entailments can come about, and that lexical meanings at least are mentally fixed by speakers in the lexicons of their languages. Although it is meaning that generates entailments, Russell, and the logicians following him, are not interested in the question of the cognitive reality of meaning in natural language. Not even when logicians begin to look at the semantics of natural language is any interest shown in the cognitive roots of meaning, as it is believed that standard twentieth-century logic and the clinically regimented but literally mindless worlds to which it applies already provide the model for semantic analysis and description. The performance of natural-language speakers in regard to deriving entailments is measured against the standards of modern logic which have been set in advance. The logic has become primary and language has to meet its standards. In actual fact, however, meanings are fixed first and the metaphysics of truth and logic follows suit, depending on how the meanings have been fixed.

The mere fact that Russell describes the mind as being 'as purely receptive in inference as common sense supposes it to be in perception of sensible objects' should raise the alarm, since one knows that the mind is anything but 'purely receptive'. On the contrary, the mind actively pre-shapes the ways in which humans perceive and interpret the world.

Although Russell, in the passage quoted above, eliminates the mind from logic, he does not yet eliminate the mind from ontology. Later on in the book he attributes to 'thought' the power to create beings, including propositions:

> *Being* is that which belongs to every conceivable term, to every possible object of thought—in short to everything that can possibly occur in any proposition, true or false, and to all such propositions themselves. Being belongs to whatever can be counted. If *A* be any term that can be counted as one, it is plain that *A* is something, and therefore that *A* is. '*A* is not' must always be either false or meaningless. For if *A* were nothing, it could not be said not to be; '*A* is not' implies that there is a term *A* whose being is denied, and hence that *A* is. Thus unless '*A* is not' be an empty sound, it must be false—whatever *A* may be, it certainly is. Numbers, the Homeric gods, relations, chimeras and four-dimensional spaces all have being, for if they were not entities of a kind, we could make no propositions about them. Thus being is a general attribute of everything, and to mention anything is to show that it is.

Existence, on the contrary, is the prerogative of some only amongst beings. To exist is to have a specific relation to existence—a relation, by the way, which existence itself does not have. This shows, incidentally, the weakness of the existential theory of judgment—the theory, that is, that every proposition is concerned with something that exists. For if this theory were true, it would still be true that existence itself is an entity, and it must be admitted that existence does not exist. Thus the consideration of existence itself leads to non-existential propositions, and so contradicts the theory. The theory seems, in fact, to have arisen from neglect of the distinction between existence and being. Yet this distinction is essential, if we are ever to deny the existence of anything. For what does not exist must be something, or it would be meaningless to deny its existence; and hence we need the concept of being, as that which belongs even to the non-existent. (Russell 1903: 449–50)

What Russell defends here, following the Austrian philosopher Alexius Meinong, is the ontological reality, though not the actual existence, of virtual objects, and thus the possibility of referring to them and uttering true or false propositions about them—a far cry from the 'programme of elimination of particulars' put into motion by himself in his celebrated article 'On denoting' of 1905 and later continued by Quine.

Russell (1905*a*) takes the further step of removing the mind, or rather, mentally constructed objects, from ontology. There he embraces the 'existential theory of judgment'—that is, purely extensional logic and semantics—that he so strongly rejected in the passage just quoted. Curiously, he does so without any comment or explanation, as if he has always been of this new opinion and as if that opinion is so obviously correct as not to require any further argument.

This new, strictly extensional, view of logic will stay with him throughout his life,[5] yet always tarnished by unclarities. One reads, for example:

Logic, I should maintain, must no more admit a unicorn than zoology can; for logic is concerned with the real world just as truly as zoology, though with its more abstract and general features. (Russell 1920: 169)[6]

[5] Later in life, especially in his (1940), Russell did indulge in (not very substantial) musings about the psychological or epistemological conditions that make people use certain logical forms such as negation or disjunction, or existential or universal quantification. This anticipates what thirty years later will manifest itself as pragmatics. The logic and the corresponding ontology, however, remain unaffected.

[6] Strangely, as pointed out to me by Cor Burger (see note 4), Russell insists, in a letter he wrote (in German) to Meinong on 15 December 1904:

I fully agree with the view that mathematics is object-theory [which, for Meinong, meant logic; see note 9]; it even is one of the main theses of my 'Principles'. (Der Ansicht, die Mathematik sei Gegenstandstheorie, stimme ich völlig zu; sie ist sogar eine der Hauptthesen meiner 'Principles'.)

This is seriously wrong, as any competent modern logician will agree. Logic has nothing to do with the world as it really is, but only with what must of necessity be true *if* some given sentences are true—where necessity is taken in the Kantian sense of analytical necessity which holds in virtue of meanings, not in virtue of what is actually the case. Logic has no business with what is actually the case (general features or not), that being the province of practical decision making or of science, or the police, for that matter. Its business is with entailments—that is, with the *conditions* of saying in truth that something is the case given that something has been truthfully said to be the case. The issue is not what the actual, extensional world is like, but rather under what conditions possible states of affairs are accepted as truth makers. Unicorns living in Hampstead Heath are as much possible truth makers as fellows living in Magdalen College.

If Russell had meant to say that (his) logic requires the kind of extensionalist ontology he defends, no matter in what possible world, that would have made more sense. It would have described the real bone of contention in the prevailing tradition of twentieth-century logic. In modern logic and the semantics derived from it, only extensional being—that is, the being of causally determined, preferably physical, and in any case fully defined objects with their own specific identity, and subject to the principle of contradiction—can make truth. In Meinongian intensionalist philosophy, intensional or virtual being can also make truth, under strictly defined conditions. Since, as will become amply clear, in natural language virtual objects and actual objects are equally good at making truth, it makes sense to follow Meinong, and not Russell, in this respect.

Meinong is a key figure in this context, because the question of virtual being is the central issue in the entire tangle of questions we are dealing with, and because Meinong was the only philosopher of the period, and the last one till the present day, with a well-developed ontology of virtual being. He lived in a philosophical environment that gave rise to a range of, mostly Continental-European, developments in philosophy during the twentieth century. He and Husserl both studied with Brentano and thus stood at the cradle of phenomenology, and therefore also of the many subsequent, rather less disciplined, forms of philosophy and quasi-philosophy witnessed over the

Then, in a letter to Meinong of 5 November 1906, he writes

> In particular, I agree with you when you say that mathematical knowledge is independent of what actually exists and belongs properly to object-theory. (Im besonderen stimme ich Ihnen bei, wenn Sie behaupten, Mathematik sei 'daseinsfreies Wissen' und gehöre eigentlich der Gegenstandstheorie zu.)

From these statements it simply follows that logic is NOT concerned with the real world—as is indeed the standard view held by professional logicians. (For the correspondence between Russell and Meinong, see Kindinger 1965.)

past six or seven decades. (Fortunately, the study of language in the twentieth century was hardly influenced by these latter 'philosophies', which would have caused even greater damage than that done by the positivist excesses of Anglo-Saxon philosophy.)

Meinong himself, however, although often obscure and not obviously consistent, steered clear of the cloudiness that subsequent generations of Continental-European philosophers fell victim to—which is probably why he never gained much influence in a Europe steeped in Heideggerian clouds, existentialism, and postmodernism. Yet despite his relatively low profile in twentieth-century philosophy, we must recognize that the study of language will benefit considerably from his insights.

At the same time, however, it must also be recognized that there is an important difference between Meinong's way of thinking and the method of analysis applied here. Whereas we feel inclined to rank Meinong among the speculative thinkers in the traditional philosophical sense (despite his own protestations to the contrary), we have the advantage of ordinary-language philosophy developed and practised at Oxford for a few decades immediately following the Second World War. Following in the footsteps of the Oxonian ordinary-language philosophers, we try to avoid philosophical speculation, taking the facts of language and language use as empirical data that place restrictions on any form of linguistic theory, including theories of the natural ontology and the natural logic underlying the use of language. In effect this means that although we are certainly helped by Meinong's analyses in many ways, we have to part company with him in other respects. Since, however, this is not a study of Meinong's philosophy, I will forgo a detailed analysis and appraisal of his ontology and will, in general, refer to him only when it is felt that his insights contribute to what seems to be a proper analysis of the ontology underlying the use of language.

In an amazing about-face that happened somewhere during 1905, Russell, having first enthusiastically embraced Meinong's theory of nonexistent objects, then suddenly rejected it, ushering in the century of extensionalist logic and semantics. Given Russell's enormous influence, it is important to look at the details of this event and at his motives for doing what he did.

In Russell (1903) and other writings by Russell of that period, Meinong is frequently referred to with the greatest respect. A telling passage is the following:

> The work of Herr Meinong on Weber's Law, already alluded to, is one from which I have learned so much, and with which I so largely agree, that it seems desirable to justify myself on the points in which I depart from it. (Russell 1903: 162)

Yet in Russell (1905*a*) all respect for Meinong is gone. Here Meinong is depicted as a fool who holds 'a difficult view' and fails to see that he violates the first principle of ontology and logic, the principle of contradiction:

[Meinong's] theory regards any grammatically correct denoting phrase as standing for an *object*. Thus 'the present King of France,' 'the round square,' etc., are supposed to be genuine objects. It is admitted that such objects do not *subsist*,[7] but nevertheless they are supposed to be objects. This is in itself a difficult view; but the chief objection is that such objects, admittedly, are apt to infringe the law of contradiction. It is contended, for example, that the existent present King of France exists, and also does not exist; that the round square is round, and also not round; etc. But this is intolerable; and if any theory can be found to avoid this result, it is surely to be preferred. (Russell 1905a: 482–3)

Is this a fair rendering of Meinong's views? Findlay has his doubts:

Russell was early aware of Meinong's great merits, and devoted the generous, brilliant 'Meinong's Theory of Complexes and Assumptions' (*Mind*, 1904) to their exposition and partial defence. It is well known that Russell's Theory of Descriptions, that famous and historically important piece of logical analysis, was, in part at least, an attempt to provide an alternative to Meinong's doctrine of *Außersein*, of extraexistential objectivity. Russell also discussed Meinong's theory of content in the first chapter of *The Analysis of Mind*. Unfortunately Russell was far too concerned to advance from Meinong to his own notions and conclusions to bother to get Meinong quite straight, and the accounts he put into circulation of Meinongian contents as consisting of sense-data and images, and of Meinong's non-existent objects as 'subsistent', are simplifying travesties of Meinong's complex opinions. (Findlay 1963: xi–xii)

Findlay paraphrases Meinong's not always entirely perspicacious wordings in clear language as follows:[8]

Meinong bases his distinction between the content and the object of an idea in the first place on the fact that we are able to think of objects which do not exist. Of these non-existent objects he distinguishes three types: we have such objects as golden mountains, whose non-existence is merely a matter of brute empirical fact; then we have such objects as round squares, which cannot exist because they involve a contradiction; finally we have such entities as the equality between three and three, or the diversity

[7] Russell might have commented on his choice of the term *subsist* in rendering Meinong's position. In the context of this passage, *subsist* is probably best read as 'be reducible to material substance'. The problem is that Meinong himself used the German verb *bestehen* not for 'exist', as one would have expected, but for the form of being that adheres to abstractions and reifications that form a part of the real world—a form of being for which Findlay uses the verb *subsist*.

[8] For the real devotee, Findlay refers to p. 382 in Meinong's *Über Gegenstände höherer Ordnung und deren Verhältnis zur inneren Wahrnehmung* of 1899, reprinted in *Gesammelte Abhandlungen*, vol. II.

between red and green—even in ordinary language we recognize that such things may subsist (*bestehen*), but not exist 'like a house or a tree'. (Findlay 1963: 18)

The entire edifice of Meinong's ontology is a great deal larger and more comprehensive, but, as far as it goes, the above quotation appears to do justice to his views. Besides the basic entities that exist extensionally ('existieren'), Meinong also accepts HIGHER-ORDER OBJECTS ('Gegenstände höherer Ordnung'), which comprise not only sets of objects, but also abstractions or reifications such as 'the difference between A and B' or 'the prime minister's popularity with the voters'. Such objects do not EXIST but they SUBSIST ('bestehen'; see note 7) in the extensional world of existing entities, in the sense that they owe their being to entities that do exist. Apart from these existing and subsisting entities, however, there are also virtual objects or entities, like the golden mountain or the philosophers' stone, which are nonexistent ('daseinsfrei'), and thus incompletely defined, although they could have existed and would then have been fully defined. These are not part of the extensional world but merely imagined and thus part of what we call a virtual world. Impossible objects, such as round squares, are also objects, although, in Meinong's view, these have no 'being', as he requires logical and semantic consistency in thinking or saying that a state of affairs X obtains for elements or objects in X to deserve the property of being. The latter is indeed 'a difficult view', if that is what one is looking for when reading Meinong. Along with Zalta (1988: 135–8) we reject this view, as it seems highly incongruent to construe objects that have no being whatsoever.

Leaning over backwards to grant Meinong his ontology of subsistent being, Russell extends his theory of descriptions to reifications:

We can now see also how to deny that there is such an object as the difference between A and B in the case when A and B do not differ. If A and B do differ, there is one and only one entity x such that 'x is the difference between A and B' is a true proposition; if A and B do not differ, there is no such entity x. Thus according to the meaning of denotation lately explained, 'the difference between A and B' has a denotation when A and B differ, but not otherwise.... Thus out of any proposition we can make a denoting phrase, which denotes an entity if the proposition is true, but does not denote an entity if the proposition is false. E.g., it is true (at least we will suppose so) that the earth revolves round the sun, and false that the sun revolves round the earth; hence 'the revolution of the earth round the sun' denotes an entity, while 'the revolution of the sun around the earth' does not denote an entity. (Russell 1905*a*: 490–1)

For Meinong, however, the denoting phrases *the revolution of the earth around the sun* and *the revolution of the sun around the earth* both denote objects—the former a subsisting object, the latter a virtual, nonsubsisting object. This being so, it seems that one may indeed conclude that Russell's rendering of Meinong's ontological views as given in his article 'On denoting' of 1905 deserves to be called a simplifying travesty.

But the next, more important, question is whether Russell's objection 'that such objects, admittedly, are apt to infringe the law of contradiction' is justified. Again we must take sides against Russell. It may be said in Russell's defence that Meinong's wording is not always as clear as one would have wished it to be. On superficial reading it does appear as if Meinong attributes extensional properties to intensional, or virtual, objects, as when he says that the property 'golden' adheres to the nonexistent golden mountain. But what he means, obviously, is not that the golden mountain is literally made of gold, only that the property of being made of gold has been mentally attributed to it: it has been thought up as being made of gold, so that it is made of gold in the virtual world in which it figures. This appears, for example, from the following passage on the first page of Meinong's essay 'On object-theory' (Über Gegenstandstheorie),[9] published in Meinong (1904) and reprinted in Meinong (1988):

> That one cannot know without knowing *something*, or more generally, that one cannot judge, not even imagine, without judging *about something* or imagining *something*, is the most natural consideration in the world, which already provides a basic element in the analysis of these experiences....
>
> But even those who disagree with my view that feelings and desires are nonautonomous psychological facts insofar as representations (Vorstellungen) are indispensable 'psychological prerequisites' for them, will immediately agree that one rejoices *over something* or is interested *in something*, that, at least in most cases, one does not simply want or desire but one wants or desires *something*. In other words, no one will deny that this specific 'being directed at something' is so extraordinarily often part of psychological thought that it is at least reasonable to surmise that it constitutes a characteristic difference between the psychological and the nonpsychological. (Meinong 1988: 1; translation mine)

The assignment of properties to objects, whether existent or nonexistent, is always mediated by some operator of thought. As regards nonexistent objects, this holds not only for objects that happen not to exist but could have existed, but also for impossible objects like round squares, which are thought up as being both round and square, and thus stand no chance of ever being instantiated as an actual object. In this respect, a round square is just like

[9] Meinong used the term *Gegenstandstheorie* (object-theory) for what is commonly called *logic*. In his letter to Meinong of 15 December 1904 (see note 6), Russell writes: 'I have accustomed myself to using the name "logic" for what you call "object-theory", and the objections you raise on your pages 20ff. against this usage appear to me hardly decisive. But this is a mere detail.'

Russell's own impossible barber who, being a male villager, shaves all the men in the village (and therefore, in Russell's analysis of the word *all*, also himself) that do not shave themselves.

Humans often think up entities, events, or situations that are assigned incompatible properties. Lewis Carroll's stories, telling of Alice's dreams, are full of them, and so are science fiction stories, especially when they treat one to time machines. But does a recognition of this fact amount to an infringement of the principle of contradiction? Of course not. All it means is that people sometimes infringe that principle in their thinking or their imagination, thereby creating virtual 'worlds' that will never be actual. The Aristotelian principle of contradiction is primarily an ontological principle, applying to any world that can lay claim to actuality. The classic formulation is as follows:

> It is not possible for the same property to adhere and not to adhere to the same entity at the same time and in the same respect. (Aristotle, *Metaphysics* IV,3, 1005b19)

By itself, this says nothing about the human capacity to think up states of affairs that violate this basic principle, as long as no truth is claimed for any such thought (see Zalta 1988: 17). But it does say something about a sound logic, as Aristotle himself clearly saw:

> That it is impossible to affirm and deny simultaneously the same predicate of the same entity is implicit in any logical proof. (Aristotle, *Posterior Analytics* I,11, 77a10–11)

It must be admitted, meanwhile, that Meinong's account gives rise to deep ontological, logical, and linguistic questions. In particular, there is a problem with the truth conditions of sentences like *The golden mountain is not made of gold.* In a sense, this sentence is true, since nonexistent objects are not made of any material. In another sense, however, it is false, insofar as the golden mountain has been thought up as being made of gold and not of some other material. Likewise with a sentence like *Sherlock Holmes is an Englishman.* In a sense, this sentence is false, since Sherlock Holmes is a fictitious character and the predicate *Englishman* requires the actual existence of its subject-term referent for truth. In a different sense, however, it is true—in the sense in which *Sherlock Holmes is a Frenchman* is false (this question is discussed in detail in Chapter 5).

There is much more to the question of virtual reality than meets the eye. Russell's answer to this question is not only less than subtle but also lacks precisely that quality of philosophizing that he so strongly insists on, a robust sense of reality and truth. It would have been easy for Russell to protect Meinong from the allegation that his ontology violates the principle of contradiction. All it would have taken was a closer and more sympathetic reading of Meinong's texts and a modicum of logical creativity.[10] Significantly, no such allegation is found in Russell's numerous earlier publications about Meinong, written when he still admired the man. In Russell (1904), in particular, it is totally absent. But in Russell (1905*b*), written after 'On denoting' and published along with it in the same issue of *Mind*, it is there, repeated from his (1905*a*).[11] What happened to make Russell change his mind so suddenly?

My personal view is that, besides his obvious joy at having hit upon his theory of descriptions (which he himself admired greatly), it was his decision to cut through the metaphysical tangles he increasingly found himself caught

[10] Carnap comments as follows:

> Meinong's conception has been critically discussed by Russell and then rejected. Russell's chief reason for the rejection is that the impossible objects violate the principle of contradiction; for example, a round square is both round and nonround, because square. Russell is certainly right in the following respect: Within the logical framework of our ordinary language, we cannot consistently apply the conception of impossible things or even that of possible nonactual things. And, as far as I am aware, neither Meinong nor Lewis nor any other philosopher has constructed or even outlined a language of a new structure which would accommodate those entities. That such a language must be different from the ordinary one is shown by the following example: In the ordinary language we say: 'There are no white ravens and no round squares.' In the new language we would have to say, instead: 'There are white ravens; however, they are not actual but only possible. And there are round squares; however, they are neither actual nor possible, but impossible.' I have no doubt that a resourceful logician could easily construct a consistent language system of this kind, if he wanted it; he would have to lay down rules for the quantifiers deviating from the ordinary rules in a way suggested by the examples. The decisive question is not that of the technical possibility of such a language but rather that of its usefulness. Only if it can be shown to have great advantages in comparison to the ordinary language structure would it be worth considering in spite of its fundamental deviation and increased complexity. (Carnap 1956: 65–6)

This comment is a great deal subtler and philosophically more 'robust' than Russell's sudden swipe at Meinong in his (1905*a*).

[11] In his letter to Meinong of 15 December 1904 (see note 6)—that is, during what appears to have been a critical period in Russell's attitude towards Meinongian philosophy—Russell writes:

> I have so far always held that every object must somehow *have being*, and I find it hard to accept objects that have no being. In cases such as the golden mountain or the round square, one must distinguish between the *sense* and the *reference*, to speak with Frege. The sense is an object and has being; the reference, however, is not an object. The distinction between sense and reference is seen most clearly in mathematical examples. 'The positive square root of 4' is a complex sense, whose reference is the number 2.
>
> (Ich habe bis jetzt immer geglaubt, dass jeder Gegenstand in irgend einem Sinne *sein* muss, und ich finde es schwer, nichtseiende Gegenstände anzuerkennen. In einem solchen Fall wie der des goldenen Berges oder des runden Vierecks muss man zwischen *Sinn* und *Bedeutung*, um mit Frege zu sprechen, unterscheiden: der Sinn ist ein Gegenstand und hat Sein, die Bedeutung aber ist kein Gegenstand. Den Unterschied zwischen Sinn und Bedeutung sieht man am Besten bei mathematischen Beispielen: die 'positive Quadratwurzel aus 4' ist ein komplexer Sinn, dessen Bedeutung die Zahl 2 ist.)

Remarkably, Russell appears to take a term's sense as its vicarious reference in cases where there is no actually existing reference object. This is, of course, fully untenable, as was demonstrated by Quine (1953: 2): when I speak of Pegasus, I do not speak of the *sense* (what Quine calls 'the Pegasus-idea') but of the *thing itself*, even if that does not exist. This shows that, in this critical period just before he published his (1905*a*), Russell was still far from clear about the ontological status of nonexisting objects, despite his apodictic statements and his *panache* in putting forward his arguments in that publication.

up in that made him anxious to take the extensionalist road and do away with 'Meinong's unduly populous realm of being' (Russell 1944: 13), or with 'Plato's beard', as Quine put it in his famous essay 'On what there is' (Quine 1953). It was 'in the air', anyway, to dispense with what were felt to be nonessentials. In the arts, cubism was becoming popular in painting and sculpture, reducing natural forms to modelled planes. Architecture was bracing up for what were to be strictly functionalist movements like Bauhaus, De Stijl, Werkbund. In psychology, the minimalist and positivist theory of behaviourism was in the offing. In philosophy and science, positivism was already sweeping the market. Russell, I think, had his nose in the air and sniffed what was going on.

Yet Russell cut off and threw away far too much, committing the same sins as were committed by positivism and behaviourism. It was no coincidence that these and similar movements came to fruition during a period of explosive technological advance undreamt of by previous generations but accompanied by unparallelled and often tragical irresponsibility and lack of wisdom. The youthful but immature enthusiasm of the period led to oversimplification and to an overestimation of mankind's technical powers. Things that were in fact highly essential were thought to be nonessential, and thus disposed of, owing to the blinkers of youth, success, and hubris. But in all cases the chickens have come home to roost. Positivism did not last long, behaviourism ended in the 1960s and the schematization of forms pursued in the arts quickly gave way to a multiplicity of artistic currents that were less axiom-driven. Yet even though these movements were shortlived, they have had lasting effects. And once the movements in question had been quelled as their disadvantages began to weigh too heavily, these effects have, on the whole and in the long term, been beneficial (if one can say that of a movement in art). My feeling is that the same is true of the twentieth-century movement of purely extensional logic and semantics, ushered in or at least propagated by Russell at the beginning of the century, along with its 'desert landscape' philosophy. The difference is that this movement is still reigning supreme. Though I have no doubt that it is as blinkered as the other movements mentioned, I admit, at the same time, that it has also contributed significantly to our present capability of seeing contours and distinctions.

In recent years there have been a few non-mainstream attempts at reviving intension-oriented studies in a perspective that does not give primacy to extensionalism. These attempts were no doubt prompted by the empirical failure of the extensionalist approaches, evident to a small number of authors. Parsons (1980) is an early harbinger. In 1985–6 a volume on intensional ontologies and logics appeared in the *Grazer Philosophische Studien*, edited by Rudolf Haller (Meinong was philosophy professor at Graz). In that volume, a few authors, such as Castañeda and Lejewski, draw attention to the problem of intensional being. Zalta (1988: 15–18) distinguishes between two grammatically identical ways of applying a predicate to its term referents: *exemplification*, producing extensional truth, and *encoding*, producing intensional truth. Bottani and Davies (2006) is an up-to-date and extremely valuable collection of papers centred largely around the theme of the ontological status of fictional entities, harking back to the classic authors who dealt with this question. At this point, however, we will not go into the merits or demerits of these and other discussions and proposals. Questions of how to account for intensional phenomena in language and cognition are discussed in Chapters 5 and 6.

To summarize, let us revert to the sentences (2.1) to (2.4) above (repeated here for convenience). These sentences are easily comprehended, interpreted, and truth-valued by speakers of English with sufficient knowledge of history:

(2.1) The god Apollo was worshipped in the island of Delos.

(2.2) The god Poseidon was worshipped in the island of Delos.

(2.3) The god Apollo lives in Kathmandu.

(2.4) The god Apollo does not live in Kathmandu.

What are the truth conditions of sentence (2.4)? The standard answer is that the proposition underlying (2.3) must be false, which indeed it happens to be in its extensional reading in the actual world.[12] But what makes it false in its extensional reading? Here ontology must step in to provide the answer. Let us assume that the ontological status of the city of Kathmandu is secure enough and that the property of living in Kathmandu is sufficiently well defined. But how about Apollo, who is not part of the actually existing world? If talk or thought about a nonexistent entity is talk or thought about nothing, as the extensionalists have it, then in the proposition underlying *The god Apollo lives in Kathmandu* the property of living in Kathmandu is assigned to nothing. But that is not so, because the property is assigned to Apollo, who happens not to exist, and not to, for example, Santa Claus, who labours under a similar fate. And likewise for sentences (2.1) and (2.2). Sentence (2.1) is true when uttered in this, our actual world, which would be hard to explain if the property of being worshipped in the island of Delos is assigned to 'nothing'. And sentence

[12] That is, (2.4) is true and (2.3) is false in their extensional readings (the falsity of (2.3) is caused by presupposition failure and is, therefore, called 'radical falsity'). In their intensional readings, (2.4) is again true, and (2.3) is *minimally false* (false without presupposition failure), because in the mythology of which Apollo is a part he lives on Mount Olympus and not in Kathmandu. See Chapters 5 and 6 for extensive comment.

(2.2) is false when uttered in the actual world, which, again, would be hard to explain if the property of being worshipped in the island of Delos is assigned to 'nothing'.

Or, to take an example involving impossible properties, if hating to be dead for the rest of one's life, or hating to save people's lives by killing them, amounts to not hating anything, in this or any other possible world, as the Montagovians have it, then (2.5a) and (2.5b) should always be substitutable *salva veritate*, since being dead while being alive and saving people's lives by killing them are equally impossible properties, no matter in which possible world:

(2.5) a. The boy refused to be drafted into the army because, he said, he hated to be dead for the rest of his life.

b. The boy refused to be drafted into the army because, he said, he hated to save people's lives by killing them.

But (2.5a) and (2.5b) are clearly not substitutable *salva veritate*, since being dead for the rest of one's life is not the same as saving people's lives by killing them, even though both are analytically impossible (ignoring metaphorical use). Possible-world semantics has so far not been able to solve this problem.

Frege thought that sentences like (2.1) to (2.4) lack a truth-value in our world as no property is assigned or denied. But this cannot be so, because (2.1) and (2.4) are both clearly true, while (2.2) and (2.3) are both clearly false. Russell presented his theory of descriptions, replacing all referring terms with quantified expressions. But this does not work either, not only because it fails to do justice to the context-bound character of sentences, but also because it founders on cases such as sentence (2.1), which is true despite the fact that, in Russell's construal of the existential quantifier, it is false to say that there is an *x* such that *x* is a god called 'Apollo'. Montague's possible world solution has likewise fallen in the crossfire of linguistic facts, as is shown by examples like (2.5a, 2.5b). And since no new proposals have appeared since Montague's, one is entitled to say that the question is still wide open.

2.2.2 *The epistemological argument*

All we can do, in the circumstances, is swallow our pride and knock again at Meinong's door. We are not so much interested, right now, in subsisting objects (Meinong's 'higher-order objects')—such as John's similarity to his brother, or the average Englishman, or the force of your argument—because these should all be reducible, one way or another, to what are construed as actually existing objects in the world. All that is needed for such subsistent objects is a 'formula of reduction' (although I admit that I would be hard put to provide such formulas). But we are interested in Meinong's *nonexistent* individual objects, like his golden mountain or Russell's present king of France, even in impossible individual objects, such as Quine's round square cupola on Berkeley College or Russell's barber, because these are not part of physical, causality-driven reality. These are virtual objects in virtual worlds, where reifications, abstractions and 'higher-order objects' are likewise allowed, but in a virtual mode of being.

Given the facts of language and mind, in particular the fact that we assign a truth-value to sentences such as (2.1) to (2.4), we are forced to face the old Kantian EPISTEMOLOGICAL ARGUMENT, which says that humans have no choice but to construct worlds in terms of what is and what could be, even in terms of what could never be. Extensionalists like Quine and countless others find it difficult to imagine that there could be things that do not exist, but this just shows the naïveté of their ideas about what the world actually is. They think the world is the way it is perceived and construed by us humans. But that is truly naïve. They should know that from the theoretical physicist's perspective a table is not an object at all but a mass of particles best described in mathematical terms with nothing object-like to it. A table is a table only from the point of view of natural cognition, which is prewired for specific forms of (object-oriented) processing of incoming sense data. We have known since Kant that what is naïvely taken to be the perception of reality is in fact the construction of a supposed reality through a naïve 'theory' which is determined by innate traits of the species as much as it is by incoming sensory data. Fortunately, the 'theory' constructed is such that it makes for relatively reliable predictions, so that humans can save their skin amidst the dangers of this world. Moreover, and equally importantly, these naturally emerging 'theories' make humans project, through an innate sense of empathy, their own construals on fellow humans, and even on members of related species, to whom they attribute similar 'theoretical' constructions. But the hypostatized world can never be *proven* to be actually there. Kant's 'Ding an sich' is unknowable and unprovable (with the exception of oneself, because one cannot deny self-experience, as Descartes has taught us). All the naïve 'theory' can tell humans at a metatheoretical level is that it is best to make the *inductive leap* and behave in all respects as if the world so construed constitutes actual reality, complete with fellow humans who do the same. As to the real nature of what is 'out there', agnosticism seems to be the only coherent answer.

But this is not all. In constructing a world picture, human and other animals are not passively restricted to incoming signals, which are then put through an automatic construal process. On the contrary, human and other animals ask questions as soon as sensory input impinges on their mind/brain, questions of interpretation, cause, reason, manner, and so on. They are self-starters, constructing worlds beyond the support of sense data. They must be, since one cannot construct a world theory without first constructing a blueprint, which is subsequently checked for adequacy. There can be NO INTERPRETATION WITHOUT REPRESENTATION, and hence without prior virtual reality. Humans talk, think, and fantasize about their virtual worlds and situations, knowing full well that such worlds and situations are not to be construed as 'real' in the sense of 'actually being there'. But they hope to make them come true or try to prevent them from coming true, or check if they are true, or what not. Quine failed to recognize that. To cut out what to him were 'disorderly elements', he imposed a further criterion, beyond Aristotle's principle of contradiction, on 'what there is'. His criterion (already proposed by Russell half a century earlier, although less emphatically) was 'no entity without identity', shown in action in the following famous passage staged as a refutation of the imaginary philosopher Wyman, who defends a Meinongian ontology against the equally fictional professor McX, who thinks it is ideas we talk about, not things:

> Wyman's overpopulated universe is in many ways unlovely. It offends the aesthetic sense of us who have a taste for desert landscapes, but this is not the worst of it. Wyman's slum of possibles is a breeding ground for disorderly elements. Take, for instance, the possible fat man in that doorway; and, again, the possible bald man in that doorway. Are they the same possible man, or two possible men? How do we decide? How many possible men are there in that doorway? Are there more possible thin ones than fat ones? How many of them are alike? Or would their being alike make them one? Are no two possible things alike? Is this the same as saying that it is impossible for two things to be alike? Or, finally, is the concept of identity simply inapplicable to unactualized possibles? But what sense can be found in talking of entities which cannot meaningfully be said to be identical with themselves and distinct from one another? These elements are well-nigh incorrigible. By a Fregean therapy of individual concepts, some effort might be made at rehabilitation; but I feel we'd do better simply to clear Wyman's slum and be done with it. (Quine 1953: 4)

How arrogant and how fatally wrong. Who or what is this Wyman? Surely, he is a fictional character, but does that make him indistinguishable from his equally fictional opponent? Are they the same possible man, or two possible men? How do we decide? How many possible men are there in that story? Of course there are exactly two possible men in the story and they are distinct because Quine has thought them up as distinct. I feel we would do better simply to leave Quine's desert landscape and be done with it.

The rhetoric of this passage merely conceals the fact that it is nothing but a sleight-of-hand. Had Quine reflected more calmly on the Cartesian and Kantian gap between the mind and 'what there is', or rather 'what must be there', he would have realized the force of the epistemological argument: humans have no choice but to construct a world theory, and in that theory they distinguish between what is construed as actually existing and what is merely thought up, applying different criteria to the former and the latter. Instead, Quine takes a criterion that may perhaps be sensibly applied to what is taken to be the actually existing world—his 'no entity without identity'—and applies it to nonexisting, virtual worlds, which, in fact, allow not only for the merging and splitting of identities but also for all kinds of other phenomena that are incompatible with construals of the extensional world.

In most cases, however, virtual worlds are not quite as bad as that. When we are not dozing off into dreamworlds, like Lewis Carroll's Alice, but are actively setting up hypotheses about what might perhaps be the case so that certain facts are seen in an explanatory light—an activity well known among scientists and to some extent also among police detectives—we tend to make sure that what we think up stands a chance of being an actual state of affairs. When I am criticizing a doorway in a house just built (or still being planned, for that matter), I may imagine a possible fat man in it and see that he would not be able to get through comfortably if he were an actual fat man. In the unlikely event that I should also wish to imagine a bald man in the same doorway, it is up to me to decide whether I imagine them as the same man or as two different men. This point is not made in any of the existing philosophical literature on virtual objects, which concentrates entirely on possible extrapolations from the actual world, neglecting the fact that these owe their being to the creative representational powers of cognizing beings. Quine's objection is thus answered by the simple statement that the identity of virtual entities is decided upon by their inventors, since it is the inventors' thought that creates them.

But it cannot be reasonably doubted, under a correspondence notion of truth, that such thought-up entities have being, since I can utter true sentences, such as (2.1), about them. If truth is correspondence between what is said or thought on the one hand and what is the case in the world on the other, then one must include the virtual entities and states of affairs created by the human and other thinking individuals populating it. These, too, 'are there' in the world.

Quine and his followers must either maim the concept of truth or accept that what we speak and think about are elements in a world as it is mentally constructed but whose real nature is beyond the reach of any cognizing being, even a postulated highest being.[13] It is an inevitable consequence of the

[13] Unless that highest being is itself what it knows to be there—the immanent god of so many philosophies and religions. Such a being would be able to say to itself 'cogito, ergo sum', and thereby establish the actual reality of the universe.

'human condition', and likewise of the condition of all nonhuman cognizing animals, that humans and other higher vertebrates act upon sensory stimuli by constructing world representations and testing them for truth. In analysing and describing this 'condition' we face a choice: either we maim the concept of truth or we accept that humans and other cognizing animals construct virtual worlds. Maiming the concept of truth takes one away from the cognitive reality of the animals involved into a world of philosophical artifacts whose relation with the real world is at best doubtful and at worst nonexistent. By contrast, accepting the construction, and hence the reality, of virtual worlds is merely a consequence of the epistemological argument. Therefore, the choice is clear: we accept that humans and other cognizing animals construct virtual worlds which they think about and which humans also talk about, and we try not to mutilate the concept of truth. This, it would seem, is sufficient to bring water to Quine's desert.

It is not at all a question of what really exists 'out there', for no one will know anything beyond what the theory tells them, and no theory can prove existence. It is all a question of what has been constructed, whether under the operator of actual existence or under some operator of virtual being. What is needed is not Montague's programme of extensionalization of intensions but the opposite, a programme of intensionalization of extensions.

3

Propositions and truth

3.1 What is a proposition?

3.1.1 *The general concept of proposition*

Few will deny that propositions are the stuff that logic is made of and that linguistic utterances express propositions. We are not among these few, but that does not tell us what a proposition is or how it can be most profitably defined. And here we find ourselves right away in deep water.

Let us begin by stating that there can be no language without the mental faculty of representing actual or virtual situations and thus of mentally representing actual or virtual objects as possessing (or lacking) certain properties. And that is what we take a proposition to be:

> PROPOSITION:
> A proposition is the here-and-now (token) mental act of assigning a property to one or more actual or virtual objects (entities).

But what is it to 'mentally assign' a property to an object? Such an assignment can only be made if the objects in question are mentally represented in some specific form. By saying that a property is assigned to an object we mean that the object in question is mentally represented as possessing that property, and such a representation is then either true or false, in principle. The 'in principle' rider has to be added because there are complications.

The basic principle is not too complicated. Suppose there is an actually existing object **o**, mentally represented as [**o**]. Now suppose a cognizing individual *I* imagines **o** as having the property **P**. This amounts to *I*'s adding the mental label for **P** to [**o**], and if **o** actually possesses the property **P**, this addition results in a true proposition; otherwise in a false one.

Logically, or mathematically, speaking, a proposition is the assignment—in this case by the mind—of an entity (or class of entities) to a class (or class of classes of entities). The essence of a proposition is that an entity (or class of entities) is placed under a rubric that admits of other, similar entities (or classes of entities). That is, a proposition consists of two terms, a predicate and a subject term. The predicate must of necessity be at least a first-order predicate, taking as its extension a class (possibly Ø) of zero-order individual entities. It may be a second-order predicate, extending over a class of classes of individual entities, or a third-order predicate, extending over a class of classes of classes of individual entities, and so on, but the predicate must always be one order higher than the referent of the subject term. Thus, in the proposition underlying the sentence *John is an American*, the subject term *John* denotes a zero-order individual entity and *be an American* is a first-order predicate, whose extension consists of all those zero-order human entities who have American citizenship.[1]

Interestingly, when, in a letter dated June 1902, Bertrand Russell presented Gottlob Frege with his famous paradox:

Let *w* be the predicate: to be a predicate that cannot be predicated of itself. Can *w* be predicated of itself? (Van Heijenoort 1967: 124)

also formulated as:

Does the class of all classes that do not belong to themselves belong to itself?

Frege replied promptly, expressing his consternation at the discovery of the paradox, but at the same time unwittingly hinting at its solution:

Incidentally, it seems to me that the expression 'a predicate is predicated of itself' is not exact. A predicate is as a rule a 1st-level function, and this function requires an object as argument and cannot have itself as argument (subject). Therefore I would prefer to say 'a concept is predicated of its own extension'. (Van Heijenoort 1967: 127)

Frege thus effectively says that 'a 1st-level function... cannot have itself as argument (subject)', since the extension of a predicate cannot comprise itself as a member. To assign the extension of a predicate to itself is absurd and can, therefore, never be expressed as a proposition. A proposition is, in virtue of its very nature, the mental act of assigning an nth-order entity **o** to an $n+1^{th}$-order class of entities of the same order as **o**.[2]

But what if there is no actual entity or object **o** so that the mental object representation [**o**] does not correspond to anything in the actual world? Our (in principle Meinongian) view is that, by definition, REPRESENTATIONS CREATE VIRTUAL OBJECTS. A mental object representation [**o**] by definition CREATES a virtual object ***o***, which is defined by the properties that have been mentally

[1] Richard Montague had, apparently, only a limited understanding of the order-raising nature of propositions. He analysed (Montague 1973) *John is an American* as 'there is an American *x* such that John is identical with *x*'. Further comment would be uncharitable.

[2] See Seuren (1998: 334–5) for a somewhat fuller account of this remarkable episode in the history of logic, and Section 6.2.2 for a more elaborate discussion of the paradox.

assigned to it. We say that the actual object **o** INSTANTIATES the virtual object ***o*** and the corresponding mental representation [**o**], or:

MENTAL REPRESENTATION	[**o**]	creates the
VIRTUAL OBJECT	***o***	[**o**] and ***o*** may be instantiated as an
ACTUAL OBJECT	**o**	

When I draw a dog I have produced not only a figure on paper but also a virtual object that has the properties I have given it in my drawing as it is interpreted by viewers, such as the property of being a dog and, say, of having drooping ears. When my drawing is meant as a portrait of a really existing dog, then the property of having drooping ears assigned to it in my drawing is either true to life or not. But when my drawing is not intended to represent a real dog, but is just a dog-representation, we have a virtual dog, created by my drawing and fitted out with the properties I have given it. I can do a great many things to this virtual dog. I can talk or write or dream about it, I can give it a name, I can forget and remember it, I can look for it (but will then, of course, never find it) and so on. But I cannot take it for a walk or feed it, though I can represent it as being taken for a walk or being fed by me.

Properties assigned to virtual objects not only help define those objects but also lead to a truth-value. Suppose I have a mental representation [**the present king of France**]. This representation creates the virtual object ***the present king of France*** even if there is no actual object **the present king of France**. Now I can add to the mental representation [**the present king of France**] the property description of being bald, thereby mentally assigning to ***the present king of France*** the property of being bald. That assignment will result in falsity because for anyone to be bald he or she needs actual existence: virtual objects are never bald, only actual objects sometimes are. But suppose I add to [**the present king of France**] the property description of being admired, thereby mentally assigning to ***the present king of France*** the property of being admired. Now the assignment may be true, since it is not only actual objects that can be admired but also virtual objects. And likewise for properties such as being talked or written or thought or dreamed about, or being looked for, or being given a specific name and so on.

The assignment of a property P to a virtual object ***o*** can result in actual truth when the predicate F expressing P does not require the actual existence of an instantiating object **o** for truth. We then say that the predicate F is INTENSIONAL with regard to the term position in question. The intensionality of a predicate must be defined with respect to an argument term (I know of no predicate that is intensional with respect to more than one term). The two-place predicate *worship*, for example, is intensional with respect to its object term but not with respect to its subject term, because in order to worship one has to exist, but in order to be worshipped that condition does not apply. Most predicates, such as the one-place predicate *bald* or the two-place predicate *beat*, are fully extensional and thus not intensional, but some predicates are intensional with respect to one of their terms.

A mental representation [**o**] can be expressed linguistically as a definite nominal term *a*, which is said to REFER to **o** when there is an actual **o** available, and otherwise to the virtual object ***o***. Using the function sign 'ρ' for the reference relation, we stipulate that $\rho(a) = \mathbf{o}$ when an actual **o** instantiating ***o*** is available; otherwise, $\rho(a) = {}^*\mathbf{o}^*$. For example, the term *the Monster of Loch Ness* refers to the actual monster if it actually exists, but to the virtual object ***the Monster of Loch Ness*** if it doesn't. This is the REFERENCE HIERARCHY discussed in greater detail in Section 3.5.2.

In the process of speaking or writing, the term *a* originates as a mental element, which is realized as an actual phonetic (or graphic) element when an actual utterance comes about. Diagrammatically, this set of elements and their relations are represented in Figure 3.1, where the bold '**refers to**' relation marking bold arrows indicates reference to an actual object, and the nonbold 'refers to' relation with the thin arrows indicates the secondary reference relation to virtual objects.

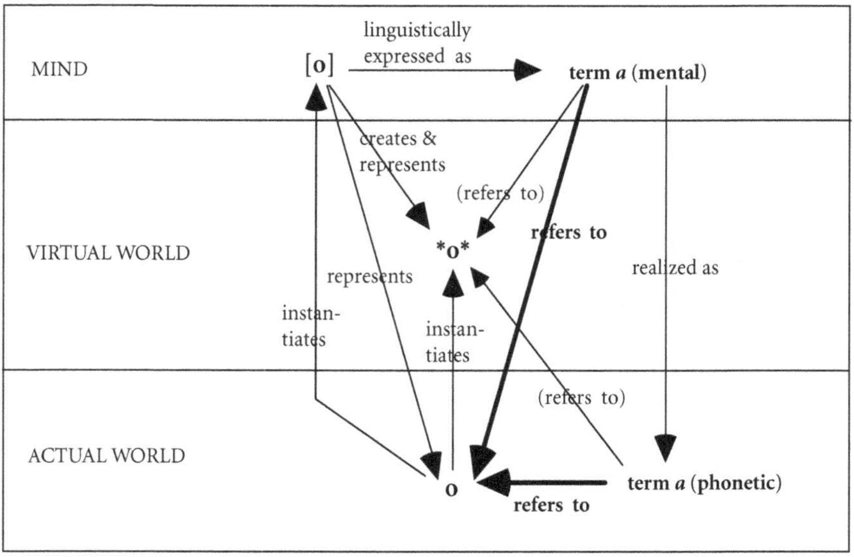

FIGURE 3.1 The various relations between [**o**], ***o***, **o** and the (mental and phonetic) linguistic term *a* referring to **o** (***o***)

It is assumed that mental object representations [**o**] have the form of DISCOURSE ADDRESSES (DAs) which accumulate the properties assigned in subsequent propositions (details of this incremental mechanism are discussed in Chapter 8 of Volume II). The mental assignment is always and necessarily made with regard to virtual objects in a virtual world called into being by the mental power of imagination and corresponding wholly, or partly, or not at all, with what is construed as actually being so in the real world. Truth or falsity results according to the conditions posed by the property in question, and thus by its linguistic correlate, the lexical predicate. This makes the proposition the primary bearer of a truth-value.

Although there can be no language without propositions, there can be propositions without language, as is the case with nonhuman higher vertebrates. Even though these have no language to show it, they must have propositions, as their behaviour cannot be understood without the assumption that they form mental pictures which they try out for truth, or avoid, or seek to turn into reality. Propositions were there long before human language came into being, even before there were humans.

As regards propositions entertained by humans, consider the following example. Given the actually existing object **Mount Everest**, I can mentally assign it the property of being on the Nepalese border (in fact, I have already done so, in virtue of writing this sentence, and so have you, in virtue of reading and understanding it). That is, I have a mental representation—that is, a DA—[**Mount Everest**] of what in reality is **Mount Everest**. This DA creates the virtual object ***Mount Everest***, to which the property of being on the Nepalese border is assigned. This is recorded by adding the description of the property of being on the Nepalese border to [**Mount Everest**], which created ***Mount Everest***. The new DA-description is now true or false, according to whether the actual **Mount Everest** is or is not on the Nepalese border. In doing this I have formed a proposition which can be rendered as something like (3.1):

(3.1) +**P(On the Nepalese border)** ([**Mount Everest**])

where **P(On the Nepalese border)** stands for the property of being on the Nepalese border, expressed by the predicate *On the Nepalese border*, and, as has been said, [**Mount Everest**] for the mental object representation creating the virtual object ***Mount Everest***, both of which are instantiated by the actual **Mount Everest**. The plus sign indicates that the property is added to the mental representation, thereby creating a proposition, which is true or false according to whether the corresponding actual mountain is or is not on the Nepalese border, respectively.

The convention of writing a proposition as

$$+P(X)([o_1], \ldots, [o_n])$$

where P(X) represents the property expressed by the predicate **X** and $[o_1], \ldots,$ $[o_n]$ the *n*-tuple of mental object representations $[o_1]$ to $[o_n]$ to whose mental description the property description P(X) is added, seems a useful way to distinguish pure propositions from their linguistic expressions, at any level of grammatical representation, while emphasizing the fact that a proposition consists in the mental act of assigning a property to one or more objects. As is elaborated in Chapter 4, a proposition *p*, when expressed in a linguistic utterance, is necessarily placed under a social force operator, with which it forms a so-called INTENT. The intent is fed into the LANGUAGE MODULE, where the appropriate lexical items are selected and cast into a first syntactic mould, the SEMANTIC ANALYSIS, or SA (also often called 'logical form'), of the sentence to be produced. The SA consists of a SPEECH ACT OPERATOR and an L-PROPOSITION, and is structured according to the syntax of a formal language of predicate calculus. The SA-structure is semantically analytic and regular, unlike the corresponding SURFACE STRUCTURE, which obeys the rules of the syntax and morphology of the specific language in which the intent is to be expressed. The correspondence between the SA and the surface structure materializes as a TRANSFORMATIONAL ALGORITHM which is part of the language module and transforms SAs into surface structures.

But let us revert to propositions proper. One single actual object can be mentally represented by two distinct DAs, creating two distinct virtual objects. This happens when the person in question is under the misapprehension, or considers the possibility, that what is in fact the same object is two different objects, as when it is thought that the morning star and the evening star are two different planets, or that Mount Everest and Chomolungma (its Tibetan name) are two different mountains. In such a case, the propositions (3.1) and

+P(**On the Nepalese border**) ([**Chomolungma**])

are two distinct (true) propositions, since the property description of being on the Nepalese border is added to two distinct DAs, resulting in two distinct property assignments.

Conversely, one single DA may in fact be instantiated by two different actual objects. This happens when one counterfactually thinks or imagines that what are in fact two different objects is the same object, as when I think that the man who addressed me on the train is the person who stole my purse, whereas in fact my purse was stolen by a mischievous urchin as I was walking down the platform. Or when I think that Smith has been murdered by one person whereas in fact there were two murderers. If one of the two murderers is insane while the other is in full control of his wits, the question arises as to the truth-value of the sentence:

(3.2) Smith's murderer is insane.

This sentence clearly expresses one single proposition, as only one mental assignment is made. But is this proposition true or false? Donnellan (1966) raises the closely related question of definite terms referring to so far unidentified but postulated objects. In the case of (3.2), this would amount to an interpretation 'whoever murdered Smith is insane'. Such sentences pose problems for any theory of truth-value assignment, not only when there was more than one culprit but also in cases where Smith wasn't murdered and, consequently, there was no murderer. Problems such as these are far from definitely settled, and it is not my purpose to settle them in the context of the present book.

3.1.2 *Some history of the term* proposition

Our use of the term *proposition* follows what many like to call 'the tradition' in that it takes propositions to be the primary bearers of truth-values, and thus as token occurrences. Even so, the tradition has not always been as clear as it could have been on what the primary bearers of truth-values should be taken to be. Although the term *proposition* has been regularly used in the history of Western thought in connection with the properties of truth and falsity, opinions on what should actually be taken to constitute a proposition have varied widely. The Dutch philosopher Gabriel Nuchelmans devoted three hefty volumes (Nuchelmans 1973, 1980, 1983) to the various concepts of proposition from Aristotle to Kant, and he could have added a fourth about concepts of the proposition since Kant. All these concepts have in common that propositions are viewed as the primary bearers of truth-values, but they differ as to what it is that bears them.[3]

The term was coined by Aristotle around the middle of the fourth century BCE. His Greek term was *prótasis*, literally 'the act of putting forward', translated into Latin as *propositio*, which passed, in the phonologically appropriate form, into all European languages (except Modern Greek, which kept the

[3] The late twentieth century distinguishes itself by some rather curious notions of proposition. One such notion, developed in model-theoretic semantics and extremely influential for a few decades, sees a proposition p as the set of possible worlds in which p, or the sentence expressing p, is true. We consider this a terminological aberration, due to the disastrous Russell–Quine 'programme of elimination of particulars', which required a collapsing of the notion of *sentence* as a type-level entity carrying type-level truth conditions with the notion of *utterance* as a token occurrence carrying an actual truth-value. It is also a break with the tradition, in that it hardly makes sense to say that a set of possible worlds has a truth-value. A similar confusion is apparent in Cappelen and Lepore (2005), where a proposition p appears to be equated with the truth conditions of the type-level sentence used to express

Greek original). The term *prótasis* occurs for the first time on the first page of Aristotle's *Prior Analytics* (*PrAn*), the book that contains his doctrine of syllogistic (*PrAn* 24a16):

A proposition (prótasis) is an affirmative or negative expression (lógos) that says something of something.

The phrase *that says something of something* is to be read as 'by which a property is assigned to one or more objects by means of linguistic expressions.' Aristotle calls the object or objects in question the *hypokeímenon* (literally 'that which underlies'; Latin *subiectum* or *suppositum*), and the property assigned the *symbebēkós* (literally 'that which has befallen to'; Latin *contingens* or *accidens*). The sentence part expressing the property is the *katēgoroúmenon* (literally 'that which is said of'; Latin *praedicatum*). Aristotle had no term for the linguistic sentence constituent we now call *subject.*

During the Middle Ages, the term *subiectum* was transferred from the object or objects denoted to the sentence constituent now called *subject* and characterized by nominative case when appropriate, whereas the term *suppositum* remained reserved, on the whole, for the thing or things to which reference is made (Seuren 1998: 121–3). Section 3.2 shows that this terminological transfer, although slow in real time, was rash in terms of linguistic analysis.

The word *lógos*, translated here as 'expression', has a large range of possible interpretations, covering the concepts 'word', 'saying', 'meaning', 'reason', 'rationality', 'thought', 'theory', 'system', and a few more. Aristotle defines it, for the purpose of logic, in his *On Interpretation* (*Int*), where he sets out his predicate calculus—that is, his theory of quantification. There we find the following:[4]

A *lógos* is a stretch of human sound that is meaningful by convention and contains one or more separate meaningful parts that may say something but not [necessarily] in the sense of an affirmation or a negation. For example, the word *human* does have

p. These authors thus avoid the unhappy possible-world terminology but at the same time they involve themselves in a conceptual tangle. In a footnote they write (Cappelen and Lepore 2005: 3):

Semantic Minimalism, as understood in this book, need not take a stand on whether semantic content is a proposition, or truth conditions, or what have you. Throughout the book we try to remain neutral by couching the issues both in terms of truth conditions and in terms of propositions.

This cavalier attitude is remarkable in a book that consists largely of a critical discussion of opinions and theories that see propositions as token occurrences with actual truth-values, not as abstract type-level objects with truth conditions.

[4] The translation is mine, but it must be admitted that the text is at least partially opaque and the published translations do not agree among each other. I have tried to strike the best possible balance between what makes good sense and what the text can be taken to say. See the discussion by Graffi (2004).

a meaning but does not say whether humans do or do not have some property. It will form an affirmation or a negation when something is added. (*Int* 16b26–32)

Then, at *Int* 17a1–7:

Every *lógos* is meaningful, not in virtue of some physical force but, as has been said, by convention. But not every *lógos* is an assertion. Only a *lógos* that has the quality of being true or false is an assertion. But this is not always the case. For example, a *lógos* can be a wish, in which case it is neither true nor false. We will disregard the others, as they are more properly dealt with in the study of rhetoric or of poetry. Our present investigation concerns assertions only. (*Int* 17a1–7)

It thus seems that Aristotle uses the term *lógos* to cover what we call 'word', 'phrase', and 'sentence', without differentiating between sentence and utterance, but in full awareness of the different types of speech act, such as assertion, question, command, wish, suggestion. At the same time he seems to identify assertions with propositions, which have a truth-value, as we see from the following densely formulated sentence:

An affirmation is an assertion of something *of* something; a negation is an assertion of something *off* something. (*Int* 17a25–26)

which can be paraphrased as something like 'An affirmation is an assertion that a property adheres to an object; a negation is an assertion that a property does not adhere to an object.'

Our view of the proposition as the mental act of assigning a property to one or more objects is Aristotelian in spirit, even though Aristotle preferred to use the term for linguistic expressions (or rather for their underlying semantic analyses), rather than for thought processes. There is no doubt that, for Aristotle, the origin of truth and falsity lay in thought processes, and we agree with him in this respect. But unlike Aristotle we think there is a point in reserving the term *proposition* for the thought processes, apart from their (underlying) linguistic expression, just as the Stoics used the term *axíōma* for the thought underlying assertive utterances. In our day, Dummett does the same. For him, 'it is the *thought* (proposition) which is what is true or false' (1964: 97). Taken in this sense, a proposition may, but need not, be expressed linguistically.

Especially in his *Metaphysics*, Aristotle attributes the quality of being true or false to thought processes. A classic statement is *Metaphysics* 1027b25:

For falsity and truth are not properties of actual things in the world (so that, for example, good things could be called true and bad things false), but properties of thought.

Yet when, at *PrAn* 24a16, he actually introduces his term *prótasis* he leans towards a verbal rather than a mental interpretation.

Curiously, although Aristotle recognized the reality of underlying cognitive processes and structures as the bearers of truth and falsity, he had no term for this cognitive basis of speech utterances. He apparently preferred to speak of linguistic entities rather than of the shadowy goings-on in cognition. And this has remained so, with ups and downs, until the present day. Our use of the term *proposition* coincides not so much with Aristotle's *prótasis* as with what the Stoics called *axíōma*, the thought expressed in an assertion, or with Augustine's *sententia*, at least in some of its uses. On the whole, there was considerable terminological confusion regarding matters of thought and language in late Antiquity and the Middle Ages (see Nuchelmans 1973 for a well-nigh complete account). One does see, in 'the tradition', attempts at making explicit the cognitive basis of language and speech. The Stoics, many medieval philosophers and, above all, mainstream nineteenth-century philosophers of language have bravely tried to penetrate into the mind by analytical means. But they never carried the day. A renewed attempt is made in the present book, in the light of the much refined, although still in many ways incomplete, notions of what makes up the human mind, developed over the past half-century.

3.1.3 *Reifications and virtual objects*

So we revert to the present and we concentrate for a moment on what can be meant by the term *object(s)* in our definition of the concept of proposition, that being a question of great magnitude. For this, we need the *epistemological argument* discussed in Section 2.2.2. Since Descartes, Hume, and Kant, philosophers have become increasingly aware of the fact that all knowledge is the result of processes of cognitive construal and construction, where the relation to what is 'out there' in the outside world, if there is anything at all, is by definition undefinable. What the thing that you and I perceive as the chair over there actually corresponds to in the 'bare' world is not known and cannot be known: Kant's 'Ding an sich' is in principle unknowable, nor is its real existence provable. It follows that absolute, provable certainty about the external world is unattainable: this is the price humans and other animals have to pay for being cognizing animals. To compensate for this basic uncertainty, humans and other animals are endowed with the natural gift to trust in the reliability of what presents itself as the 'outside world' and as 'actually there' in it. This trust is deeply entrenched in human nature. Humans consider their perception of what presents itself as being 'out there' so certain that they call it 'knowledge' and they live by it as if they had certainty. This primeval and all-encompassing 'inductive leap', leading to a world theory or world construal, is bred into all forms of animal intelligence. Apparently, it is a good theory, perhaps even the best that can be achieved on the basis of incoming sense data, since its predictions have a high degree of consistence and reliability, which makes it indispensable for survival.

But apart from this central overall aspect of all knowledge, it is clear that human cognition is capable of creating representations of complex 'objects' in the full knowledge that these 'objects' have no proper existence as individual entities in the world. Consider the following propositions:

(3.3) a. +P(**Powerful**) ([**the military-industrial complex**])
b. +P(**High**) ([**the average age of Boston taxi drivers**])
c. +P(**Positive**) ([**Jim's attitude towards environmental legislation**])

What makes these propositions peculiar is the fact that they assign properties to objects of which the individuals who make up such propositions know perfectly well that they are not objects that exist as such in the actual world, although they owe their being to actually existing objects and the properties they have. In the Meinongian terminology explained in Chapter 2, they do not *exist* but they can be said to *subsist*. In normal human thought processes, 'objects' such as the military-industrial complex, the average age of Boston taxi drivers, or Jim's attitude towards environmental legislation, are not construed as separate individual objects in the world, but as 'abstractions' of a kind often called REIFICATIONS. Reifications are abstract objects shaped by a cognitive process of selectively taking together a number of features of object representations and treating them as if they represented a single object, even though one realizes perfectly well that they do not. Reification is, although badly underexposed in the existing literature, an integral part of human cognition, which is obviously, in computer-science terminology, OBJECT-ORIENTED: the advantages of object-oriented programming were, apparently, known to Nature long before computers were even dreamt of.

It is interesting to note, in this context, that Russell's famous and uniquely influential article 'On denoting' (Russell 1905*a*) claims to provide a solution to the problem of both reifications and virtual objects, but in fact only deals with virtual objects (Russell's famous king of France) and provides no solution to the problem of reifications (his 'difference between A and B'). Consequently, whatever debates there have been around these matters have concentrated on virtual objects, while reifications have remained practically undiscussed in the literature since 1905.

The objects involved in the propositions represented as (3.4a–3.4c) are remarkable for a different reason.

(3.4) a. +P(**Englishman**) ([**Sherlock Holmes**])
b. +P(**Go missing**) ([**Little Red Riding Hood**])
c. +P(**Handsome**) ([**Apollo**])

The objects to which the properties in question are assigned are likewise creations of the thinking mind, but not as an object-oriented instrument for a more adequate handling of complex relations in the world. Rather, objects like ***Sherlock Holmes***, ***Little Red Riding Hood*** and ***Apollo*** are created as virtual individual objects, fitted out with the properties the thinking mind attributes to them. What distinguishes them from, for example, ***Mount Everest*** is the fact that the latter is instantiated by an actually existing object, the actual **Mount Everest**, while the former lack such instantiation. Their form of being remains restricted to virtual being. Like actually existing objects, they are mentally represented by the DAs in the mind, but unlike actually existing objects, they lack actual incarnation. The propositions (3.4a–3.4c), which assign the property of being an Englishman, going missing, and being handsome to virtual persons, are strictly speaking false (*pace* Chapter 5), because it takes actual existence to be an Englishman, to go missing, or to be handsome.

Virtual objects have been a sore point in twentieth-century Anglo Saxon philosophy, which banned them completely, saying, in effect, that when we talk about Sherlock Holmes we talk about nothing because there is nothing in the actual world that corresponds to any figure called *Sherlock Holmes.* As was argued in the previous chapter, we take modern Anglo Saxon philosophy to be fundamentally mistaken in this. Any ontology, whether made by philosophers or naturally emerging in the human mind, is the result of a process of construal of sense data and storage in memory. The way humans construe sense data is CREATIVE in the sense that they mentally construct individuals and states of affairs that may or may not match with what are taken to be actually existing counterparts in the actual world. That is, the human mind has the power to represent things that may or may not exist. It follows that the objects involved in propositions are not necessarily objects construed in naïve natural human ontologies as having actual, causally determined, physical and individual existence, but are often merely thought-up, virtual objects.

3.1.4 *The type–token distinction: eternal versus occasion sentences*

A further unclarity that has to be removed in the present context concerns the distinction between TYPES and TOKENS. The terms were introduced by the American philosopher Charles Sanders Peirce (1839–1914), but the distinction itself has been known at least since the German philosopher Gottfried Wilhelm Leibniz (1646–1716), who spoke of the IDENTITY OF INDISCERNIBLES. What is at issue is the fact that humans (and other animals), in their construal of the external world, identify unique token occurrences of objects and events in terms of types. Perception and understanding of the world crucially depend on the interpretation of token occurrences in terms of types: without that, it would be impossible to detect any systematicity in the world.

It is by definition impossible for a token to be repeated, since every repetition will produce a different token, but types can be repeated in the sense that a new token belonging to the same type may be produced. When Tim says that he has the same car as Tom, he will, normally speaking, not mean that he and Tom are co-owners of the same vehicle, but rather that his car belongs to the same type as Tom's car, where the degree of 'sameness' may be vague or may vary according to the context at hand. Or, to revert to the realm of language, when I ask you to repeat what you just said, you might, uncooperatively, reply that that would be impossible, since you would utter different sounds. But if you are in a more responsive mood, you will produce a token of the same type.

Linguistic utterances are unique tokens or token occurrences, but, no matter how garbled, they are interpreted in terms of a sentence type which is defined, in principle, by the lexicon and the grammar of the language in question but, in cases of garbled or improvised utterances, may also be set up ad hoc. Propositions are also token occurrences, but in virtue of the type-level machinery of any given language, which provides type-level predicates and type-level structures for linguistic expressions, they can, if that is the subject's decision, be cast in the mould of a type-level semantically explicit L-proposition, which is transformed by the grammar module into a corresponding type-level surface structure, which can, in the end, be realized as a token utterance. The resulting utterance is again a token occurrence, but the corresponding sentence, whether as a surface structure or as a semantic form, is a type, constructed according to the instructions of the lexicon and the grammar of the language in question. Token propositions are thus channelled through a type-assigning device so that they can be produced as token utterances.[5]

This view reflects a nonstandard perspective. In the view considered standard, there is no distinction between sentence types and utterance tokens: statements or sentences are considered to be both type and token. Truth and

[5] This is, in fact, what 'processing' in a general sense amounts to: tokens are fed into a type-level processing machinery that categorizes and transforms, and produces new tokens.

falsity are taken to adhere to statements or sentences regardless of context and situation. The fact that language fails to comply with this view is taken to mean that it is logically defective. A logically perfect language should, in the view favoured by the majority of modern logicians, contain only L-propositions that are independent of context and situation—Quine's *eternal sentences*—and ban all *occasion sentences.* This view is fundamentally misguided, as is argued by Strawson in the first pages of his seminal article (1950). In principle, (assertive) sentence *types* are not bearers of truth-values but (assertive) utterance *tokens*, expressing underlying propositions, may be considered secondary bearers of truth-values, the primary bearers always being the propositions expressed. We revert to this topic in Section 1.5 of Volume II.

Most of modern logic, and much of modern semantics and philosophy of language, is characterized by a PROGRAMME OF ELIMINATION OF PARTICULARS (Quine 1960), whereby occasion sentences are reduced to eternal sentences, not only for the purpose of logic but also, disastrously, for the purpose of semantics. A central feature of this programme is that definite determiners, such as the article *the*, which establishes token reference, are 'translated' into a form of quantification, which is by definition independent of token reference. This programme aims at an obliteration of the distinction between utterance tokens and sentence types, so that sentence types can now be taken to be the bearers of a truth-value. The underlying mental token process of creating a proposition by assigning a property to one or more objects is mostly left out of the picture altogether. We consider this programme to be a serious infraction of the most basic principles of human language. In our view, Quinean eternal sentences are merely marginal cases, where the dependency on context and state of affairs is reduced to zero or near-zero.

3.1.5 *Anchoring for interpretability and keying for truth testing*

Normally speaking, propositions do not arise out of the blue. They normally arise in a specific context. More precisely, as is shown in Section 3.2, they normally arise as a result of a wish to provide an answer to a question explicitly asked or implicitly anticipated in a given discourse. We call this context-dependency the (CONTEXTUAL) ANCHORING of propositions. Propositions are also, by definition, ABOUT, or TRAINED ON, certain objects at a given time in a given state of affairs. This aboutness relation we have decided to call (INTENTIONAL) KEYING. Propositions are by definition contextually anchored and intentionally keyed. By contrast, L-propositions and surface structures are, taken by themselves as linguistic objects, nonanchored and nonkeyed, since they are types, not tokens. They can only be said to be anchored and keyed to the extent that, in actual speech or sublinguistic thought, they express a given proposition. Nonanchored (and hence nonkeyed) L-propositions and surface structures are said to have a MEANING, whereas well-anchored L-propositions and surface structures also have an INTERPRETATION.

The contextual anchoring of propositions is reflected in various ways in their linguistic expression, for example (but not uniquely) by the use of linguistic means for definite reference, sometimes in the form of anaphora. When I say to you, here and now, as an isolated utterance, *The man was drunk*, it is impossible for you to say whether this utterance is true or false. But when I say, here and now:

> Last night I went for a walk. As I crossed Newton Road, about where the cinema is, I saw a man and a woman. The man was drunk.

then, provided you know where Newton Road and the cinema are, you will be able to make sense of the question whether *The man was drunk* is true or false. This is because the utterance of the sentence *The man was drunk* is now properly anchored in a given context—that is, linked up with what has been stored in a special discourse memory containing the information conveyed by preceding utterances and having open access to available world knowledge—and also properly keyed to an identifiable state of affairs by means of the temporal and other references made.

Then there is the fact that many predicate meanings have one or more built-in PRECONDITIONS, which restrict the usability of the predicates in question to certain contexts, on pain of incoherence. For example, the predicate *be closed* has the precondition that the object denoted by its subject term must have been open, for an object cannot be closed unless it has been open first. Likewise for *get divorced*, which denotes something one cannot do unless one has been married first. This means that such predicates cannot be used in a discourse where the preconditions cannot be satisfied, either because the context contains information that is incompatible with the information added by the sentence in question or because one or more of the objects the predicate is applied to are of the wrong kind. Thus, when it has just been said or implied that John was never married, an utterance like *John has got divorced* is uninterpretable in the sense of destroying the coherence of the discourse in question. And sentences like *Persuasion has been closed* or *The property market has got divorced* will, normally speaking, be uninterpretable unless the discourse contains information that makes it possible for persuasion to have been open or for a property market to be married. Such contingencies are, of course, a little far-fetched, but we know that discourse and context can do a great deal to make them seem quite natural. Without an accommodating discourse, such sentences are said to suffer from a *category mistake*. We will not enlarge on this question here, as it is discussed more fully when we come to metaphor in Section 9.5 and to presupposition theory in Chapter 10 of Volume II.

The question of what actually constitutes intentional keying in terms of cognitive functioning, or how intentional relations come about in general, is, although perfectly legitimate, still without a satisfactory answer. All we can do is take note of the still largely mysterious yet absolutely real fact, signalled and analysed by late nineteenth-century philosophers such as Meinong, Brentano, and Husserl, that cognizing animals, humans in particular, zoom in on certain objects, events, and situations in the world and are capable of making other individuals share that focusing. Intentional focusing, or keying, is like consciousness, in that it cannot be denied that the phenomena in question are real, yet it has so far proved impossible to capture them in terms of any formal theory or modelling. This cannot mean, of course, that therefore such phenomena must be left out of account. Psychologists implicitly recognize that the notion 'consciousness' is, as yet, not open to formal modelling, yet they feel free to operate with it whenever that is felt to be useful. We do the same with regard to intentionality phenomena, including phenomena of reference.

It is important to realize that whatever mechanisms underlie the process of fixing reference are not linguistic or even semantic in nature, but psychological. For some strange reason, authors with a formal semantics background seem to find this hard to accept, even when they look in detail at the complexities of reference and keying. Cappelen and Lepore (2005), for example, are willing to supply propositions with indexical values for such parameters as speaker, addressee, time and place of speaking, more or less à la David Lewis (1970), but otherwise insist on maximizing the notion of a proposition as something not intimately linked up with and integrated into available knowledge and ongoing cognitive processes—that is, the old Quinean notion of 'eternal sentence'—thereby keeping semantics 'minimal' and shifting the load of all kinds of semantic phenomena onto the shoulders of the pragmatists.[6]

This is strange and a little out of touch with reality. For what is wrong with accepting (a) that the use of language is intimately bound up with the rest of cognition, and (b) that this means that there are many things we do not, as yet, understand? Isn't it normal for science to deal with phenomena that are not, or not yet, understood? And isn't it abnormal for science to impose an *a*

[6] These, in turn, tend to 'radicalize' their semantics in the sense that they try to reduce semantics as much as possible to utterance interpretation, minimizing the role of the system that does the processing in terms of type-level elements, structures, and procedures. This development of 'radical pragmatics' is discussed in Section 1.3.3.2.

priori mathematical schema on the phenomena studied? Yet this is what these formal semanticists do, following in the footsteps of Descartes, for whom good science consisted in deduction from self-evident, preferably mathematical, notions.[7] It seems clear that this 'weltfremd' attitude, still current in varieties of formal semantics, is due to the logico-mathematical origins of this branch of studies. History, however, is not on its side. History has, for very good reasons, rejected such an *a priori* notion of science and has given preference to a methodology that forces scientists to submit to the tyranny of empirical evidence, of facts and figures.

There seems to be no good reason why an appeal to psychological factors and processes should be considered anathema. Psycholinguistic research (for example Clark and Wilkes-Gibbs 1990; Clark 1992: 21–7) has traced, sometimes in meticulous detail, the many different ways and means used in actual linguistic interaction to establish or maintain reference relations, and it is obvious that contextual anchoring is an important factor in the establishment of reference: NO KEYING WITHOUT ANCHORING. It is thus clear that reference is not compositional in the sense current in formal semantics, but is heavily dependent on the sharing by speaker and listener of a common fund of situational and world knowledge, as well as on shared preceding discourse. This fact has not been recognized in the standard forms of formal semantics literature, where it is still preferred to write intentionality and reference out of the semantic script.

3.2 Confusion around subject and predicate: topic–comment structure

At this point something must be said about an important problem to do with the internal structure and the context-dependency of propositions. The problem was discovered, but not recognized as such, around the middle of the nineteenth century and subsequently dominated virtually all theorizing about language until about 1930, when the linguistic world lost interest and turned to the more structural matters of phonology, morphology, and syntax. Only recently has it been revived, in the context of studies on text structure and information packaging. In the literature on linguistic theory, however, it is still largely ignored, despite its central importance.

The problem finds its origin in the fact that, as a rule, a new utterance tends to be presented as an answer to a question that has arisen, explicitly or

[7] See Seuren (2004*a*: 123), where the same critique is formulated with regard to Chomsky's Minimalist Program in linguistics.

implicitly, from what has been said before (Seuren 1985: 295–304). That being so, it is the form of the question that determines, at a certain level of analysis, the form of the answer. For example, suppose I say to you *I met somebody when I was in London last week.* Now you may ask, or I may anticipate that you will want to know, who it was that I met in London last week. So as to oblige you, I may then say *I met my old class-mate Harry.* To show the structural correlation between your (explicit or merely anticipated) question and my answer, we may analyse your question as something like 'the one you met in London last week is WHO?' and my answer as something like 'the one I met in London last week is MY OLD CLASS-MATE HARRY'. A full formal representation of the discourse thus has to contain a representation of my first sentence *I met somebody when I was in London last week*, followed by a representation of the (explicit or implicit) question *The one you met in London last week is WHO?*, followed by a representation of the answer to that question *The one I met in London last week is MY OLD CLASS-MATE HARRY.* In the literature on discourse structure, it is normally said that in such a case *my old class-mate Harry* is the new information provided, usually called the COMMENT, whereas *the one I met in London last week* denotes the information that was already given, constituting the TOPIC of the (implicit) question.

Or take another example. Suppose the context has given rise to the question *Where does coffee grow?* (to hark back to an example in Steinthal 1860). Now we may say that the proposition underlying this question is:

+P(WH?) ([**the place where coffee grows**])

where [**the place where coffee grows**] is the mental representation of the actual thing focused on (Aristotle's hypokeímenon) and +P(WH?) occupies the position where a property is to be filled in. The answer to that question will then express a proposition that specifies the property asked for:

+P(**Africa**) ([**the place where coffee grows**])

expressed as the L-proposition:

Be_v Africa (the x [coffee grows in x])

where the predicate Be_v assigns a value to the parameter expressed by the topic—in the case at hand *the place where coffee grows.* In the corresponding surface sentence, *Coffee grows in AFRICA*, the Be_v-predicate *Africa* will then receive main nuclear accent.

Similarly for *What did John do to the letter?* followed by the answer *John SIGNED the letter*, or *What did John sell?* followed by *John sold THE CAR*, or *Who sold the car?* followed by *JOHN sold the car*, as in the L-propositions

(3.5a–3.5d). We note that, according to the definition given, the answers in (3.5c) and (3.5d) express different propositions even though their surface structures are identical (apart from intonation): *John sold the car.*

(3.5) a. Be_V WH? (the x [coffee grows in x])
⇒ Be_V Africa (the x [coffee grows in x])
b. Be_V WH? (the x [John did x to the letter])
⇒ Be_V sign (the x [John did x to the letter])
c. Be_V WH? (the x [John sold x])
⇒ Be_V the car (the x [John sold x])
d. Be_V WH? (the x [x sold the car])
⇒ Be_V John (the x [x sold the car])

It is thus the L-propositional predicate, not necessarily the surface predicate, that receives main nuclear accent. (Clearly, the grammar will have to specify how and in what shape the L-propositional topic–comment structure turns up as a linguistic surface structure, but that is not our concern right now.)

This shows that, in many instances of linguistic utterance, the contextual anchoring of propositions gives rise to a curious disparity between, on the one hand, the propositional subject–predicate structure which is, normally speaking, determined by a discourse-driven question–answer game and, on the other, the corresponding linguistic expression in the form of a grammatical surface sentence containing a grammatical predicate with grammatical argument terms. The disparity consists in the fact that the propositional predicate very often does not correspond to the grammatical predicate, and the propositional subject term does not correspond to the grammatical subject term but rather incorporates an entire propositional structure with one element abstracted out for comment in the form of the propositional predicate.[8]

Only in a limited class of nontypical cases can one say that propositional structure may be taken to lack a topic–comment structure and to stand in a relation of direct correspondence to surface grammatical structure. This occurs, for example, when a newsreader announces *The king has fired the entire cabinet*, or *Arsenal has beaten Manchester United.* Utterances of that nature simply state a new fact, which may give rise to subsequent (implicit or

[8] The disparity in question is the main topic in Gardiner (1932), a lucid text which, unfortunately, was written at a time when the formal notions necessary for a clear analysis were not yet available. Gardiner assigned the topic–comment distinction to 'speech', but the grammatical predicate–argument distinction to 'language'. At the time, this was a very enlightening point of view. Nowadays, however, we tend to attribute both the topic–comment and the predicate–argument(s) distinction to 'language', which is now, as a whole, seen to be sensitive to contextual restrictions—a view that could not have been developed during the 1920s.

explicit) questions so that a question–answer game gets off the ground. Such utterances may be described, in theatrical terms, as FEEDING LINES or FEEDERS. They have also been called 'hot-news sentences'. In such cases the problem of the disparity between L-propositional and surface form does not present itself.

Let us have a brief look at the history of the disparity problem (see Seuren 1998: 120–33 for extensive discussion). As has been said, it was discovered in the second half of the ninteenth century and gave rise to a widespread debate that lasted well into the twentieth century and dominated practically all theoretical thinking about language during that period. About 1850, well before there was any talk of topic or comment, a number of (mostly German) scholars observed that if one associates the grammatical subject term with the object or objects the proposition is about and, analogously, the grammatical predicate with the property assigned, then one is in trouble, because in many if not most cases the grammatical structure of a new utterance does not reflect at all what one understands as being the (new) property assigned to the (given) object or objects. Steinthal formulated this problem as follows:

> One should not be misled by the similarity of the terms. Both logic and grammar speak of subject and predicate, but only rarely do the logician and the grammarian speak of the same word as either the subject or the predicate.... Consider the sentence *Coffee grows in Africa.* There can be no doubt where the grammarian will locate subject and predicate. But the logician? I do not think the logician could say anything but that 'Africa' contains the concept that should be connected with 'coffee grows.' Logically one should say, therefore, 'the growth of coffee is in Africa.'... Grammatical form is therefore a completely free, subjective product of the popular spirit. (Steinthal 1860: 101–2; translation mine)

We would, of course, no longer agree with the last sentence, which betrays a basic lack of understanding of what syntax amounts to—a misconception that persevered until well into the twentieth century (and still stains the theory known as Cognitivism; see Section 1.3.3.1). But apart from that, the observation that *Be in Africa* should be seen as the predicate and *the growing of coffee* as the subject term, in terms of the original Aristotelian or Stoic concept of proposition, is both acute and accurate. It gave rise to the enormous subject–predicate debate that raged in linguistics and in the philosophy of language from about 1860 till the 1930s, when it petered out, mainly, one is inclined to think, for lack of a clear solution. Scholars like Meyer-Lübke, Von der Gabelentz, Wegener, Lipps, Stout, Wundt, Mathesius, Gardiner, and others added to Steinthal's observations and proposed solutions. The idea was, on the whole, that sentences should be considered to have a double structure. On the one hand, there is a grammatical structure where a predicate is distinguished that takes a subject term and often also a direct and/or indirect object term plus any number of adverbial adjuncts. On the other hand, there is a 'psychological' or 'logical' structure, which specifies the property 'really' assigned to the 'really' given object or objects and thus consists of the two main components, subject and predicate, at what for us is the semantic or L-propositional level of analysis.

Although the observations were sharp and correct, one tended to get lost in conceptual and terminological confusion, mainly owing to the lack of clarity of the notions and the criteria involved. The only substantial theoretical advance, we may say in hindsight, was made by a few scholars, in particular Wegener, Lipps, and Stout, who proposed that the 'psychological' or 'logical' structure should be taken to be bound up with discourse structure. A telling passage, where the terms *subject* and *predicate* are taken in the psychological, not the grammatical, sense, is the following, taken from Stout:

> The predicate of a sentence is the determination of what was previously indeterminate. The subject is the previous qualification of the general topic or universe of discourse to which the new qualification is attached. The subject is that product of previous thinking which forms the immediate basis and starting-point of further development. The further development is the predicate. Sentences are in the process of thinking what steps are in the process of walking. The foot on which the weight of the body rests corresponds to the subject. The foot which is moved forward in order to occupy new ground corresponds to the predicate.... All answers to questions are, as such, predicates, and all predicates may be regarded as answers to possible questions. If the statement, 'I am hungry' be a reply to the question, 'Who is hungry?' then 'I' is the predicate. If it be the answer to the question, 'Is there anything amiss with you?' then 'hungry' is the predicate. If the question is, 'Are you really hungry?' then 'am' is the predicate. Every fresh step in a train of thought may be regarded as an answer to a question. The subject is, so to speak, the formulation of the question; the predicate is the answer. (Stout 1909, vol. 2: 212–14)

As a result of this new insight, the Prague scholar Mathesius (1928) proposed his theory of topic and comment or 'functional sentence perspective', a proposal which has since been promoted by the Prague school of linguistics and has led to important but formally not very well sustained developments in pragmatics and discourse studies later on in the twentieth century.

The relevance of this complex question in the present context is the fact that the original Aristotelian concept of proposition is now seen to reflect a discourse-bound, topic–comment structure rather than a grammatical predicate–argument structure. Yet the linguistic tradition, starting with the ancient Alexandrian grammarians, simply identified the part of the proposition that is keyed to the thing or things spoken about with the part of the sentence that is in the nominative case, the rest being treated as the predicate, which may contain object terms and adjuncts. This identification is now seen to have been too rash, since it is simply not true that the grammatical subject always denotes the object to which a property is assigned in the underlying mental proposition.

Nevertheless, the grammatical subject–predicate analysis also became the analysis adopted in the logical tradition of the Middle Ages and later. For many centuries this tradition analysed propositions as consisting of a (possibly complex) predicate and a subject in the *grammatical* sense of these terms—that is, with the subject in the nominative case and the verbal element in the predicate agreeing in person, number, and, where appropriate, gender with the subject term (the language used was, of course, Latin). During the twentieth century, the grammatical subject–predicate analysis was enriched in the sense that linguists began to analyse sentences, and logicians began to analyse L-propositions, as consisting of an *n*-ary predicate plus a number of terms and adjuncts. Although this analysis, based on grammar, constituted an important enrichment of logic, it also meant a further distortion of the original Aristotelian subject–predicate concept of proposition, which is essentially a topic–comment structure and differs, therefore, from the grammatical structure of surface sentences.

In fact, Aristotle found himself in a dilemma, probably without being aware of its precise nature. On the one hand, his concept of proposition implied an occasion-bound token occurrence of mentally assigning a property to one or more given objects: 'a proposition is an affirmative or negative expression that says something of something', as we saw in Section 3.1.2. On the other hand, however, he could only develop a logic on the basis of type-level semantic properties of the token occurrences involved. One now understands more clearly why he decided to ban all occasion sentences, with their token-specific properties, from his logic and take the first steps on the road that was continued by Russell and Quine two millennia later. But, as has been said, this is not a road we can take.

The result has been that the original Aristotelian discourse-bound topic–comment notion of the proposition, which, therefore, requires contextual anchoring, was lost and was replaced with a static, discourse-independent *n*-ary analysis in both grammar and logic. Does this matter? The grammarians tend to say it does not, because, they say, the topic–comment distinction is a matter of main nuclear accent and hence of intonation, not of grammar.

But they are wrong. That the topic–comment distinction is also a matter of grammar and semantics can be seen, for example, from cases of reflexivization. Take the following little dialogue between a father and his young son, who is crying because he has just hurt his knee:

(3.6) Father: Well-educated boys don't cry.
Son: **I** didn't educate me, YOU did!

If, as it is reasonable to assume, reflexivization of nonsubject pronouns expresses the fact that the predicate has been used in the reflexive mood ('self-hate', 'self-wash', 'self-admire', and so on), one will wish to have a principled account of why the predicate cannot reflexivize, *salva significatione*, in the son's answer but requires that the pronoun *me* occurs as a fully referential expression: the reflexivized sentence ***I*** *didn't educate myself,* YOU *did!* does not express what the boy wanted to express. The difference is naturally expressed by means of a paraphrase in terms of the cleft construction 'the one who educated me is not ME, but YOU', which differs from 'the one who self-educated is not ME, but YOU', where the predicate *educate* has been reflexivized. If this paraphrase relation is given the status of explanation, all that remains is for the grammar to show how the two underlying cleft constructions are turned into their corresponding surface forms. This shows that it is reasonable to assume that main nuclear accent reflecting topic–comment structure does play a role in defining what makes for grammatically well-formed expressions of given underlying propositions.

Further indications of the grammatical relevance of topic–comment structure are not hard to come by. In Turkish, for example, yes/no questions are characterized by the suffix *-mı*—that is, *mi, mü, mı* or *mu*, according to the laws of vowel harmony—attached to the L-propositional comment–predicate of the question. Thus, Turkish has the following variants:

(3.7) a. Mehmet kitab-ı al/AL-dı-**mı**?
Mehmet the book$_{OBJ}$ buy/BUY$_{PAST\text{-}QUESTION}$
Did Mehmet buy/BUY the book?

b. MEHMET-**mi** kitab-ı al-dı?
MEHMET$_{QUESTION}$ the book$_{OBJ}$ buy$_{PAST}$
Did MEHMET buy the book?

c. Mehmet KITAB-I-**mı** al-dı?
(Mehmet the BOOK$_{OBJ\text{-}QUESTION}$ buy$_{PAST}$)
Did Mehmet buy THE BOOK?

Lexical choice may be influenced by topic–comment structure. Consider the difference between German *aber* and *sondern*, both meaning 'but', yet *sondern* can only be used to correct a previous comment, as shown in (3.8a, 3.8b). A similar phenomenon is found in Malay and in the Surinamese Creole Sranan

(Seuren 1981, 1983), both of which have special contrastive negations, *bukan* and *a-no*, respectively, to deny a comment, as shown in (3.8c, 3.8d):

(3.8) a. Dieses Haus ist nicht schön, aber billig.
This house is not beautiful, but it is cheap.

b. Dieses Haus ist nicht schön, sondern billig.
What this house is, is not beautiful but cheap.

c. Bukan dia yang datang. Abang-nya.
not he who came. Brother-his.
It is not he who came, but his brother.

d. A-no bigi yu oso. A doti.
is-not big your house. It dirty.
What your house is, is not big but dirty.

Phenomena such as those illustrated in (3.6)–(3.8) show that any adequate linguistic analysis of sentences must take into account their underlying topic–comment structure.

Semanticists and logicians, when asked the same question, tend to answer the way the grammarians do, though for a different reason. Topic–comment structure, they say, does not matter, because it does not affect the truth-value of the sentence in question and is therefore logically irrelevant. Sometimes an appeal is made to the device of LAMBDA ABSTRACTION, which produces formulae that single out one particular argument term from an *n*-tuple of arguments and incorporates the remaining arguments into a complex lambda-predicate. Lambda-abstracted formulae, it is said, are logically equivalent to their non-lambda-abstracted counterparts. For example, instead of writing the binary structure (3.9a), one may write the lambda-abstracted (3.9b) or (3.9c). In (3.9b), the lambda-predicate is λx [Sell (x, the car)] or 'be someone who sells the car'. In (3.9c), the lambda-predicate is λx [Sell (John, x)] or 'be something John sells'. In neither case is there any difference in truth conditions:

(3.9) a. Sell (John, the car) (John sold the car)

b. λx [Sell (x, the car)] (John) (John is one of those who sold the car)

c. λx [Sell (John, x)] (the car) (the car is one of the things John sold)

Which form a speaker selects depends, according to the logicians, on pragmatic factors. As they see it, it is nothing to do with logic or semantics.

This, however, is again inaccurate. Just as the grammarians are tripped by, for example, reflexivization or the Turkish question morpheme, the logicians are tripped by EMOTIVE INTENSIONAL CONTEXTS. Ignoring topic–comment structure (or reducing it to lambda abstraction) steers clear of disaster as long as the logical system is kept strictly extensional. But it goes wrong when intensionality is admitted into the system, since truth-conditional differences do occur when different topic–comment versions are embedded under verbs of emotive propositional attitude, such as *anger* or *surprise*, as one sees from the following examples:

(3.10) a. It surprised Ann that JOHN (and not Kevin) had sold the car.
 b. It surprised Ann that John had sold THE CAR (and not the motorcycle).

(3.11) a. It angered Ann that JOHN (and not Kevin) had sold the car.
 b. It angered Ann that John had sold THE CAR (and not the motorcycle).

The consequences of this fact are momentous. It means, first, that semantic theory will have to explain how the (a)-sentences may be true while the (b)-sentences are false and vice versa. This is important enough in itself, as it adds to Frege's discovery that co-referring terms cannot be substituted *salva veritate* in intensional contexts (see Section 6.1). Here we see that different topic–comment structures cannot be substituted *salva veritate* in emotive intensional contexts. The question is why that should be so. And the answer will have to be that discourse structure is indispensable to semantic analysis, since what causes the emotion in question is expressed by the newly added information, the comment, not by the topic representing given information. But apart from the problems that the observation made in (3.10) and (3.11) poses for semantic theory, logic itself will have to take topic–comment structure into account as soon as it ventures beyond its traditional territory and tries to explore the logic of propositional attitude sentences.

To revert to the observations made by Steinthal and others regarding a second structural layer of topic–comment structure, it is worth mentioning that they are strongly confirmed by the semantic behaviour of negation. This is something our learned nineteenth-century colleagues failed to notice, yet it is remarkable. Consider the negation of Steinthal's sentence:

(3.12) Coffee does not grow in Africa.

In the normal interpretation, this sentence does not deny that coffee grows, or that there is coffee, but rather that it is *in Africa* that coffee grows, leaving the fact that there is coffee and the fact that coffee grows unaffected. This follows from the analysis in terms of topic–comment L-propositional structure, since (3.12) is then analysed as:

(3.13) Not [Be_V Africa (the x [coffee grows in x])]

which says nothing about the reality of the 'object' referred to by the subject term, namely the place where coffee grows, but only about the applicability of the property denoted by the predicate *Africa* to that 'object'.

We thus take it that sentences have, in principle, two layers of structural analysis. There is first a grammatical *n*-ary analysis in terms of a predicate and its *n* argument terms, with possible adjuncts added. If I may be allowed to speculate for a moment, I would say that this kind of structure probably came first in the evolution and development of human language, as it is grammatically most deeply embedded in the structure of sentences and semantically simple in that it does not have to carry the weight of contextual anchoring. It is also the most obvious to the analytical eye and, probably for that reason, it is the structure that is used in standard logical analysis. Then there is the topic–comment structure, which is in fact the structure that the original Aristotelian, or the original Stoic, concept of a proposition is based on. I would guess that this analytical layer is probably younger in the evolutionary history of human language, as it appears to have been incorporated into the grammar in secondary ways, largely, but not exclusively, by means of an intonational overlay (main nuclear accent) over the already fixed syntactic structure. If this is indeed the case (but I may well be wrong on this), syntactic clefting constructions must also be younger than the canonical predicate–argument structure.

It may be objected that the very first manifestations of language were almost by definition strongly tied up with specific, concrete situations. True, but this only means that *deixis* (in both verbal and gestural forms), as well as *ad hoc interpretation mechanisms*, must have been prominent in very early language. What is meant here is that the technical machinery for the *systematic* anchoring of utterances to preceding context, such as topic–comment structure, presuppositions, and anaphora, was probably developed way after the emergence of 'plain' syntactic structure. But who is to tell?

3.3 Truth, lexical satisfaction conditions, and valuation spaces

3.3.1 *Truth*

We adhere in principle to the Aristotelian concept of truth as correspondence, applied both to mental propositions and to well-anchored L-propositions interpreted in terms of a given referential key. As is shown in Seuren (1998: 12–15), Aristotle applies the predicates *true* and *false* to both thoughts and sentences, but it is clear that he sees thoughts as the primary bearers of these truth-values. At *Metaphysics* 1027b25, as was shown in Section 3.1.2, he speaks of truth and falsity as properties of thought. And again at *Metaphysics* 1051b1–16, where truth and falsity are regarded primarily as properties of thought, and only secondarily of spoken sentences (Aristotle's terminology of 'objects being united' or 'divided' is to be read as objects belonging or not belonging to the extension of a predicate):

Now truth and falsity depend, in the case of objects, upon their being united or divided; he who thinks of what is divided that it is divided, or of what is united that it is united, is right; but he whose thought is contrary to the real condition of the objects is in error. This being so, when do 'truth' and 'falsity' apply or not apply? We must be clear about the meaning of these terms. You are not white because we truly think you are, but it is because you are white that we speak the truth when we say you are.

However, at *Metaphysics* 1011b26 truth and falsity are defined as properties of what is *said*, not what is *thought*:

We begin by defining truth and falsehood. Falsehood consists in saying of that which is that it is not, or of that which is not that it is. Truth consists in saying of that which is that it is, or of that which is not that it is not.

But, leaving this ambiguity aside, we see that, for Aristotle, truth consists in a *correspondence* between what is thought/said and what is the case in the world, and falsity in a noncorrespondence.

An important corollary to the Aristotelian analysis of the concept of truth is brought along by the fact that, for occasion sentences, propositions (or the L-propositions expressing them) cannot be true without the mind being trained on specific objects and states of affairs—that is, without keying. Since keying requires anchoring and since both keying and anchoring are mental processes, it follows that the mind is an indispensable factor in the truth relation, at least for occasion sentences. Truth, though still correspondence between what is thought/said and what is the case in the world, is now seen to require mental processes for it to come about. The extensionalists' ideal of a mindless logic is thus seen to be illusory. More is said about this aspect in Section 3.4.1, in connection with the Principle of Cognitive Independence, which is the dominant viewpoint in modern logic.

A question, only very partially answered by Aristotle, is: what does this correspondence consist in? When we want an answer that is as formally precise as possible, we had better stick to the *verbal* concept of truth as correspondence between what is said and what is the case, since notions about the structure of thoughts are still vague and indeterminate, whereas notions about the structure of sentences and their corresponding L-propositions are much more precise, at least nowadays. Therefore, when correspondences are to be formally defined, L-propositions are more useful than propositions. This view was expressed as follows by the German philosopher Wolfgang Stegmüller, who used the term 'judgement' for what we call 'proposition' (Stegmüller 1957: 16–17; translation mine):

> The concept 'true' is used as a predicate. But as a predicate of what? ... In principle there are two possibilities. 'True' can be regarded as a predicate of sentences as linguistic structures, or as a predicate of that which is expressed by these sentences, i.e. of judgements. At first sight it might seem that the decision should go in favour of judgements. For sentences are there for the judgements, and not the other way round. Sentences are merely linguistic tools; what matters is the judgements expressed in them. Yet it is not advisable to take the judgement option, since one gets immediately involved in a mass of philosophical discussions that have repeated themselves endlessly for a long time without ever leading to any satisfactory solution and thus to agreement among the investigators involved. When the sentence, as a linguistic expression, is taken as the point of departure, such discussions ... can be avoided. We see here a factor that has dominated the whole development of the newer logic. Whenever logical analysis did not take the linguistic expression as its point of departure, three factors would invariably play up: 1. One got immediately involved in a preliminary discussion on whether this kind of investigation was really logical, and not one that belonged to a different discipline such as psychology. ... 2. The very reality of the phenomena concerned was often called into question. ... 3. The decisive argument in this context is, however, that only those investigations that sought a connection with linguistic expressions have led to really positive results. The whole development of symbolic logic, and indeed of the modern logic of science, has been possible only because it was a sentence logic and not a judgement logic taking the 'cognitive content' of a sentence as its object of enquiry.

On the whole, Stegmüller's advice is sound and is followed up in the present book whenever we intend to speak of truth in a precise way. This should, however, not be confused with the view, dominant in modern logic and modern formal semantics, that the determination of truth is entirely a matter of linguistic form. In Section 3.4.1, it is argued extensively that cognitive input is essential for the determination of truth and falsity. But the extent to which, and the form in which, cognitive input is used in determining truth-values is guided and limited by linguistic form.

What needs to be specified now is what we mean by linguistic form. What we do not mean by it is the surface form of natural language sentences, in all their flexible, stylistically varied manifestations. Aristotle already saw that some sort of regimentation of surface linguistic form is necessary for a systematic determination of truth-values, and thus for the development of a sound logic. Modern logic goes a little further and posits that truth determination is best served by using the language of predicate calculus (LPC), a product of the cumulative effort of Frege, Peano, and Russell, rather than natural language surface structures, to define truth and truth relations.

It was through the work of James McCawley and others (including the present author), during the late 1960s and the early 1970s, that it became clear on empirical grounds that, indeed, LPC embodies, in main outline, the best available hypothesis about the form that should be taken to underlie, both semantically and syntactically, the surface structures of natural language sentences, their 'deep' structure. In fact, we hold that, when an utterance is produced, a mental proposition (Stegmüller's 'judgement') is first cast, at a purely mental level of representation, into the mould of an underlying linguistic structure in some variety of LPC. The term we normally use for this level of representation is 'L-proposition', but we also speak of 'semantic analysis' or SA. L-propositions are mapped onto surface structures of sentences by the grammar of the language.

We take it that the general format of L-propositional form is the following:

(3.14) $F(a_1, \ldots, a_n)$

where F is the predicate and $a_1, \ldots, a_n$ stands for the appropriate number *n* of terms to which the predicate F is applied. If, as is the case for the majority of L-propositions, the L-proposition expresses a discourse-bound proposition with a topic–comment structure, then F stands for the comment and the one term *a* for the topic. The grammar of the language in question will normally have means for transforming the topic–comment structure into either an *n*-ary surface structure with main nuclear accent or a cleft construction.

Given the general format given in (3.14), we can define truth as follows:

(3.15) **Definition of the concept 'truth of an L-proposition $F(a_1, \ldots, a_n)$':**
A properly anchored and keyed L-proposition $F(a_1, \ldots, a_n)$ is true just in case the *n*-tuple of objects $< \rho(a_1), \ldots, \rho(a_n) > \in [\![F]\!]$—that is, just in case the series of objects denoted by the terms $a_1, \ldots, a_n$, in that order—satisfies the conditions posed by the predicate F.

We use the symbol 'ρ' for the cognitive-intentional function of referential keying: for any term *a*, $\rho(a)$—the ρ-value of *a*—is the object (of whatever kind) referred to by the term *a*. The notation $[\![F]\!]$ is used in the standard way: it stands for the *extension* of the predicate F—that is, the set of objects (or, for *n*-ary predicates, the set of *n*-tuples of objects) that satisfy the conditions posed by F for the attainment of truth.

It is possible to apply this form of definition to underlying mental propositions. When we do that, we can say:

(3.16) **Definition of the concept 'truth of a proposition +P(X) ([o_1],..., [o_n])':**
A proposition +P(X) ([o_1],..., [o_n]) is true just in case the *n*-tuple of objects <o^*_1,..., o^*_n > satisfies the conditions posed by the property P(X), where **o*** is the actual object **o** if there is one, and otherwise the virtual object ***o***, mentally created and represented by [**o**].

This, however, is less helpful than the definition of truth of an L-proposition given in (3.15), because, as has been said, the language of L-propositions is well defined and open to logical computation, whereas we do not have a well-defined 'language' of mental propositions, let alone a logical calculus grafted onto it.

3.3.2 *Satisfaction conditions for the predicate* true

How do we specify the conditions posed by a predicate F for the attainment of truth, the SATISFACTION CONDITIONS of F? In principle, one wants the satisfaction conditions of a predicate F (Frege's 'Sättigungsbedingungen') to be specifiable in terms of conditions to be fulfilled by any object or *n*-tuple of objects to which F is applied in an L-proposition P for P to be true. In practice, however, it turns out to be extremely hard to specify satisfaction conditions of predicates, owing to their apparently whimsical behaviour in different contexts. Yet for the moment we stipulate, as a matter of principle, that it is the satisfaction conditions of F that define the MEANING of F. The satisfaction conditions of F are formally presented as conditions on membership of the extension of F, or [[F]], as defined earlier.

The question of what metalanguage or technical terminology is to be used in specifying the satisfaction conditions of lexical predicates has not, so far, found a satisfactory answer. In practice, a makeshift device is used for approximating such specifications with the help of some more or less regimented variety of English, although one knows that this method is basically flawed. Lexicographers are familiar with the problem at the practical end. They know how hard it is to get satisfaction conditions right. Frequent appeals must be made to human experience, impressions, subjective evaluations, functional use, and in particular to the cognitive phenomenon known as prototypicality, which co-determine the meanings of lexical predicates, as is amply illustrated in Chapters 8 and 9.

Such complications, however, hardly arise in the case of the predicate *true*, at least when it is taken in its literal sense, without any polysemous extensions (as in *true to life*, *the true love of his life*, *a true diplomat*, and so on). For cases like (literal) *true* we can safely adopt the descriptive device of satisfaction conditions, as set out earlier. The notation used for one-term predicates is as in (3.17a); for two-term predicates the notation is as in (3.17b):

(3.17) a. $[\![\text{PRED}]\!] = \{ x \,|\, x \ldots \ldots \ldots \}$

b. $[\![\text{PRED}]\!] = \{ <x,y> |\, x \ldots\ y \ldots \ldots \}$

The conditions to be fulfilled for anything, or for any *n*-tuple of objects, to 'deserve' the predicate in question are formulated within the curly brackets. The formula $\{ x \,|\, x \ldots \ldots \ldots \}$ is to be read as 'the set of all *x*'s such that *x* . . . ', with the conditions filled in for the dots. For the one-term predicate *house*, for example, the formula may read 'the set of all *x*'s such that *x* is a permanent building intended for inhabitation by a limited number of persons.' Although it is doubtful whether the format of (3.17a, 3.17b) is sufficiently flexible to render the manifold varieties in which lexical meanings manifest themselves, it would seem that it will do for the limited purposes of the technical predicates of logic and metalogic, such as the predicate *true*.

How do we capture the predicate *true* in the format of satisfaction conditions as illustrated in (3.17a, 3.17b)? We can 'translate' (3.15) into that format as follows, which has the advantage that the language of L-propositions is formally well defined ('F $(a_1,\ldots, a_n)$' ranges over well-anchored and well-keyed token occurrences of any given L-proposition):

(3.18) $[\![\text{True}]\!] = \{ F\ (a_1,\ldots,a_n) \mid < \rho(a_1),\ldots \rho(a_n) > \in [\![F]\!] \}$
(The extension of the predicate *true* is the set of well-anchored and well-keyed token occurrences of any L-proposition F $(a_1,\ldots, a_n)$ such that the *n*-tuple of objects referred to by $a_1,\ldots, a_n$ is an element in the extension of the predicate F.)

Or, if one wants a definition of *true* to reflect the fact that truth and falsity are primarily properties of propositions, we can take *true* to be a predicate over propositions (see (3.16) for the symbol **o***):

(3.19) $[\![\text{True}]\!] = \{+P(X)\ ([\mathbf{o}_1],\ldots, [\mathbf{o}_n]) \mid <\mathbf{o}^*_1,\ldots, \mathbf{o}^*_n>$ satisfies the conditions of P(X)}
(The extension of the predicate *true* is the set of well-anchored and well-keyed token occurrences of any proposition +P(X) $([\mathbf{o}_1],\ldots, [\mathbf{o}_n])$ such that the *n*-tuple of objects $<\mathbf{o}^*_1,\ldots, \mathbf{o}^*_n>$ satisfies the conditions posed by the property **P(X)**.)

3.3.3 *Valuation spaces*

Definition (3.18) requires that the objects satisfying the predicate *true* are well-anchored and well-keyed token occurrences of L-propositions, whose predicates must be satisfied for truth to arise. On the basis of this a further step can be taken, which brings the predicate *true* in line with the operators of propositional calculus when these are treated as predicates, as is done in Section 2.3.4 in Volume II. This method makes it possible to treat *true*, together with the operators of propositional calculus, as a TRUTH PREDICATE—that is, in a format specific to predicates that take well-anchored and well-keyed token occurrences of L-propositions as term referents and whose satisfaction conditions are expressible in terms of mathematical set theory.

To see what is meant, an appeal to VALUATION SPACE MODELLING is of great help. The notion of valuation space (VS) for any given well-anchored and well-keyed token occurrence of an L-proposition was introduced by Van Fraassen (1971). Informally, the VS for any given (well-anchored and well-keyed) L-proposition P, or /P/, is the set of situations in which P is true. A situation *sit* is a possible state of the world (a more technical definition is given in Section 2.3.3 in Volume II) Since there always is one specific privileged situation sit_{act}, which is the ACTUAL state of the world, it makes sense to say that a token occurrence of an L-proposition P is true in some situation *sit*, but not in the actual situation sit_{act}. The universe of situations U is defined as the set of all possible situations.

It would be wrong to think that what are called situations here amounts to the possible worlds in formal, model-theoretic semantics. In the approach developed here, situations play a very different role from possible worlds in model-theoretic semantics. Our (sets of) situations are, to begin with, mathematical constructs used to define the properties of logical systems and to compare and evaluate logical systems with regard to each other. Apart from that, however, (sets of) situations may well have some psychological reality and as psychologically real objects they may well be endowed with some degree of formal precision, provided they stay within certain definable limits of naturalness. In taking this view, we may leave open the question of what precise 'mental model' does the work (Johnson-Laird 1989). In model-theoretic semantics, by contrast, the possible worlds are an integral part of the ontology, containing the things spoken about.

The concept of valuation space can now be informally described as follows:

(3.20) VALUATION SPACE (VS):
The valuation space /P/ of a well-anchored and well-keyed L-proposition P is the set of situations *sit* such that, for each $sit \in$ /P/, P is true in *sit*.

The VS of an L-proposition or of a set of L-propositions P is thus defined by the truth conditions of P.

This makes it possible to specify the meaning of the predicate *true* as follows, where *X* ranges over L-propositions *modulo key* and, therefore, /X/ over the valuation spaces of L-propositions (or sets of L-propositions), again *modulo key*:

(3.21) $[\![\text{True}]\!] = \{ X \mid sit_{act} \in /X/\}$
(the extension of the predicate *true* is the set of all L-propositions X such that the actual situation sit_{act} is a member of the VS of X)

Definition (3.21) is, of course, of little use when one wants to find out whether a specific, given proposition is true or false, but it is of great use in the specification of metalogical relations such as entailment, equivalence, contrariety, or contradiction, and in the semantic specification of truth predicates, such as the operators of propositional calculus, whose satisfaction conditions are best expressed in terms of valuation-space relations, as is shown in Section 2.3.2 in Volume II.

3.4 The role of cognition in the emergence of truth-values

3.4.1 *The untenability of the Principle of Cognitive Independence*

This leaves us with the question of how and under what conditions truth-values of L-propositions come about. This question is of particular importance because there exists a great deal of unclarity, among semanticists, pragmatists and philosophers of language, about the role of cognition in the coming about of truth-values, opinions ranging from a totally truth-conditions-free pragmatics-based utterance semantics to a pragmatics-free semantics fully determined by truth conditions. I shall refrain from discussing the wide range of opinions and arguments, mainly because if I did, I would never get round to presenting my own views. What I will do is take the most conservative, standard view of twentieth-century logic as my starting point and see where it breaks down and is in need of modification or repair.

The standard view in logic, and in philosophy in general (but much less in semantics and pragmatics), is that the truth-value of a well-anchored and well-keyed L-proposition P is a metaphysical property of P, adhering to P regardless of the knowledge state of the person producing or comprehending P. Thus, when I produce here and now, sitting in front of my computer in Nijmegen, a token of sentence (3.22), then that token utterance is true or false, regardless of what I know about the weather in New York at the moment of typing:

(3.22) It is raining now in New York.

This much seems incontrovertible, backed as it is by robust intuitions we have about the notions of truth and falsity. Yet one must be careful not to eliminate altogether the role of cognition in the emergence of truth-values. In fact, close observation-cum-analysis quickly shows that, in all sorts of ways, cognition is indispensable in establishing the precise links between type-level lexical and clausal meanings on the one hand and those elements and aspects in any given situation that are crucial for truth-testing on the other. Let us have a closer look.

When one asks logicians what the role is of people, or rather their minds, in the emergence of truth-values, some will say that this is not the concern of logic. This is all right for a logician whose interest goes no farther than mathematical logic *per se*, and we can, therefore, leave mathematical logicians alone from now. But we cannot leave alone those logicians who apply their logic to language. Some of these will answer that the human mind, one way or another, must be equipped with some kind of machinery, probably a form of logical model theory, that checks the world and decides upon a truth-value for any well-formed string of words or morphemes of the given language, the implication being that truth-value assignment is a self-contained, compositional calculus taking as input only a well-defined language and a well-defined state of affairs. This attitude was, to a large extent, shaped by Bertrand Russell, who rejected any 'intrusion of a psychological element' and held that 'the mind . . . is as purely receptive in inference as common sense supposes it to be in perception of sensible objects' (Russell 1903: 33).

I believe one can say that this restricted view of how truth and falsity are determined has done great harm to semantic theory during the twentieth century. So let us give it a name and a definition, in order to be in a better position to show its shortcomings. We say that in standard modern logic the following principle holds:

> *Principle of Cognitive Independence of truth-value assignments: (PCI)*
> The truth-value of any well-anchored and well-keyed L-proposition P depends exclusively on the linguistically defined, type-level meaning of P on the one hand and the state (*cum* history) of the actual world at the moment of P's token production on the other. The mind acts as a rubber-stamp machine doing the processing but not making any contribution of its own.

This purified, unworldly view has been of great use to standard mathematics, which is likewise purely formal, never vague, and whose dependency on mental contingencies is easily factorized out. But it has also been applied, especially since the 1960s, to the study of linguistic meaning, whereby it was assumed that linguistic meaning, like mathematical meaning, is independent of mental contingencies and nonvague. This assumption has, however, proved unwarranted over the past quarter-century. Not only do most predicate meanings in natural language impose contextual restrictions, called preconditions, which generate presuppositions. They are also often vague and/or they incorporate all kinds of purely cognitive (often evaluative) conditions, besides the conditions to be satisfied by the objects themselves to which the predicate is applied. We revert to this point in greater detail in Chapters 8 and 9, but it has to be mentioned here, because of its basic importance.

As it stands, PCI is in any case untenable. It was shown earlier in this chapter that any L-proposition must be anchored in context for sense and keyed to a specific verification domain for truth or falsity. As was said in Section 3.1.4, Quine's eternal sentences are marginal cases, where the dependency on anchoring and keying is reduced to zero or near-zero. In principle, therefore, we maintain that without anchoring and keying there is no proposition and hence no truth-value, though there may be a type-level linguistic structure—an L-proposition or surface structure—that is meaningful but has no truth-value (Strawson 1950). Since both anchoring and keying are mental processes required for the coming about of a proposition, it is impossible to maintain that the mind plays no role other than as a rubber-stamp in the assignment of truth-values. Clearly, the unfortunate attempt by Russell and Quine to reduce all sentences to eternal sentences and to eliminate from the semantic analysis of sentences all 'particulars' that are in need of a key was inspired by the desire to maintain PCI. But the cost to be paid for this is severe linguistic inadequacy.

Definite reference is a clear case in point. Most uttered sentences contain definite terms referring to specific objects or sets of objects to which, truly or falsely, a property is assigned. For the reference relation to be successful it is necessary that the means be available to identify the object or objects in question (Clark and Wilkes-Gibbs 1990). In most cases, these means can only be provided if the mind is in a contextually and referentially restricted information state. Reference clearly requires specific anchoring and keying.

One knows that reference gives rise to multiple problems. Sometimes a definite term in an L-proposition, though well-anchored, may lack a proper referential key. This may happen, for example, when you and I know that Magee has five sons and I utter (3.23), with the definite term *Magee's son*. Then (3.23) may be true or false for me, provided I know which son I mean, but for you, the listener, the truth-value of (3.23) is held in abeyance, pending further cues, because I have not made it clear which of Magee's five sons I want to refer to.

(3.23) Magee's son works as a postman.

In that case the keying condition is not fulfilled for you, the listener, because you cannot reconstruct a proposition, and are thus not in a position to establish a truth-value. But it may also be the case, for example, that I, the speaker, know of only one son of Magee's while you, the listener, may or may not know that Magee has five sons but you know which of Magee's sons I know. Or I may just have seen one of Magee's sons delivering letters and I utter (3.23) to my neighbour with whom I am having a chat across the fence. In such cases, the listener knows which of Magee's five sons I am referring to when I utter (3.23), so that there is a proposition, hence a truth-value, both for the speaker and for the listener. Referential keying, in other words, is heavily dependent on speaker's and listener's knowledge, whether it be encyclopedic or background or situational knowledge.

A similar tale can be told about anchoring. Suppose I start a discourse by uttering (3.24a) and subsequently add (3.24b). Then my discourse is incoherent because (3.24b) makes no sense in the context of (3.24a): there is no known connection between towns and masks. A connection can be forged, but only when more context is added. Given that (3.24b) makes no sense against general world knowledge, it does not express a proposition and thus must do without a truth-value. By contrast, (3.24c) makes perfect sense, because it is fairly normal for an Italian town to have a cathedral:

(3.24) a. There is a town in Sicily called Milazzo.
b. Its mask was made in the twelfth century.
c. Its cathedral was built in the twelfth century.

Now it so happens that Milazzo is a very small town that does not have a cathedral, though, of course, it has churches. Nearby Messina does have a cathedral, which was indeed built in the twelfth century, but Milazzo does not have one. The question is now: does (3.24c) express a proposition and if so, what is its truth-value?

For Russell, as one knows, (3.24c) does express a proposition and thus does have a truth-value, namely false. Russell (1905*a*) analysed sentences like (3.24c) roughly as (3.25), turning (3.24c) from an occasion sentence into an eternal sentence by treating the definite description *its cathedral* as a propositional function under the existential quantifier (the 'uniqueness clause' of Russell's analysis has been disregarded here):

(3.25) $\exists x$ [Cathedral-of-Milazzo(x) $\wedge$ Built-in-12th-century (x)]

By contrast, Frege (1892) and Strawson (1950, 1952, 1954, 1964) held that (3.24c) lacks a truth-value because a property ('be built in the twelfth century') is assigned to a nonexisting object and thus to nothing—which means that no property assignment has taken place and no proposition, hence no truth-value, has come about.

In my view, making contextual sense is a necessary condition for an L-proposition or its corresponding surface sentence to express a proposition, though it is not always a sufficient condition. On my reckoning, (3.24c) does express a proposition and does have a truth-value, namely (a variety of) false. I thus go along with Russell, but only up to a point, since I reject his analysis as given in (3.25) and defend the more traditional analysis, also defended by Frege and Strawson, in which surface definite descriptions are likewise definite descriptions in semantic form—that is, at L-propositional level. For me, as for Frege and Strawson, (3.24c) has the semantic form (3.26) (leaving out tense):

(3.26) Be-built-in-12th-century (the x [Cathedral-of-Milazzo(x)])

But unlike Frege and Strawson, I maintain that (3.24c) and (3.26) do express the assignment of the property P (Be built in the twelfth century) to an object, not an *actual* but a *virtual* object, the thought-up cathedral of Milazzo, which does not exist but has virtual being in virtue of its having been thought up. Sentence (3.24c) thus does express a proposition with a truth-value. And since the predicate *be built* (unlike, for example, the predicate *be worshipped*) presuppositionally requires for truth that the referent of its subject term be an actually existing object, and since (3.24c) fails to satisfy that condition, (3.24c) is not just false but *radically false*, that being the truth-value reserved for cases of presupposition failure. As is illustrated in Figure 3.1, a definite term that occurs in a well-anchored L-proposition and fails to refer to an actually existing object necessarily refers to a virtual object and is, therefore, automatically well-keyed.

In a general sense, therefore, contextual anchoring and referential keying are mental processes that pose conditions for the emergence of a truth-value. This means that we have no choice but to reject PCI as being irreconcilable with natural language. For if sentences normally require anchoring and keying to have a truth-value, it follows that the machinery that does the anchoring and the keying, *in casu* the human mind, must be at some suitable point in the development of a discourse or context and must be intentionally focused on—that is, keyed to—a particular state of affairs for the truth-value assignment to take place successfully.

Further manifestations of the cognitive dependency of meaning—and, therefore, of truth-value assignment, are found with regard to lexical meanings. QUALIA predicates, as is shown in Section 8.2.2, depend for their satisfaction entirely on the sensations corresponding to the qualia properties. In other cases, the satisfaction conditions of predicates depend on what is taken to be NORMAL in a given context. Consider, for example, the predicate *many.* If it is normally so that out of an audience of three hundred taking part in a TV quiz nobody gets through to the maximum amount of prize money, then, when in one session three participants get through to the highest prize, it is true to say that there were many highest prize winners in that session. But if only three out of three hundred people voted for me to be the representative of the cleaning staff in the institute that I work in, then, one fears, it is false to say that many people voted for me.

Or consider cases of what is called 'dynamic filtering' in Section 9.6.3, found all over the lexicons of natural languages. For example, the conditions for the predicate *flat* to be satisfied differ considerably when it is applied to a tyre, a road, or a mountain. And a definite term like *the office* will have different interpretations according to whether it stands as a subject term to the predicate *be on fire* or, for example, *have a day off.* In the former case the phrase *the office* is interpreted as referring to a building. In the latter case it is interpreted as referring to a group of people. Likewise for predicates of possession, expressed in words such as *have, with, without,* or as a genitive or a dative case. As shown in Section 9.6.3, possession predicates typically require an appeal to world knowledge for decisions about truth or falsity. The lexical meaning description of possession predicates must be taken to contain an open parameter referring the language user to his or her knowledge base.

Cases of what is often called 'viewpoint' are likewise telling (see Section 9.7 for more discussion; cp. also Seuren 1998: 402–3). Consider the case of a circular fountain and a tree which are located at a distance of, say, twenty yards from each other. Now to decide whether sentences such as (3.27a) or (3.27b) are true it must be known from what angle the fountain and the tree are being looked at by some 'homunculus' or 'mental viewer' in the mind:

(3.27) a. There is a tree behind the fountain.
b. There is a tree to the left of the fountain.

If the mental viewer is standing in the continuation of the line running from the tree to the fountain, (3.27a) is true and (3.27b) is false. But if the homunculus is standing at right angles with that line, with the fountain to the right, then (3.27a) is false and (3.27b) is true. Clearly, unless one is describing what is seen on a postcard, the viewpoint is determined by previous discourse, and hence by a cognitive process.

Viewpoint differences are likewise to be discerned in cases that show an orientation towards or away from a speaker's mental position, as with predicates of the *come/go*-type:

(3.28) A few men were waiting to be shown into the office. The door opened and one of them went in. After two minutes another man WENT IN / CAME IN.

The difference between the two versions is, although clear, hard to express in other than metaphorical terms. With *went in*, the 'camera' (or the homunculus) has, so to speak, remained outside the office. With *came in*, it has 'moved along' with the first man and is now inside the office. Such phenomena are clearly semantic, in that they reflect meaning differences, even though they do not lead to different truth-values.[9]

Other lexical restrictions look more like being based on criteria of prototypicality (see Section 8.8). Suppose, for example, that John is a man with no hair on his chest. Then, when one says that his chest is bald, that would perhaps not be false in a strict sense, but it would certainly be inappropriate, the reason being, probably, that the predicate *bald* prototypically but not truth-conditionally, requires the top of the head when applied to humans. In similar fashion one observes that what we call 'red wine' is called the equivalent of 'black wine' in Modern Greek. We may say in English that a person, when frightened, turns pale, but not that his or her shoulders are 'pale' when not sun-tanned. Examples like these, of course, come in the thousands.

Lexically fixed prototypicality also has a logical counterpart. For example, when a man says that he is eating out because *all his wives are ill*, then that may be taken to be true in a strict, logical sense,[10] but not in the ordinary sense which would make one understand that he has a collection of wives. This example probably hinges on an element of prototypicality in the linguistic meaning of the logical word *all*.

In view of all this it seems that we are forced to conclude that the Principle of Cognitive Independence lacks empirical backing and seems to have been added to the philosophers' rule book merely to avoid complications in the theory of truth. The fact that it is widely considered sacrosanct in circles of logicians is part of what one might call the mythology of logic. Meaning in natural language is a great deal more complex than what we see of it in standard logic. The standard, logically coloured concept of meaning is highly idealized and streamlined, which creates the impression that one may

[9] *Come/go* predicates do not necessarily require that the speaker/writer's mental position coincides with the speaker/writer. Depending on the language, it is also possible for the speaker/writer to take the position of the listener/reader. In Italian, for example, it is both normal and polite to use the *come*-variant when speaking or writing to a person elsewhere, as in, for example: 'Intendo di *venire* a Roma in Ottobre' (I am planning to *come* to Rome in October), when the addressee is in Rome but the speaker/writer in, say, London.

[10] In modern standard predicate calculus it is even true when the man has no wife at all. In traditional Aristotelian-Boethian logic it requires the real existence of at least one wife.

adequately operate with a strictly bivalent system of simple truth and simple falsity. Since, however, natural language is very much more complex and more varied, there is room for the idea that the logical system of language requires a more finely grained set of truth parameters than just True and False. This is one of the questions explored in this book.

3.4.2 *Reference and the OSTA Principle*

All human (and many forms of nonhuman) perception and interpretation processes appear to be guided by an overarching deeply ingrained functional principle forcing the perceiving subject to do the utmost to make sense of incoming signals. This 'drive for sense' is well-nigh irresistible and can only be turned off after specific training. It is hardly surprising, therefore, that the same is found in the processes involved in the interpretation or construal of linguistic utterances. We call this the *principle of optimization of sense, truth, and actuality*, or OSTA. This principle, which is appealed to repeatedly in the text to follow, compels listeners, first, to try and make optimal sense of incoming messages, no matter how garbled or deformed, then, to do so in such a way that any proposition asserted or presupposed can be taken to be true and finally, to seek truth preferably vis-à-vis the actual world, rather than any virtual world.[11]

It would seem that the three stages of sense, truth, and actuality should be taken to become operative in the order given. First, optimal sense is sought. When there are still questions left, truth is preferred to falsity. And finally, when even then interpretation is not fully determined, reference to actually existing objects is preferred to reference to intensional objects. OSTA is probably also best taken to involve a feedback process from the assumption of truth and/or actual reality back to optimal sense and optimal selection of reference objects. It is obviously premature, given the present state of our knowledge, to try to capture OSTA in terms of a formally defined algorithm (a partial formalization is given in Section 9.5 in Volume II). Yet this does not make the principle less important, not only for a proper understanding of what enables humans to comprehend and interpret utterances, but also, in a more general sense, for cognitive psychology as a whole.

[11] Theo Janssen kindly pointed out to me that a version of OSTA already existed in medieval philosophy of language. Marmo (1995) discusses the concept 'bonitas intelligentis' (the charity of the interpreter) introduced by an anonymous Praguian author (Marmo 1995: 175):

> [I]t is admissible that an equivocal term is determined to mean one of its contents 'de bonitate intelligentis'. The goodness of the 'interpreter' consists in ascribing one meaning to an equivocal term so that no contradiction or non-sense is introduced in the proposition where it occurs.

First and foremost, OSTA optimizes sense. That is, it makes us use all the rules in the book to force sense on utterances, if at all possible.[12] It selects readings and reference relations that fit in best with shared situational, contextual, and world knowledge. Thus, OSTA makes the listener select the Swiss banker mentioned in (3.29) as the antecedent for the expression *the fifty-year-old bachelor* in (3.29a) and *the man* in (3.29b), whereas the expression *the arresting officer* in (3.29b) is identified with the unmentioned agent of the passive predicate *was arrested* in (3.29):

(3.29) Yesterday evening a Swiss banker was arrested at Heathrow Airport.
  a. The fifty-year-old bachelor declared that he had come to Britain to kidnap the queen.
  b. The arresting officer declared that the man had come to Britain to kidnap the queen.

It also makes one see better how metaphor works, as in example (3.30) (for further comment, see Section 9.5):

(3.30) And the train which had picked them at sunrise out of a waste of glaciers and hotels was waltzing at sunset round the walls of Verona.
(E. M. Forster, *Where Angels Fear to Tread*)

In a strictly literal but dumb interpretation, this sentence fails to make sense, since trains are incapable of dancing: the predicates of dancing have a categorial precondition requiring animate beings as reference objects of their subject term. OSTA, however, enables the listener to make sense of (3.30) by considering the train in question, for the purpose of the text at hand, as an animate being capable of dance, which, subsequently, opens a treasure of further possibilities to vary on this theme.

What exactly is meant by *sense* is still a largely unexplored question. Perhaps one should follow Sanford and Garrod, who maintain that the notion of sense is intrinsically linked up with the notion 'scenario' in that the proposition contained in any utterance must be taken to describe a situation that is, or can be, integrated into a cognitive network of relations in such a way that this integration has certain effects on the cognitive network in question (Sanford and Garrod 1998). An utterance of, for example, *Your discord's health will be*

[12] One way to see the power of OSTA is by observing the way young children react to silly questions. As their capacity for lateral thinking is still underdeveloped, they will take any silly question seriously. The other day, for example, a five-year-old boy told me that he had an elder half-brother. I then jokingly asked him whether it was his brother's left or right half that was his brother. He stood thinking for a minute and then replied that his half-brother was on his mother's side—indeed a sensible answer to a truly silly question. What this boy did was reply to a possibly related but sensible question, reflecting the way he applied OSTA.

subscribed to in terms of stock taking is unlikely to fulfil such a requirement and will, consequently, leave the listener/reader perplexed.

It is, of course, true that the term 'sense' is part of the phenomenological phraseology of human experience rejected in Section 1.3.3.1 as being unscientific. For that reason it is important to emphasize that the use of this term implies a research programme reducing the intuitive experience it denotes to general, empirically testable principles of discourse-bound interpretation and of cognition in a general sense. For the time being, the use of the term *sense* is largely intuitive and heuristic, and the analysis aims at making this notion explicit—a common procedure in the human sciences.

Then, OSTA optimizes truth in that there appears to be a definite preference for truth—or, rather, the likelihood of truth—in selecting reference objects. Three examples will suffice here. Seuren (1985: 317–18), gives the example of two professors of English in the same department, one a Dutchman and the other a British national. Now *he* in the second sentence of (3.31) selects as its reference value the British professor (who, in addition, is said to be male), not the Dutch professor, apparently because truth is preferred to falsity. Whatever is said about the man in subsequent discourse is then taken to apply to the person from Britain, not the Dutchman:

(3.31) There is a professor of English in this department. *He* is British.

A similar, and perhaps better, example is given in Clark (1992):

A listener must often take account of what is being asserted about the referent. Suppose two men are walking down the street, one very fat and the other very thin, and you say [(5)], [(6)], or [(7)] to a companion as you nod in their direction:

[(5)] That man weighs too much for his own good.

[(6)] That man weighs too little for his own good.

[(7)] That man is my neighbor.

In [(5)] and [(6)], the only way your companion can find a demonstrative relation *F* is by using what you are asserting about the referent; you would assert [(5)] of the fat man and [(6)] of the thin man, but not vice versa (unless you were being ironic). In [(7)], where he can find no reasonable relation, he is uncertain. (Clark 1992: 94)

Here, the speaker and his companion have a mental representation of the fat man and of the thin man. That is, they have two discourse addresses (DAs), one for each of the two men they see walking down the street. What Clark is saying is that an utterance of [(5)] will automatically make the listening companion select the DA to which [(5)] can be incremented in such a way as to produce truth, and similarly for [(6)], while [(7)] will lead to uncertainty regarding the selection of the proper DA, and hence to referential perplexity, precisely because the listener has no way of knowing whether [(7)] is true or false. The likelihood of truth, in combination with the optimization of sense, is thus instrumental in fixing reference.

The third example is an instance of what is called 'implicit intensionalization' in Section 5.3. Implicit intensionalization typically occurs with fictional characters (see also Linsky 1967: 28). Consider sentence (3.32):

(3.32) Sherlock Holmes is an Englishman.

In a literal interpretation, this sentence is false—radically false, owing to presupposition failure, since to be an Englishman one has to have actual existence, which Sherlock Holmes, despite his great fame, has always lacked. Yet we want to be allowed to consider (3.32) true, since in Conan Doyle's stories, which are part of the world's cultural heritage, there is no doubt at all about Holmes's nationality. So, in order to make (3.32) true, listeners allow the strictly extensional predicate *Englishman* to apply to the intensional object Sherlock Holmes, in virtue of their knowledge of the Conan Doyle stories about him. In doing so, they place the sentence under an implicit intensional operator 'in the Conan Doyle stories', thereby intensionalizing the predicate *Englishman* for the occasion, allowing it to take an intensional object, and not only actually existing objects, as the reference value for its subject term. That is, listeners seek an interpretation in which (3.32) is true, ignoring the interpretation in which it is false.

Finally, OSTA optimizes actuality. To see what is meant by this, one should first realize (as is argued in Section 6.2.2) that any referring definite term necessarily has a virtual object corresponding to it, since referring requires anchoring in a given discourse domain, and discourse domains are mental representations, and representations create virtual objects by definition. These virtual objects may fail to have a corresponding counterpart in the actual world. For such cases the misnomer 'reference failure' is standardly used, which is then taken by some to lead to falsity and by others to the lack of a truth-value. That so-called reference failure may lead to truth appears from cases like (3.33), which is true in the actual world with its actual history:

(3.33) The god Apollo was worshipped by the Greeks on the island of Delos.

Here, the definite term *the god Apollo* fails to refer to an actually existing object, but it still refers to the virtual object denoted by that name. And since that virtual object was indeed worshipped by the Greeks on Delos, (3.33) makes for a true utterance.

The optimization of actuality now means that when there is an actual object available for making a proposition true, then that object is preferred to its virtual counterpart. Thus, an utterance of (3.34), with the definite noun phrase *the cat*, may be made true by an actual cat but also by a merely thought-up and contextualized virtual cat.

(3.34) The cat frightens the little boy.

Yet when an actual cat is available to make (3.34) true, the interpretation machinery settles for that as the reference value, not for its virtual correlate. Only when no actual reference object is available, must the machinery be content with the virtual reference object. It is in this sense that we say that OSTA optimizes actuality.

The functionality of OSTA should be obvious. Optimal sense is sought on the grounds that messages are best taken to have been sent for some good reason. Truth is sought because information that is taken to be true is more conducive to the construction of a coherent and meaningful world view than information taken to be false. And actual reality is sought because the primary necessities of life revolve around tangible rather than merely thought-up objects and states of affairs.

3.5 Existential import is determined by the lexicon

3.5.1 *Extensionality and intensionality of term positions*

We conclude this chapter by giving an account of what causes so-called existential import (ExIm) in the language of modern predicate calculus (and hence in grammatically underlying L-propositions). Existential import is the entailment schema that allows one to infer the actual existence of an entity or object in an L-proposition *P*, on the assumption that *P* is true.[13] There is a general, but superficial, belief that ExIm is licensed by both definite and existentially quantified terms. In the case of definite terms, ExIm is often treated as a presupposition of actual existence, induced by definite determiners, such as the article *the*. This belief is supported by examples like (3.35a) and (3.35b), which entail the actual existence of both the (or a) thief and of my wallet:

(3.35) a. That thief has stolen my wallet.
b. A thief has stolen my wallet.

[13] The notion of entailment is dealt with in detail in Chapter 1 in Volume II, where entailment is defined as an inescapable inference from the truth of a given sentence or set of sentences, based on the meaning of the sentences involved.

Those who are better informed, however, know that this general belief is not only superficial but also false.[14] This appears from sentences such as (3.36a, 3.36b):

(3.36) a. That mermaid has enthralled the whole village.

b. A mermaid has enthralled the whole village.

Clearly, (3.36a) and (3.36b) can be true without there being an actually existing mermaid, which is sufficient to show that the general belief is false. Sentence (3.36a) shows that the definite determiner, *that* in this case, does not induce ExIm and is, therefore, not the source of existential presuppositions. Sentence (3.36b) shows that the existential quantifier ∃ in modern logic, as applied to natural language, cannot by itself induce ExIm, for if it did, (3.36b) would entail the actual existence of at least one mermaid, which it does not. We assume, of course, that the logical analysis, in terms of modern predicate calculus, of (3.35b) and (3.36b) are entirely analogous. In terms of standard modern predicate calculus (3.35b) and (3.36b) are rendered as (3.37a, 3.37b), respectively:

(3.37) a. $\exists x\,[\text{Thief}(x) \wedge \text{Have Stolen}(x, \text{my wallet})]$

b. $\exists x\,[\text{Mermaid}(x) \wedge \text{Have Enthralled}(x, \text{the whole village})]$

Since (3.35b) does and (3.36b) does not carry ExIm, it follows that ExIm cannot be induced by the existential quantifier ∃ as defined in modern predicate calculus. Therefore, when logic is applied to natural language, the existential quantifier must be stripped of its existential force and be curtailed to an entailment of mere being. This corollary is important because it is still almost universally held, among semanticists and philosophers of language, that the natural language counterparts of the existential quantifier induce an entailment of actual existence.

The question thus arises: what determines ExIm? The answer is relatively simple: ExIm with regard to a term *t*, whether definite or quantified, is determined by the satisfaction conditions of the predicate G with regard to *t*. When G requires the actual existence of the referent of a term *a*, we say that G is EXTENSIONAL with regard to *t*. Most predicates are extensional with regard to all of their terms. Extensionality of a term position is, therefore, the default case. But nondefault cases abound. When G does not require the actual existence of the referent of a term *t* but allows for, or perhaps even requires, virtual

[14] Parsons clearly is among the better informed. He presents the following case (Parsons 1980: 7):

(i) There are winged horses—Pegasus, for example.

Like (3.36b), sentence (i) shows that there are existentially quantified sentences without ExIm.

objects as reference value, we say that G is INTENSIONAL with regard to the term position of *t*. The validity of ExIm with regard to a term *t* thus depends entirely on the question of whether *t*'s predicate is or is not extensional with regard to the term position of *t*. In all cases but the predicates of existence (discussed in Section 6.2.2), the requirement of extensionality or intensionality with regard to a term position is laid down in the PRECONDITIONS of the predicate in question, thus giving rise to a PRESUPPOSITION OF EXISTENCE, or the absence thereof.

A few examples will suffice. Transitive predicates like *hit*, *steal*, *touch*, *eat*, and so on, are lexically defined as being extensional with regard to both their terms, as is illustrated by (3.35a, 3.35b). *Enthrall*, as in (3.36a, 3.36b), is lexically defined as intensional with regard to its subject term, since fictitious entities may, though nonexistent, still have the power to enthrall, but it is extensional with regard to its object term. In this respect, *enthrall* differs from, for example, *bewitch*, which is extensional with regard to both its terms, since, it seems, only actually existing entities can be attributed the power to bewitch anything or anybody. *Love* is intensional with regard to its object term, as one may well love a nonexistent being, although the one who loves must really exist. Likewise for predicates like *think about*, *speak about*, *laugh about*, *be afraid of*, and so on, which are all intensional with regard to their objects term but extensional with regard to their subject term. The predicate *be called* '*N*', where *N* is a proper name, is intensional with regard to its subject term, since it is not only actually existing but also virtual entities that bear names.[15]

This takes us to the predicates of existence, such as *exist* or *imaginary*, which have caused so much trouble and confusion in modern philosophical literature. Both these predicates are defined as being intensional with regard to their subject term. When the ρ-function gets through to an actual object, *exist* yields truth and *imaginary* yields falsity; and vice versa when this function stops at a virtual object. This ensures that no presuppositions are involved, so that no presupposition-cancelling negation is required for negative sentences with these predicates to produce truth. In this respect, existence predicates differ from other predicates with ExIm, which require actual existence

[15] This, in fact, solves the problem Saul Kripke had no answer to (Kripke 1972, 1980): how can proper names be attached to virtual entities if they are so-called 'rigid designators'? The answer is that proper names are not 'rigid designators' but simple noun phrases of the type 'the x such that x is F', where F is a predicate. Thus, the NP *the butterfly* is analysed as 'the x such that x is a butterfly'. Analogously, *Nixon* is analysed as 'the x such that x is called "Nixon"'. The predicate *be called 'Nixon'* leads to truth or falsity in virtue not of physical but of social reality, as do predicates like *be guilty/innocent*, *be rich/poor* and many others.

presuppositionally—the existential presupposition that is widely, but wrongly, regarded as being induced by definite determiners.

3.5.2 *The reference hierarchy and a corollary for predicate logic*

The mechanism described above crucially depends on what we call the 'reference hierarchy' for definite terms, which is described as follows. The ρ-value of a definite term is fixed in a stepwise procedure, guided by the so-far largely unformalizable process of intentional keying. Every interpretable definite term has a VIRTUAL ρ-value, or else it cannot be used in an interpretable (well-anchored) L-proposition. Some interpretable definite terms also have an ACTUAL ρ-value instantiating their virtual ρ-value. When an actual ρ-value is available for a given term, the interpretation machinery settles for that as its final ρ-value. (Thus when I say that a cat is worshipped in some place, and the proposition is made true by an actual cat, then all subsequent talk about the cat worshipped in that place is talk about the actual cat, not its virtual counterpart.) But when no actual ρ-value is available, there is always, by definition, a virtual ρ-value, or else the L-proposition in which the term in question occurs is uninterpretable (unanchored). In the absence of an actual ρ-value, the reference process automatically stops at the virtual ρ-value, which is always there. This reference hierarchy is a direct consequence of the overarching, inbred and highly functional OSTA principle, introduced in Section 3.4.2, guiding the interpretation or construal of utterances.

When a predicate F under which the term in question occurs in a given position is extensional with regard to that position and the ρ-value is restricted to a virtual object, then falsity ensues and vice versa, as has been stipulated. In the light of this machinery, it is at least misleading to speak of 'reference failure' in the absence of an actual ρ-value, (a) because there always is the virtual ρ-value and (b) because absence of an actual ρ-value may well lead to truth, as has been shown. Real reference failure—that is, absence of a ρ-value for a definite term—inevitably leads to uninterpretability and hence to there not being any underlying proposition at all, as in cases where sentences or their underlying L-propositions are presented to illustrate a point of grammar or of semantics.

This has an important corollary for predicate logic. Standard modern predicate logic violates natural intuitions in that the universal quantifier yields truth when the class quantified over is null. By contrast, in traditional predicate logic, usually crowned with the name 'Aristotelian', an inconsistency arises when the class quantified over is null, making this logic faulty in a strictly extensionalist ontology admitting only of actually existing objects. Yet this logic corresponds more closely to natural intuitions. In Section 10.7 in Volume II it is shown that this conflict is solved in favour of traditional logic, since, by definition, the class quantified over cannot be null: if there is no class of actual objects to quantify over, there always is a class of virtual objects as values for the quantifier's variable—on pain of uninterpretability.

It is unnecessary, and also wrong, to add a (pre)condition of existence to the universal quantifier, thereby creating an existential presupposition for all universally quantified sentences. No presupposition of existence induced by the universal quantifier is needed, since the acceptance of virtual objects as possible reference objects will do. What is needed is a specification of whether the sentence's main predicate G is extensional or intensional with regard to its term positions. In cases where G is extensional with respect to a term *t*, G induces a presupposition of actual existence of ρ(t), the reference value of *t*.

Thus, a sentence like (3.38a) is to be judged false on grounds of presupposition failure, since the predicate *be married* has the precondition that the ρ-value of its subject term be an actually existing person—unless (3.38a) is interpreted under a tacit intensional operator such as 'in the stories', as in (3.32) above, in which case it is also false, but not on presuppositional grounds. By contrast, (3.38b) is true, because the predicate *fictitious* (see Section 6.2.2) does not carry such a precondition and because the actual world has never been embellished by an actually existing mermaid:

(3.38) a. All mermaids are married.

b. All mermaids are fictitious.

Much more remains to be said about this intriguing topic, but it is better to proceed step by step and not to unload everything in one go.

4

Speech acts: socially binding force

4.1 The emergence of speech act theory

So far we have been speaking about propositions, L-propositions with their logic, and surface sentences, and among these we have distinguished the category of assertions or statements, considering them, along with most of the logical and philosophical tradition, to be expressions of propositions. The tradition, considering only assertions and taking these to be expressions of propositions, has mostly taken it to be the essential function of language to describe how things are, or, in the more trendy twentieth-century formulation, to 'convey information' about the world. This, however, cannot be the full story, since an assertion is more than just the expression of a proposition. It has been known, of course, since Aristotle that there are other sentence types beyond assertion. As was pointed out in Section 3.1.2, Aristotle (*Int* 17a1–7) distinguished assertions from, for example, wishes, but he felt that such distinctions were more properly dealt with in the study of rhetoric or poetry (nowadays one would say 'pragmatics'). And that is largely the way matters stayed until quite recently.[1]

It was not until the late nineteenth century that some began to see that sentence type distinctions involve more than rhetoric, poetry, or pragmatics. A very early reference is found in the linguistics literature. Sadock (1975: 16–17) quotes the American linguist William D. Whitney (1827–94), who writes:

When, now, we come to speak in our own persons, we change *ich behaupte, dass du mich liebst*, 'I maintain that thou lovest me', into *du liebst mich*, 'thou lovest me', the assertion of the assertion being usually a quite unnecessary formality; *ich will wissen, ob du mich liebest*, 'I wish to know whether thou lovest me', becomes *liebst du mich*,

[1] The Greek grammarian Apollonius Dyscolus (2nd cent. CE) is sometimes credited with having spotted performative verbs as an implicit part of semantic structure. This is, however, somewhat overdrawn. In his *On Syntax*, Apollonius writes (Householder 1981: 219) '...it is clear that *I call you a thief* signifies "I assert by means of the speech which I am uttering that you have done the act of stealing" '. But there is no sign of Apollonius having seen the significance of performative-verb usage or of him having detected the speech-act quality of utterances. All this quote represents is a casual remark, whose significance obviously escaped the ancient author.

'lovest thou me?', the wish to know being intimated by arrangement and tone; and *ich verlange, dass du mich liebest*, 'I require that thou lovest me', is changed into *liebe du mich*, 'love thou me!' the desire or demand being expressed by arrangement, tone and appropriate verbal form. (Whitney 1870: 205)

This interesting observation, however, was never picked up by the linguistic world, which did not wake up to the distinction between implicit and explicit performatives until it had been proposed by the Oxford philosopher John Austin.

Frege made a distinction, in various places, between the SENSE (Sinn) and FORCE (Kraft) of a sentence, intending the term *force* to capture whatever it is that brings about the socially binding effect of speech acts. But, as amply demonstrated by Dummett (1973: 295–363; 1981: 306–7), Frege had no clear idea of how to distinguish between what we have called a proposition and its being incorporated into a speech act of assertion, question, command, and so on.[2] Nor, as is pointedly shown by Dummett (1973: 311–12), had other philosophers of the day, in particular Russell, who in his *Principles of Mathematics* of 1903 got hopelessly entangled in the distinction between the uttering and the asserting of a truth.

An important further step was taken by the American philosopher Clarence I. Lewis (1883–1964). In his remarkable book *An Analysis of Knowledge and Valuation* of 1946 we read:

A proposition is a term capable of signifying a state of affairs. To define a proposition as an expression which is true or false, is correct enough but inauspicious, because it easily leads to identification of the proposition with the statement or assertion of it; whereas the element of assertion in a statement is extraneous to the proposition asserted. The proposition is something assert*able*; the *content* of the assertion; and this same content, signifying the same state of affairs, can also be questioned, denied, or merely supposed, and can be entertained in other moods as well.

For example the statement, 'Mary is making pies,' asserts the state of affairs, Mary making pies now, as actual. 'Is Mary making pies?' questions it; 'Oh that Mary may be making pies,' expresses it in the optative mood; and 'Suppose that Mary is making pies,' puts it forward as a postulate. When we say 'If Mary is making pies, then pies are being made by Mary,' we consider it and affirm that it has a certain logical consequence. And if we state, 'Either Mary is making pies or we shall have no dessert,' we likewise entertain it without assertion but affirm its being one of two alternatives.

[2] The term *speech act* is used indiscriminately to refer to an act of speaking or of writing.

If we wish to disengage this common content from any particular mood of its entertainment, we might do so—in a manner more precise than ordinary language commonly affords—if we should have symbolic devices indicating these various moods of entertainment; e.g., '+*p*' for assertion of '*p*', '*Hp*' for the postulation of it, '! *p*' for the mere greeting of it as a presentation of sense or imagination, '? *p*' for putting it in question, '*Mp*' for entertainment of it as consistently thinkable or possible. and so on.[note 8] And the common content, here represented by '*p*', would be something expressible in the manner of indirect discourse, e.g., 'that Mary is making pies now', or by a participial phrase, 'Mary making pies (now)', which can be asserted, questioned, and entertained in all these different ways, and which signifies the state of affairs which they all concern.

Note 8: Exact logic has not as yet much concerned itself with these various moods of entertaining propositions. Assertion is recognized, and postulation—though postulation is usually dealt with inconsistently and confused with assertion. Also it begins to be understood that the imperative or hortatory mood has its own logical principles, and that the so-called modal statements, of possibility and necessity, demand separate consideration. If these matters were to be adequately treated, we should, of course, expect some attempt at economy, the reduction of some moods to expression in terms of others. (Lewis 1946: 48–9)

In the Preface, Lewis makes the following attribution to Sheffer:

The manner of interpreting propositions, in Chapter III, is the same in essentials—though the terminology and some details are different—as that which Professor H. M. Sheffer has for some years presented to his classes. (Lewis 1946: xi–xii)

We may thus infer that the first serious steps toward a theory of speech acts were taken by the Harvard philosophers Sheffer and Lewis during the early 1940s. What they concentrated on was the *structural* distinction, in the semantic representation of sentences, between a speech act component (operator) on the one hand and the propositional content on the other (see also Seuren 1969: 117–18).

The importance of Lewis's analysis lies mainly in his pointing out that the grammars of natural languages do not allow for utterances expressing just a propositional content, like *Pablo driving the car*, meant to express the mere proposition +P(drive)([Pablo],[the car]). *Pablo driving the car* is not a well-formed sentence and the same goes for similar expressions in most languages. The best such structures can achieve on the way towards sentencehood is as *structurally unattached noun phrases*, such as *Pablo driving the car* in (4.1a). The *which* in (4.1b), with the antecedent *her watching TV*, shows that mere propositions can be represented as a distinct nominal constituent at some level of representation. The *which* in (4.1c), with the antecedent *watching TV*, shows the same for bare VPs:

(4.1) a. What? Pablo driving the car? That's impossible![3]
b. She was watching TV, which is quite unusual.
c. She was watching TV, which is quite unusual for her.

Such examples show that expressions of mere propositions may at most achieve the status of, possibly unattached, nominal constituent, at some level of grammatical or semantic representation, but full sentencehood is, one fears, unattainable.

The breakthrough in speech act theory, however, came with the Oxford philosopher John L. Austin (1911–60), who started publicizing, but not publishing, his speech act theory from about 1950 onward. Austin, who was well acquainted with Frege's works—he translated Frege's *Grundlagen der Arithmetik* into English (Frege 1974²)—as well as with Bertrand Russell's views, took a less structural and more philosophical stance. He observed that statements such as (4.2a–4.2f), that is, statements in the present tense and with the verb in the first person singular (or, rarely, plural), and with a verb belonging to what he took to be the class of PERFORMATIVE VERBS, have the curious property of being made true by having been pronounced under the right conditions. They are not contingently true or false in the manner of synthetic truth, that is, in virtue of a structural correspondence with an existing state of affairs. Nor are they necessarily true or false in the manner of analytic truth or falsity, that is, in virtue of their meaning. Rather, it makes sense to say that if they are true, they are so in virtue of the fact that they make themselves true by having been uttered 'felicitously', that is, by the right person on the right occasion, and intended to be taken seriously with respect to the occasion at hand.

(4.2) a. I (hereby) promise to pay you fifty pounds by Monday.
b. I (hereby) name this ship 'Queen Mary'.
c. I (hereby) bequeath my watch to my brother James.

[3] It is interesting to note, in this context, that the subject term of such mere-proposition constituents takes oblique case (*him, her, me, them*) in English. In Dutch and German, for example, this is not so. These languages assign nominative case to mere-proposition subjects, as shown in the following German example:

(i) Er (*ihn) früh aufstehen? Niemals!
he_{NOM} (*him_{ACC}) early get-up? Never!
Him getting up early? Never!

Such examples show that the principle, much vaunted in circles of Chomskyan generative grammar, that it is the finiteness ('Infl') of the finite verb that assigns nominative case cannot be correct.

d. I (hereby) bet you sixpence it will rain tomorrow.

e. I (hereby) bid you goodbye.

f. I (hereby) open the meeting.

Austin called such statements PERFORMATIVE UTTERANCES, in the sense that performative utterances make themselves true by the mere fact that they are uttered felicitously. Although he apparently leaned towards the view that such statements have a truth-value, in that he says that they make themselves true in virtue of having been uttered felicitously, he never comes down clearly on the question of whether it makes sense to call performative utterances true or false. I follow Lewis (1970), Isard (1975) and others who say that the best result is obtained when one takes the view that such statements do have a truth-value in that they are, exceptionally, made true by the mere fact of having been uttered felicitously. The felicity conditions have thus become part of the truth conditions of the sentences underlying such statements. When the felicity conditions are not fulfilled, such statements are legitimately taken to be false.

Austin then realized what Whitney had seen before him, namely that every linguistic utterance, or speech act, usually functions as a performative utterance, but in a grammatically implicit way. Thus the performative utterance (4.3a) can be reformulated less ponderously as (4.3b), (4.4a) as (4.4b), (4.5a) as (4.5b), and so on:

(4.3) a. I (hereby) assert that today is Monday.

b. Today is Monday.

(4.4) a. I (hereby) order you to leave.

b. Leave!

(4.5) a. I (hereby) ask what time it is.

b. What time is it?

Austin called (1962: 32) the (a)-sentences EXPLICIT PERFORMATIVES and the (b)-sentences IMPLICIT PERFORMATIVES or ILLOCUTIONARY ACTS.

Here, however, Austin hit upon a number of quandaries from which he was unable to extricate himself. One problem (Austin 1962: 147–63) was that the class of performative verbs turned out to be ill-defined. Many verbs, such as *blame, admit, criticize, argue, conclude, predict, estimate,* 'waver' between descriptive and performative use (Austin 1962: 85), since sentences like those in (4.6) can be interpreted both performatively and descriptively:

(4.6) a. I blame him for having ruined my chances.
b. I admit that what I said was tactless.
c. I argue that corruption must not be tolerated.
d. I predict that there will be snow tomorrow.

Similar difficulties arise with verbs like *refer* or *open*. May one say that *refer* is a performative verb because I can say *I hereby refer to you*, or that *lock* is a performative verb when I have a device in my car that is uniquely sensitive to my voice and the sounds used when I utter the sentence *I hereby lock the car*, so that it reacts by locking my car?[4]

Another difficulty lies in the grammatical form of performative utterances (Austin 1962: 57), which may vary from totally explicit to partially explicit to totally implicit and even, in some cases, to no utterance at all but just a gesture or a movement. Roman slaves, for example, could be freed in a ceremony of manumission, whereby the owner touched the slave with a rod (the *vindicta*). While doing this, the owner could say (in Latin, to be sure) anything like the following:

(4.7) a. I hereby declare you a free man/woman.
b. You are hereby a free man/woman.
c. You are free now.

But he could also remain silent, as the legal force of the ceremony lay in the touching with the rod, not in the words. Or when a sergeant asks for volunteers for a dirty job, a soldier may offer himself by uttering, say, (4.8a) or (4.8b), but he may also stay silent and simply take a step forward from the rank:

[4] Compare the famous passage in Linsky (1967: 121):

If we assume that whenever in an assertion something is mentioned by name by a speaker he is referring to that thing certain very paradoxical conclusions can be deduced. It would follow that when I write in my paper 'I am not, of course, referring to Ludwig Wittgenstein' I would be referring to Ludwig Wittgenstein. But if someone were asked to show where in my paper I had referred to Ludwig Wittgenstein it would be absurd for him to point to the statement in which I say, 'I am not referring to Ludwig Wittgenstein.' The same would be true of the statement in which I say, 'I am referring to Ludwig Wittgenstein.' In both cases I would have used Wittgenstein's name. Therefore, to mention someone by name is not necessarily to refer to him. And consider this example. Suppose the porter at Magdalen College asks me whom I am looking for. I answer, 'Gilbert Ryle.' Would anyone say I had referred to Gilbert Ryle? But if I say, in the course of a talk, 'I am not referring to the most important of present-day philosophers,' I would then and there be referring to Ludwig Wittgenstein; though in saying as I just did, 'I would then and there be referring to Ludwig Wittgenstein,' I could not be said to have referred to Ludwig Wittgenstein. And this is so notwithstanding the fact that Ludwig Wittgenstein is the most important of present-day philosophers. This, then, is the paradox of reference. In saying 'I am referring to Ludwig Wittgenstein' I am not referring to Ludwig Wittgenstein.

This passage is certainly thought-provoking, even if one may have reservations about some of its aspects (one may well, for example, disagree, as I do, with Linsky's view that Ludwig Wittgenstein is 'the most important of present-day philosophers').

(4.8) a. I (hereby) volunteer for the job.
b. I'll do it.

Yet, despite such problems and unclarities, Austin's work on speech act theory triggered a whole new development in the philosophical and linguistic study of language—a development so important that one may well wonder why the 'discovery' of the performative aspect of speech acts was not made earlier. Quite rightly, the German philosopher Wolfgang Stegmüller vents his vicarious embarrassment at the fact that the speech act factor was discovered so late in the history of Western thought:

> It really is a shame. In fact, it is an embarrassing shame for all those who, during the past 2500 years, have occupied themselves, one way or another, with languages, that they did not make J. L. Austin's discovery much earlier, the essence of which can be summarized in one short sentence: 'By using linguistic utterances we can perform a wide variety of acts.' It is particularly curious that even after the 'linguistic turn' in modern philosophy it should have taken several decades until a philosopher made the discovery that there are such things as *speech acts*. (Stegmüller 1986: 64–5; translation mine)

4.2 Speech acts create social reality

What Stegmüller failed to mention, however, is the SOCIALLY BINDING force (in the Fregean sense) of speech acts. This aspect of speech act theory has not been totally neglected in the literature, but it has definitely been underexposed. To the extent that it is discussed, one usually finds the terms *pragmatic* or *social effects* used to refer to the socially binding force of speech acts. Yet given the other, more legitimate, uses of the term *pragmatic*, such terminology is misleading to say the least. (It is much more useful, and less misleading, to speak of 'pragmatics' only with regard to the vicissitudes of well-keyed and well-anchored utterances in situations of use.) Moreover, discussions of the socially binding force of linguistic utterances are usually incomplete, as the authors mostly fail to mention the speaker's binding commitment as regards the truth of the proposition expressed that comes with assertions. Sadock, for example, speaks of speech acts as 'establishing automatic social effects' (1994: 404), but he limits this to 'requirements for action', that is, imperatives and questions. He might have mentioned that assertions also create an automatic social effect, namely the speaker's personal guarantee of, or vouching for, the truth of the proposition expressed.[5]

[5] See also the discussion further down in the present section of Dummett (1973) and Fodor (1983).

We will now devote some specific attention to the socially binding force of speech acts, not only because it is in itself of fundamental importance, but also because it will help to unravel some questions that have so far not been cleared up satisfactorily.

We start by stipulating the PRINCIPLE OF SOCIAL BINDING, which we consider to be axiomatic and defining not only for human language but also for all other systems of human and nonhuman conscious communication:

> PRINCIPLE OF SOCIAL BINDING
> Every serious linguistic utterance *u* has a FORCE in that it creates a socially binding relation consisting in either a COMMITMENT (of varying strength) on the part of the speaker, or an APPEAL (of varying strength) issued to the listener(s), or the INSTITUTION OF A RULE OF BEHAVIOUR with regard to the proposition expressed in *u*, or an APPELLATION, as in vocatives ('Hey, you over there!'). The social partners with regard to whom the speech act is valid form the FORCE FIELD of *u*.

This principle implies that those who are bound are PERSONS in the sense in which this term is used in moral philosophy and in the theory of law—that is, as holders of rights and duties and hence of responsibility and dignity. Persons are accountable and every serious speech act creates an accountability relation, no matter how trivial or insignificant, between speaker and listener. The fact that the principle of social binding is axiomatic and defining for human language (and, in fact, for any form of communication by means of signs) means that language is primarily an instrument for the creation of accountability relations and not for the transfer of information.

The notion of force field is defined for speech acts as a whole, just like the notion of anchoring. Keying is not defined for overall speech acts but is restricted to propositions—as shown in Figures 7.5 and 7.6. It is important to note that, by definition, the socially binding force inherent in speech utterances cannot take effect until the thought has been turned into an actually perceived and interpreted utterance, in the form of speech or writing, since there is no commitment or engagement without the uptake by one or more comprehending listeners or readers.

The socially binding forces of commitment, appeal, the setting of rules of behaviour and appellation have, between themselves, been channelled into at least seven universal linguistic categories, namely:[6]

[6] According to Sadock (1994: 403): '... most languages make nonpropositional distinctions among at least three basic types, clustering around the central features of statement making, question asking, and order giving.' I believe that the three basic types cluster around the central features of commitment,

(a) ASSERTIONS, also called statements or declaratives;
(b) QUESTIONS, also called interrogatives, directed at the listener;
(c) COMMANDS, also called imperatives, directed at the listener;
(d) EXHORTATIONS, directed at oneself, one's own group, or at a third party;
(e) PERMISSIONS/OBLIGATIONS, setting rules for behaviour (modals);
(f) IMPRECATIONS, especially wishes and curses.
(g) VOCATIVES, when a name or predicate is used to address or refer to someone.

When a speaker utters an ASSERTION *A*, (s)he vouches for—that is, takes on a commitment with regard to—the truth of, the proposition expressed in *A*.[7] When a speaker utters a QUESTION *Q*, (s)he either appeals to the listener to specify the truth-value of the proposition expressed in *Q* (yes/no question) or to fill in the value of a variable (WH-question). When a speaker utters a COMMAND *C* (usually in the form of an imperative), (s)he appeals to the listener, with force of authority, to make the proposition expressed in *C* true. When a speaker utters an EXHORTATION *E*, (often using a conjunctive or a 'let us' type of imperative), (s)he appeals to him/herself, or to a group to which (s)he belongs, or to a third party, to make the proposition expressed in *E* true. When a speaker utters a PERMISSION or imposes an OBLIGATION *R* (mostly using a modal verb), (s)he does so in virtue of his or her power to impose rules of behaviour with regard to the proposition expressed in *R*. When a speaker utters an IMPRECATION *I* (using a conjunctive or optative mood), (s)he appeals to an unspecified higher authority to make the proposition expressed in *I* true. And when a speaker utters a VOCATIVE (s)he may use a name (*Hey you, Simon*, that is: 'I hereby call you "Simon" ') or an epithet (*Hallo, love*, that is: 'I hereby call you "love" ').[8]

Languages also contain EXPRESSIVE EXCLAMATIONS like *Wow*, *Damn*, *Oh dear*, *My word*, *Now there.* It is probably best to deny them the power of

appeal, behavioural rule-setting, and appellation. They are linguistically channelled into at least the seven (but perhaps more) sentence types, specified in (a)–(g).

[7] A large number of languages worldwide have a grammatical category, obligatory for all assertions, of EVIDENTIALITY, qualifying the speaker's commitment as regards the source of the information contained in the proposition expressed: direct perception, report, speaker's inference or deduction and so on (see Aikhenvald 2003). As observed by Joseph (2003: 324), this category quickly turns into a 'cover one's rear' stance, in the sense that the speaker takes only a qualified responsibility for the assertion, adding the reservation 'don't take me up on this: I've merely been told'.

[8] So-called epithet pronouns, such as *the little runt* in:

(i) I'm looking for the professor but I can't find *the little runt.*

are best accounted for by assuming an appellative speech-act operator: 'I can't find him, and I hereby call him "a little runt" ' (see also Section 4.5.2).

establishing relations of social accountability. They are conventionalized ways of expressing surprise, pleasure, pain, admiration, disgust, and the like, but they appear to be linguistically marginal. For one thing, they are, normally speaking, grammatically simple and can often do without a propositional component: if languages have a 'grammar of exclamations' it is a very elementary one. Moreover, even though in most cases expressive exclamations follow the phonological system of the language in question, they are not subject to any condition of felicitousness to be effective or 'valid'. In fact, they are often highly infelicitous! And finally, exclamatory vocalizations that are expressive of pain, pleasure or fear are frequently found among animal species which have no socially functional system of signification that may in any way be regarded as an evolutionary precursor of human language (see Section 4.4). Expressive exclamations, in other words, may be seen as a very primitive form of vocalization. They have only come within the range of human language on account of their phonological regimentation and their (partial) linguistic conventionalization.

As was prominently observed by Austin, there is, within the general categories of social binding created by the linguistically defined sentence types of assertion, question, command, exhortation, behavioural rule-setting, imprecation, and vocative, a wide range of socially binding effects brought about in ways that are not reflected by any specific grammatical category. These come in two varieties. First, there are the so-called INDIRECT SPEECH ACTS, which have the grammatical form of one of the seven grammatically defined sentence types, but function in a different way. Examples are:

(4.9) a. Can you pass me the salt? (at the dinner table)
b. Why don't you sit down? (to a visitor)
c. You can go now. (to a student who has been reprimanded)
d. I suggest you call her now. (CEO to assistant; reply: Is that an order?)
e. I'll have the lamb chops, please. (in a restaurant)[9]

[9] Bach and Harnish (1979: 199) consider the addition of *please* to a sentence that is not grammatically a request or an order ungrammatical, giving as examples (asterisks theirs):

(i) *Why do you please pass the salt?
(ii) *You never please pass the salt.

I do not believe this to be correct. The particle *please* should be seen as a polite admonition by the speaker to grant whatever is literally or nonliterally requested, in the context given. It thus presupposes that there is a request inherent in the utterance at hand. Sentence (i) is then interpretable, for example, as a polite admonition to answer the question *Why do you pass the salt?*, and (ii) makes perfect sense, for example, in a context where the speaker reminds the addressee of the rule never to pass the salt—though in both cases one would prefer the *please* at the end of the sentence, and with emphatic accent.

which have the grammatical form of a yes/no question, as in 4.(9a), or a WH-question, as in (4.9b), or an assertion, as in (4.9c–4.9e), but normally function, in situations like those indicated, as a request to pass the salt, as an invitation to sit down, as permission to leave, as an order to call her now, or as a choice for lamb chops, respectively.

There is a sizeable but inconclusive body of literature on the status of indirect speech acts. One thing that does result from that literature is that the line between indirect speech acts and Austin's so-called PERLOCUTIONARY ACTS is fluid. A perlocutionary act is defined as a speech act whose communicative function is to be derived by means of sensible social reasoning, as when I say *It's cold in here*, hoping that you will take the hint and turn the heating up or shut the window, as the case may be.[10]

I shall not go into the sometimes convoluted arguments of philosophers and pragmatists as to the status of indirect speech acts, but I will follow the hard theoretical line argued for, among others, by Lewis (1970), Isard (1975), and Bertolet (1994). This theoretical hard-line position implies that (4.9a) is a yes/no question, (4.9b) a WH-question and (4.9c–4.9e) assertions, and that their indirect-speech-act character is accounted for by an appeal to social, not linguistic, competence. Thus considered, indirect speech acts and perlocutionary acts are of a kind, differing only in the extent to which the intended practical result is reflected in the words used: indirect speech acts are more direct expressions of the result intended than perlocutionary acts.

The second variety, more interesting in the present context, is the class of EXPLICIT PERFORMATIVE UTTERANCES, as exemplified in (4.2a–4.2f). These are normally defined as utterances that make themselves true by the mere fact of having been uttered felicitously. This, however, is a definition that calls for further refinement, precisely because it neglects the socially binding force which is clearly meant to be included. Appealing to the Principle of Social Binding formulated above, we take it to be a constitutive element of all speech acts,

Please is thus a close cognate of the particle *sorry*, which expresses a polite request for forgiveness for either the speech act at hand or whatever else is happening or has happened as a result of the speaker's actions.

[10] Indirect speech acts, including perlocutionary acts, are often subject to social and/or linguistic convention, which has to be learned in order to participate adequately in a society. Sentence (4.9b), for example, when translated into a language other than English, is not normally interpreted as an invitation to sit down, but literally—that is, as a way of asking for the reason why the addressee does not sit down. Or, to take a different example, the proper way, in certain circles, of informing a man that his fly is undone is to say *You are not properly dressed*. From here it is but a small step to ritualized utterances, such as *I wish you joy* to grant membership of a club or society. Idioms now come into sight as well. Thus, *to bring the house down* is, to a competent speaker of English, less dramatic than it sounds. And in colloquial British English at least, the expression *And Bob's your uncle* usually does not mean what it says, but is typically used to say that, once a simple procedure has been gone through, the desired result follows automatically.

whether of the common-or-garden implicit, or of the more ornate explicit kind, that what they bring about is some sort of socially binding relation which is always determined by the grammar and semantics of the language in question.

What distinguishes explicit performative utterances from their more ordinary implicit counterparts is (a) that they only come as grammatical declaratives (assertions), never as imperatives or interrogatives and (b) that, as declaratives, they can be used to perform a much wider range of socially valid acts than just assertion, question, command, exhortation, rule-setting, or imprecation. They can be used to open or close a meeting, name a ship, issue a warning or threat, place a bet, make a bequest, claim property, and so on. Explicit performative utterances are the primary, although not the only, means used in human societies for establishing socially binding relations or states of affairs, often of an institutional or legal nature. They are indispensable in any form of organized society.

One reason why the socially binding force of speech acts, and of explicit performatives in particular, needs to be emphasized is that one also encounters quasi-performative utterances that do not produce a socially binding but some other effect. One may, of course, choose to define the notion of explicit performative utterance without imposing the condition of social binding—disregarding the condition that every serious utterance needs a FORCE FIELD that defines the boundaries of its social validity—but such a definition would fail to do justice not only to Austin's notion of performative utterance but also to a central aspect of human language. When I say, for example, *I am hereby producing sound* or *I am hereby speaking English*, or, as in the case mentioned above, *I hereby lock the car*, I have made the utterance true by saying it out loud (and felicitously), but we do not want to include such utterances in the class of explicit performative utterances, even though, in uttering them, I create the state of affairs in which I am producing sound, or speaking English, or locking my car. All I have done is produce an assertion that stands little chance of being false and whose truth-commitment is, therefore, trivial.

Or suppose some sorcerer had the magic power to set people on fire simply by saying *You are hereby on fire.* Such an utterance should not be taken to constitute a (half-explicit) performative utterance, even though the speaker performs an act by using a linguistic utterance and makes the utterance true by uttering it, giving it little chance to be false. And whether or not a monk who has taken a vow of perennial silence produces an explicit performative utterance in the sense intended when he says *I am hereby sinning*, or *I am hereby breaking my vow*, thereby creating his sinning or vow-breaking, will depend on whether one takes sinning, or vow-breaking, to be violations of socially binding commitments (they will be if God is regarded as a social partner).

Explicit performative utterances thus create the linguistically standard socially binding relation of an assertion. In virtue of producing his or her utterance seriously in the force field at hand, the speaker vouches for the truth of the proposition expressed. Yet this truth-vouching is no longer the primary socially binding effect intended. The socially binding effect that is primarily intended in explicit performatives is expressed not in the assertive speech act operator but in the embedded proposition, the function of the assertive speech act operator having been reduced to triviality. Explicit performatives thus create, besides the commitment to truth, a linguistically nonstandard but culturally defined socially binding relation defined by the proposition expressed. This further illustrates the statement made above to the effect that explicit performative utterances are token occurrences of sentences whose felicity conditions have become part of their truth conditions: failure to satisfy the felicity conditions results in falsity. Here again, we stand by the theoretical hard-line policy referred to above.

It is thus proper to say that (4.2a) is an explicit performative utterance resulting in the promise to pay the addressee (who is necessarily, but not necessarily exclusively, part of the force field) fifty pounds by Monday, since that promise is a socially binding relation with regard to the state of affairs described by the proposition expressed in (4.2a) and created by the mere felicitous uttering of (4.2a). And likewise for (4.2b–4.2f). In (4.2f), for example, which is an explicit performative utterance resulting in the institutional fact of the meeting being open, the fact of the speaker opening the meeting makes (4.2f) true and is realized only if (4.2f) is uttered felicitously, which shows that the felicity conditions are part of the truth conditions of this explicit performative. Sentence (4.3a) is an explicit performative utterance resulting in the assertion that today is Monday, and since the fact that this assertion has come about by the felicitous uttering of (4.3a) makes (4.3a) true, it follows that the felicity conditions of (4.3a) are part of its truth conditions. And similarly for all other such cases.

Similar phenomena occur with imperative speech acts. When I say *Consider the possibility that John is still alive*, my utterance, in virtue of its being understood, automatically makes any keyed-in listener consider the possibility that John is still alive. Here the linguistically codified imperative function has been reduced to triviality since what is, playfully or politely, presented as an obligation is, in fact, an inevitable result. What counts is the effect envisaged, namely that the addressee considers the possibility that John is still alive. The command itself stands little chance of being disobeyed.

Or consider the more complex case of *Open Sesame!*, as used in the Arabian Nights stories. This has the linguistic form of a command but it can only function as a command if taken metaphorically, as if Sesame were a social partner. The imperative function of *Open Sesame!* is a form of play-acting since, in the thought-up reality of the story, Sesame is neither a free agent nor a social partner but is only treated as if he (it) were one. In terms of this twice-fictional situation—a metaphorical addressee within a fictional story—*Open Sesame!* is still an imperative speech act, but the listener/reader knows that it stands little chance of being disobeyed, in the story.

Although Austin himself was clearly aware of the socially binding force of both explicit and implicit performative utterances, it looks as if he felt uncomfortable with it. That at least is the impression one gets if one considers the lack of clarity, the hedges and the excess of stylistic flourish in the following passage, where Austin argues that if one wants to maintain that to marry is simply 'to say a few words' and betting is 'simply saying something', then it must be taken into account that the words in question are not just a true or false declaration of an inner intent, but create a socially binding relation once they are spoken seriously (Austin 1962: 9–10):

> So far, well and good. The action may be performed in ways other than by a performative utterance, and in any case the circumstances, including other actions, must be appropriate. But we may, in objecting, have something totally different, and this time quite mistaken, in mind, especially when we think of some of the more awe-inspiring performatives such as 'I promise to . . .'. Surely the words must be spoken 'seriously' and so as to be taken 'seriously'? This is, though vague, true enough in general—it is an important commonplace in discussing the purport of any utterance whatsoever. I must not be joking, for example, nor writing a poem. But we are apt to have a feeling that their being serious consists in their being uttered as (merely) the outward and visible sign, for convenience or other record or for information, of an inward and spiritual act: from which it is but a short step to go on to believe or to assume without realizing that for many purposes the outward utterance is a description, *true or false*, of the occurrence of the inward performance. The classic expression of this idea is to be found in the *Hippolytus* (l. 612) ,[[11] PAMS] where Hippolytus says
>
> ἡ γλῶσσ᾽ ὀμώμοχ᾽, ἡ δὲ φρὴν ἀνώμοτος,
>
> i.e. 'my tongue swore to, but my heart (or mind or other backstage artiste) did not'. Thus 'I promise to . . .' obliges me—puts on record my spiritual assumption of a spiritual shackle.

[11] The *Hippolytus* is a play written by the Greek tragedian Euripides in 428 BCE.

It is gratifying to observe in this very example how excess of profundity, or rather solemnity, at once paves the way for immorality. [[12] PAMS] For one who says 'promising is not merely a matter of uttering words! It is an inward and spiritual act!' is apt to appear as a solid moralist standing out against a generation of superficial theorizers: we see him as he sees himself, surveying the invisible depths of ethical space, with all the distinction of a specialist in the *sui generis*. Yet he provides Hippolytus with a let-out, the bigamist with an excuse for his 'I do' and the welsher with a defence for his 'I bet'. Accuracy and morality alike are on the side of the plain saying that *our word is our bond*.

The overall purport of Austin's argument is that the effect of the words by which one marries, promises, or bets consists in a socially binding relation, no matter whether the words were spoken sincerely or insincerely. He thus duly stresses the legal force of the spoken word as against the speaker's real intention, which may or may not be sincere.

The creation of a socially binding relation or state of affairs within a social force field is a defining condition not only of explicit performative utterances but also of implicit performatives. A serious utterance of, say, *Give me that book* is every bit as socially binding as its explicit counterpart *I hereby order you to give me that book*. The difference is that the former is a grammatical imperative and creates a command, whereas the latter is a grammatical assertion and still creates a command—besides creating a trivial truth commitment. According to the definition of explicit performative utterances, which says that such utterances make themselves true by the mere fact that they are uttered felicitously, the former is not an explicit performative utterance, just a speech act of the imperative type, while the latter is an explicit performative utterance creating a command. We now see that all speech acts, whether of the implicit or the explicit performative kind, are performative in that they create a socially binding relation or state of affairs.

In fact, we see that the primary function of language is not 'communication', in the sense of a transfer of information about the world, but social binding, that is, the creation of specific interpersonal, socially binding relations with regard to the proposition expressed in any utterance or speech act. It will be clear that this kind of social binding is a central element in the social fabric that is a necessary requirement for human communities. Looked at from this angle, linguistic theory is in the jurisdiction not only of psychology, but also of the theory of law and justice: speech act theory and the theory of law and justice are closely connected in virtue of the intrinsic socially binding effect of utterances.

[12] The original text has 'immodality', a word that appears not to exist in the English language.

If 'communication', in the sense of a transfer of information about the world, is not the primary function of language, is it its secondary function? I think the correct answer is that there are many nonprimary functions of language, the transfer of information about the world being one of them, and that these nonprimary functions cannot be ordered hierarchically. An important nonprimary function of language is the speaker's demonstration of social rank or status or the expression of the speaker's desire to belong to a particular social group. That speakers are almost always unaware of the fact that this is what they are doing does not make it less of a fact. Labov's classic study (1966) on the pronunciation of the phoneme *r* in New York City English and many other works by Labov and other sociolinguists on similar phenomena, have shown conclusively that subconscious mechanisms of demonstration of membership of, or of association with, a given group, class, or rank are powerful determinants of human linguistic interaction. They also play a crucial role in linguistic change.[13]

Phonology and phonetics are not the only area where such phenomena are manifest. Grammar and lexicon often play a role as well. Lexical choice, for example, is crucial in the following, not entirely fictitious, little exchange between a Dutch TV interviewer (DI) and an American celebrity (AC) who is visibly annoyed at having to spend a few minutes on what he sees as an unimportant TV network:

(4.10) DI: May I ask you a few questions, sir. I am from The Netherlands and I am sure the public in Holland will be interested in what you have to say.

AC: Now which is it, The Netherlands or Holland? I want to be clear on that first.

DI: Let's settle for Holland. But come to that, shall we say America or The United States, sir?

[13] We cannot, obviously, in this place propose a fully fledged theory of the factors underlying processes of linguistic change. Suffice it to say that it looks very much as if there are conflicting tendencies in the development of languages. There are functional tendencies resulting in simplification and greater semantic transparency. But there are also tendencies resulting in greater complexity and a decrease in semantic transparency. The latter are often traceable to the wish of socially dominant 'reference' groups to make it hard to manifest linguistic membership. Such groups are apt to introduce new lexical distinctions, morphological exceptions, idioms and the like, just to make it difficult to speak like them. Other factors may also be involved. It seems, for example, that the criterion of functionality works out differently for speakers and listeners. Speakers like syntactic and morphological coding, as coding makes for compactness and thus saves energy in the production process. Listeners like semantic transparency, often the opposite of coding. Yet flectional morphology, one form of coding, has the advantage of reducing ambiguity and is thus useful to the listener. Linguistic change thus somehow has to find its way in this complex game of conflicting interests.

Now AC gets even more annoyed and the interview fizzles out, neither party seeing much point in going on. What has caused the breakdown? The answer is obvious: AC wants to assert what he feels to be his superiority by suggesting to DI that the latter is inconsistent in his lexical selections and by forcing him to make a lexical choice, but DI retorts by doing the same to AC, which results in a conflict of social ranking and blocks further fruitful interaction.

But let us revert to the socially binding force in human language. We then see that the linguistic, semantic, and pragmatic literature after Austin shows no improvement. Hardly anywhere does one find an awareness of the fact that language is primarily an instrument not for the transfer of information about the world but for the creation of socially binding relations. When one looks at the formal-semantics literature, one sees that it contains a certain body of literature on speech acts, in particular questions and imperatives (e.g. Hintikka 1974; Hamblin 1976, 1987; Karttunen 1977; Stalnaker 1978; Hoepelman 1981; Groenendijk and Stokhof 1984). This literature, however, concentrates on problems of the delimitation of sets of possible worlds, mostly in the context of information-state modelling—that is, a formal specification of what exactly is asserted, asked, or ordered when an assertion, question, or order is uttered and sometimes also of what may constitute an adequate answer to a question posed or an adequate compliance with an order given or a request made. But little or nothing is said about what making an assertion, asking a question, or issuing an order actually amounts to in terms of human interaction and human relations. That sort of question is, if mentioned at all, relegated to what is seen as the nonformal study of pragmatics, which is allowed to work with impressionistic criteria.

Stalnaker, for example, concentrates on the context-changing effect of assertions—an aspect that is discussed more fully in Volume II—and takes that to be 'its essential effect' (1978: 323), saying that 'the essential effect of an assertion is to change the presuppositions of the participants in the conversation by adding the content of what is asserted to what is presupposed'. On the next page, he wonders (1978: 324): 'Would it not be more plausible to characterize assertion as trying to get the audience to accept that the speaker accepts the content of the assertion?' and proceeds to show that 'this Gricean twist is not required'. But there appears to be no awareness of the fact that the 'essential effect' of an assertion is the speaker's guarantee that the proposition expressed is in fact true.

To take a further example, Hamblin describes (1987: 167–99) a particular predicament involving imperatives, as in:

(4.11) I am obliged to order you to do D, and I hereby do so. But my private advice to you is not to.

This predicament is, obviously, closely related to Austin's example, quoted above, of Hippolytus saying 'my tongue took an oath, but my mind remained unsworn', and also to the Moore paradox, which will be discussed in a moment. Hamblin's analysis remains focused on criteria of logical consistency. He introduces (1987: 179) a Gricean rule telling speakers not to issue inconsistent orders or make inconsistent promises, but that is considered to be an external pragmatic matter, which we 'can go on, if we wish, to discuss' (1987: 197). No awareness is shown of the fact that the semantics of speech acts is part of a theory of, and not just small talk about, socially binding relations, including, among other things, a theory of law and justice (in so far as violations of speech act commitments may give rise to conflicts and may provoke sanctions).[14]

The neglect of the socially binding aspect in formal semantics is matched by a similar neglect in philosophy, where the psychological state of having an intention is placed in the centre of interest. To the extent that the nature of a speech act is gone into in the philosophical literature, the emphasis is invariably on the intention aspect of speech acts, that is, on what the speaker HAS IN MIND or INTENDS to do when producing a speech act, and not, or hardly, on what is ACTUALLY ACHIEVED when the speaker produces a speech act. It is this basic blindness to the socially binding aspect that has given rise to a number of discussions regarding insincere speech acts, mostly involving a variant of the Moore paradox shown in (4.12a) below.

A glaring example is John Searle's (1969) widely quoted analysis of promises. In his section 3.1 (1969: 57–61), Searle describes the conditions that must be fulfilled for a speech act to constitute a 'sincere and nondefective promise'. In the subsequent section 3.2 (1969: 62), he elaborates the notion of an insincere promise. His section 3.1 starts with the following sentence (Searle 1969: 57):

> Given that a speaker *S* utters a sentence *T* in the presence of a hearer *H*, then, in the literal utterance of *T*, *S* sincerely and non-defectively promises that *p* to *H* if and only if the following conditions 1–9 obtain:

Here one might wish to observe, first, that *T* should be uttered not just 'in the presence of a hearer *H*' but should be actually directed at *H*, since when *S* is speaking in a dream or is soliloquizing, no promise comes about, regardless

[14] Historically, this state of affairs is to be seen in the light of the dominant position of logic, which is by definition carried out on L-propositions as linguistic expressions of mental propositions, regardless of any socially binding speech act element. And formal semantics, which is directly derived from modern logic and logical model theory, has simply inherited the focus on the propositional aspect of sentences, at the expense of the semantically equally important aspect of social binding, which is not of a logical but rather of a socially obliging, moral, or even legal character.

of whether someone happens to be overhearing *S*'s utterances. Secondly, the qualification 'sincerely and non-defectively' in the definiendum quoted above turns out to be an unwanted element, since, as Searle concedes in his section 3.2 on insincere promises, an insincere promise is also a valid and binding promise. Let us look at Searle's analysis in some detail.

The conditions 6 and 7, as formulated by Searle, run as follows (1969: 60; 'A' stands for the act with respect to which the promise is made, mentally represented by proposition *p*):

6. *S intends to do A.*

The distinction between sincere and insincere promises is that, in the case of sincere promises, the speaker intends to do the act promised; in the case of insincere promises, he does not intend to do the act. Also, in sincere promises, the speaker believes it is possible for him to do the act (or to refrain from doing it), but I think the proposition that he intends to do it entails that he thinks it is possible to do (or refrain from doing) it, so I am not stating that as an extra condition. I call this condition the *sincerity condition.*

7. *S intends that the utterance of T will place him under an obligation to do A.*

The essential feature of a promise is that it is the undertaking of an obligation to perform a certain act. I think that this condition distinguishes promises (and other members of the same family such as vows) from other kinds of illocutionary acts. Notice that in the statement of the condition, we only specify the speaker's intention; further conditions will make clear how that intention is realized. It is clear, however, that having this intention is a necessary condition of making a promise, for if a speaker can demonstrate that he did not have this intention in a given utterance he can prove that the utterance was not a promise. We know, for example, that Mr Pickwick did not really promise to marry the woman because we know he did not have the appropriate intention. I call this the *essential condition.*

In condition 7 we read that 'having this intention is a necessary condition of making a promise', so that an insincere promise is NOT a promise, despite the distinction made in condition 6 between sincere and insincere promises. The undesirable mutual interference of the notions of intention and sincerity stands out more clearly in his section 3.2 (Searle 1969: 62):

A promise involves an expression of intention, whether sincere or insincere. So to allow for insincere promises, we need only to revise our conditions to state that the speaker takes responsibility for having the intention rather than stating that he actually has it. A clue that the speaker does take such responsibility is the fact that he could not say without absurdity, e.g., 'I promise to do *A* but I do not intend to do *A*'. To say, 'I promise to do *A*' is to take responsibility for intending to do *A*, and this condition holds whether the utterance was sincere or insincere. To allow for the possibility of an insincere promise, then we have only to revise condition 6 so that it states not that the speaker intends to do *A*, but that he takes responsibility for intending to do *A*, and to avoid the charge of circularity, I shall phrase this as follows:

6a. *S intends that the utterance of T will make him responsible for intending to do A.*

Thus amended [...], our analysis is neutral on the question whether the promise was sincere or insincere.

One wonders, of course, why the sincerity condition was introduced in the first place, if, in the end, it turns out not to contribute to the definition of what constitutes a promise. The reason seems to lie in Searle's blinkered focusing on the notion of intention and his blindness to the linguistically defined force of utterances. Searle's view that 'a promise involves an expression of intention, whether sincere or insincere' is radically wrong. Making a promise is assuming an obligation by the speaker as a result of a speech act of promising, whether the speech act is of the explicit type or is made in the form of an indirect speech act. As such it is nothing to do with the speaker's INTENTION and everything with the speaker's actual utterance, made of his own free will and directed at an audience: there is no obligation until the words have been spoken, as Austin rightly stresses. The speaker's intention with regard to a promise made is to do with the speaker's reliability, not with his making a promise. Clearly, when a speaker is known to be unreliable, or when it is known that, for external reasons, a promise, though validly made, cannot be kept, the promise will not be invalid, but merely worthless—just as money issued by a bankrupt government is still legal tender, although it may be well-nigh worthless.

The same applies, *mutatis mutandis*, to other speech acts. When an order is issued but it is clear at the same time that the one who gives the order is unwilling, or unable, to enforce it, or that the addressee is unable to carry it out, the order is not invalid but merely worthless, or futile, and therefore ineffective (except, of course, in cases of nonliteral use, like *Go to hell!*, which have a nonfutile conventionalized secondary interpretation).

The famous MOORE PARADOX should be seen in the same light. The classic example is (4.12a), keyed to the time of speaking (and not, for example, to one or more past times, as in (4.12b), which makes good sense as a report about one's own occasional delusions):

(4.12) a. The cat is on the mat, but I don't believe it.

b. Sometimes, the cat is on the mat, but I don't believe it.

The mental jarring caused by sentences such as (4.12a) has given rise to a substantial but inconclusive body of philosophical literature, most of which appeals to a 'logic of belief' and revolves around the notion of the intended sincere assertion of a given belief. The heart of the matter, however, appears to be the fact that although (4.12a), keyed to the time of speaking, is a VALID speech act, it is INEFFECTIVE. When a speaker vouches for the truth of a proposition and at the same time openly expresses a disbelief in its truth, there is reason to doubt the speaker's moral integrity or even his mental sanity. The commitment to the truth of *p* loses all force when conjoined with the commitment to disbelief in the truth of *p*. The solution to the Moore paradox thus seems to lie in a combined theory of the *validity* of speech acts and their being *effective* on account of a presumption of the speaker knowing what vouching for the truth of a proposition amounts to in terms of social responsibility and social consequences, rather than in a theory of truth or in an analysis of speakers' intentions taken to underlie the use of the predicates *assert* and/or *believe*.

The use of sentences like (4.12a) or Searle's *I promise to do A but I do not intend to do A* undermines the implicit social pact upon which all speech acts are based and thus the effectiveness of the speech act in question. In general, speech acts are valid on account of their having been uttered under appropriate conditions. Their validity is part of the semantics of the language in question since it is bound up with the system of the language at type level. Speech acts are effective in virtue of the presumption of the speaker being a properly functioning social being. The effectiveness of a speech act appears to be a matter of PRAGMATICS, as it is bound up with the circumstances in which sentences are uttered. For a speech act to be effective, it must (a) be performed *in a valid way*, that is, uttered under appropriate conditions, and (b) be supported by a *presumption of social competence*. When condition (b) is unfulfilled, the speech act may still be (semantically) valid, but it is (pragmatically) futile and thus not effective.[15] If one wants a 'logic' for this, it should be a 'logic' of the validity and effectiveness of speech acts, not a 'logic of belief'.

Like Austin, Dummett came close to appreciating the socially binding force of speech acts. Yet he failed to apply that notion to assertions:

A command has definite consequences: disobedience to a command given by a person in authority confers on that person a right to punish, or at least reproach, the person commanded. For this reason, the utterance of a sentence of a certain form, unless special circumstances divest this act of its usual significance, in itself constitutes the giving of a command. It is irrelevant what intentions the person speaking may have

[15] The distinction between validity and effectiveness must be drawn sharply. A climatologist can say, for example, without any form of inconsistency or absurdity:

(i) The earth will, therefore, explode within ten years from now, though I hope I am wrong on this.

had: it is, for instance, possible to give a command in the hope that it will be disobeyed, and that one will therefore have an excuse for punishing the person commanded; or again, one may simply be carrying out the duty of transmitting the orders given by a superior. An assertion has no such definite consequences. (It of course has consequences in the wider sense that a difference may be made to the course of events by the making of an assertion: it is essential to the understanding of assertions that we know what it is to act on an assertion. But it does not have consequences of such a kind that its conventional significance consists in a tacit agreement to bring about, and accept, such consequences in given circumstances involving the assertion.) Assertions take place against the background of a custom of uttering them with the intention of saying something true (together with all those conventions which determine the sense of the assertion, i.e. the conditions under which it is true). (Dummett 1973: 301–2)

Dummett's comment on commands seems perfectly adequate, but what he says about assertions must, unfortunately, be rejected. To make an assertion is to vouch for the truth of the proposition expressed in it, which is every bit as socially binding as the issuing of an order or the performance of any other kind of speech act. To vouch for—that is, to put oneself on the line with regard to—the truth of a proposition which turns out to be false may well have serious consequences for the one who has done the vouching. These consequences will normally vary according to whether the false information was given in good faith or in bad faith, the latter case constituting an instance of lying.

That assertions should 'take place against the background of a custom of uttering them with the intention of saying something true' may be so, but it is not germane to a description of what constitutes an assertion. Dummett has fallen victim here to a view that has been common among philosophers for a long time, namely that comprehending an assertion is not only a matter of knowing what the world will be like when the assertion is true (its truth conditions), but also a matter of believing that the assertion is true.

An outstanding example of this misconception is found in Fodor (1983), where one reads, in the context of what it is to interpret a visual stimulus:

Now about language: Just as patterns of visual energy arriving at the retina are correlated, in a complicated but regular way, with certain properties of distal layouts, so too are the patterns of auditory energy that excite the tympanic membrane in speech exchanges. With, of course, this vital difference: What underwrites the correlation between token utterances and distal layouts is (roughly) a convention of truth-telling. In the root case, the convention is that we say of x that it is F only if x is F. Because that convention holds, it is possible to infer from what one hears said to the way that the world is. (Fodor 1983: 45)

A note is attached here:

> Strictly speaking, I suppose, a convention must be something one can adhere to if one chooses; so perhaps the principle at issue is not 'Say only what is true' but rather 'Say only what you believe.' General adherence to the latter injunction will license inferences from utterances to how the world is, given the assumption (which is, anyhow, in all sorts of ways epistemologically indispensable) that much of what people believe is true. (Fodor 1983: 132)

Apart from the fact that Fodor identifies token utterances with assertions, a common but illicit narrowing of focus, he also puts forward the—I am sorry to say—absurd view that the construal of utterance meaning is based on a 'complicated but regular' (yet 'rough') correlation between certain aspects of the real world and the sounds arriving at the tympanic membrane. Whether or not it is correct to say that much of what people believe is true or that there is a convention of saying only what one believes to be true, is largely immaterial for the construal of meaning and the comprehension of utterances, even if the OSTA principle is taken into account. (Literal) utterance meaning is captured by reconstructing the mental proposition in the speaker's mind together with the modality of the socially binding force put into action by the speaker's utterance, according to the rule system of the language in question.

Seeing a ship at sea is sufficient to make us know for sure that there is a ship at sea, or hearing Joan laugh is sufficient to make us know for sure that Joan is laughing, in virtue of the processes of perception and interpretation of sensory data, epistemologically dicey as they are. Seeing smoke is sufficient to make us know for sure that there is fire, or seeing footsteps in the sand is sufficient to make us know for sure that someone has passed there, in virtue of our knowledge of causal laws, epistemologically dicey as these may be. And being told in earnest that the moon is made of green cheese is sufficient to make us know for sure that the speaker vouches for the truth of the proposition '+P(made of green cheese)[the moon]' in virtue of our competence in the English tongue and our social competence, epistemologically dicey as these may be. But being told in earnest that the moon is made of green cheese is definitely NOT sufficient to make us know or believe that the moon is actually made of green cheese. Only if there is a presumption of the speaker's factual reliability and moral integrity can we be confident that the moon is truly made of green cheese (see also Section 8.1).[16]

[16] The misconception at hand has a long history, going back at least to Descartes, who, in his *Le Monde, ou Traité de la Lumière* (The World, or Essay on Light), puts sensory perception, the

4.3 Truth conditions and liability conditions

The main conclusion resulting from the preceding pages is that every speech act or linguistic utterance made in earnest establishes the speaker's socially binding position-taking with regard to the proposition expressed in the utterance. This implies that every utterance signifies a mental event consisting in the speaker's decision to define his or her socially binding position with regard to the proposition expressed. Let us call that mental event the INTENT underlying any utterance or speech act. As was said in Section 3.1.1, we take it that an intent consists of two main components: (a) the BINDER, which defines the kind of socially binding force involved (assertion, question, command, exhortation, imprecation) in the force field at hand, and (b) the (discourse-bound) PROPOSITION. Correspondingly, it makes sense to postulate that any linguistic form expressing an intent, in other words any sentential utterance, must, at a semantic level of analysis, be taken to consist of two main components, (a) a SPEECH-ACT OPERATOR expressing the linguistically defined force of the utterance, and (b) the L-PROPOSITION, symbolizing the proposition with respect to which the speech-act operator operates. Together they form the SEMANTIC ANALYSIS (SA) of the sentence used to express the intent in question.

A proposition, looked at from the barest possible point of view, leaving out all possible affective aspects that may occur and may codetermine lexical choices in its linguistic expression, may be said to be defined by its truth conditions: when the objects intended do indeed have the property mentally assigned to them in a proposition *p*, then *p* is true; otherwise *p* is false. The truth or falsity of a proposition *p* thus depends on the nature of the objects involved and the satisfaction conditions of the mentally assigned property.

interpretation of natural signs and the interpretation of conventional (linguistic) signs into one epistemological basket:

You know well that words, which have no resemblance to the things they signify, nevertheless enable us to understand them, even without our taking note of their sounds or their syllables. In fact, it often happens that, having heard a stretch of discourse, we have understood its sense perfectly well but are unable to recall in what language it was pronounced. Well now, if words, which signify nothing other than by human convention, are sufficient to make us understand things to which they have no resemblance, why could not Nature also have established a certain sign that should make us have the sensation of light, even though this sign should have nothing in itself resembling this sensation? And is this not the way she has established laughs and tears, to allow us to read joy and sadness on the faces of people? (Adam and Tannery 1909: 4; translation mine)

A few lines later, Descartes concedes that this view is open to objection, but immediately brushes this fact aside, saying that he does not wish 'to lose time over such disputes'! Even so, the analogy between perceptual and linguistic sensory input 'will make quite a career in seventeenth and eighteenth century theories of perception (e.g. in those of Berkeley and Reid)' (Meijering 1981: 113).

A similar notion must now be developed for the intents of assertion, question, command, exhortation, rule-imposing, and imprecation. An ASSERTION is satisfied when the proposition expressed is true. A YES/NO (or POLAR) QUESTION is properly answered by the correct specification of the truth-value of the proposition expressed. A WH-QUESTION is properly answered by the correct identification of the entity questioned. A COMMAND is properly followed up by bringing about the state of affairs described in the proposition. An EXHORTATION is properly followed up when the speaker, his/her group, or the third party intended, actually do what is specified in the proposition expressed. A PERMISSION is properly implemented both when the addressee does and when (s)he does not do what is specified in the L-proposition expressed, while an OBLIGATION is properly implemented by the addressee's actually doing it. And an IMPRECATION is properly fulfilled when the state of affairs specified in the embedded proposition becomes real, through whatever force.

What we need now is a single term that covers all these different kinds of fulfilment vis-à-vis the person or persons with respect to whom the speech act in question is binding (the force field). I have chosen the term LIABILITY CONDITIONS for the purpose at hand. One reason why this term seems appropriate is that it implies the notion of force field: there can be no liability unless there are persons with respect to whom the liability holds. An intent is thus semantically defined by its liability conditions.

Just as truth conditions provide the (barest possible) semantic definition of not only the mental occurrence of a proposition but also of its linguistic expression, first as an L-proposition at the level of Semantic Analysis or SA, and finally of the corresponding surface structure, in the same way liability conditions provide the (barest possible) semantic definition of, first, the mental occurrence of an intent, and then also of its linguistic expression as an SA, which comprises both a speech-act operator and an L-proposition, and finally of the corresponding surface structure.

It is now clear that for assertions, which express a vouching for the truth of the proposition p at hand, the liability conditions coincide with the truth conditions of p: when p is true, the underlying intent, and therefore its liability condition, is satisfied. For assertions the specific force field does not matter, at least if it does not matter for truth and falsity who speaks and who listens. For the other sentence types this is not so, as has been shown. There, the liability conditions involve truth conditions, but the liability conditions do not coincide with the truth conditions, as is the case with assertions. This sheds some light on the age-old confusion between assertions and propositions.

4.4 The socially binding force more primitive than the proposition

The socially binding force is constitutive not only of speech utterances but also, more generally, of all nonlinguistic forms of signalling, whether between humans, between animals, or between humans and animals, and whether improvised or conventionalized. The binding that comes with linguistic utterances is not unique to human language, but is an intrinsic and general property of all forms of signalling that occur among human and other cognizing beings. This applies to traffic signs, conventionalized or improvised gestures, signals produced according to an agreed code and the like, but also to forms of communication between humans and animals, and probably even to communication among animals. It is not possible for a human being (or a cognizing animal) wilfully to produce a sign or give a signal without the sign or signal being placed under an operator that establishes a socially binding relation of commitment, appeal, or appellation.

Road signs, for example, are never representative of a mere proposition but always, necessarily, constitute a legally binding prohibition, admonition, injunction, or truth commitment with regard to some implicit proposition.[17] The same goes for animal signs. Bird song, as is well known, often signifies a territorial claim, which is very much part of social reality within the species concerned. Warning cries are found all over the animal kingdom.[18]

There is, moreover, the highly interesting fact, known to all parents but hardly studied academically, that young children, from the moment they begin to understand speech utterances, clearly possess an already strongly developed sense of justice, which manifests itself on two main parameters: a sense of equal rights among equals and a natural expectation that others, especially adults, will keep their trust and especially their word. Since it is empirically impossible that such young children have been taught these principles of justice, it must be assumed that they are innate—that is, a result of a genetically determined programme of personal development. Fully domesticated animals, such as dogs and horses, equally clearly manifest a

[17] Road signs may even have internal structure involving a, possibly complex, proposition, as in the case of the well-known sign forbidding overtaking of one car by another. Those signs usually consist of a disk representing two juxtaposed cars, both seen from the back, one in red and one in black, and indicating that any car finding itself in the position corresponding to the red car on the disk will be in contravention of the law. Sometimes such signs are placed underneath or next to another sign representing a heavy-transport vehicle, a bus, and/or a car with caravan. The latter sign then acts as a semantic operator restricting the prohibition to the designated categories of vehicles. Then again, the operator sign can be 'lexicalized' by replacing the red car image in the standard sign with a red image of a heavy-transport vehicle, where the image stands not only for actual heavy-transport vehicles but also for buses and any other vehicle with a weight exceeding some legally fixed limit.

[18] Vervet monkeys have specific warning cries for threats posed by eagles, leopards, and pythons (Seyfarth *et al.* 1980).

sense of justice, which, for them, amounts to an expectation of consistency and integrity in behaviour and attitude on the part of the persons in their environment.

It thus seems plausible to assume that the socially binding force in the intent of linguistic and nonlinguistic conventional signs that are uttered or otherwise displayed is at least as ancient a part of the human genetic endowment as the internal, linguistic structuring of the propositional part. And it may well be more ancient. I am prepared to venture the speculation that it is this socially binding aspect that gave the initial spur to the coming into being of human language. Even if it is hazardous to try to imagine what an early hominid society without language would have been like, it does not seem too far-fetched to assume that such societies would have benefited enormously from the availability of systematic means for the expression of socially binding forms of position-taking with regard to an infinite variety of specific, structurally specified propositions.

If I had no language but were otherwise in full possession of my wits, my sense organs and my ability to plan and execute actions, including vocalizations, I am sure I would, in all sorts of ways, develop means of signalling the way I want to define my position in an interpersonal cooperative network with regard to some proposition, using a combination of gesturing, touching, and vocalization. I would, for example, make it clear, by pointing at myself, at a nearby tree and at a nearby axe, while nodding my head and imitating the movement made when one uses an axe, or what not, that I am taking upon me the task of felling that tree. That is a commitment of the kind we call a promise, with regard to the proposition 'me felling that tree'. Clearly, if I then fail to do what I promised, I will be open to a charge of unreliability and may incur sanctions. Or, when I see my mate looking for his axe, I may make a sound to attract his attention and subsequently point at the deer hide nearby, making it clear to him that his axe is underneath the hide. This amounts to my guaranteeing the truth of the proposition 'your axe is underneath that hide'. And again, if it subsequently turns out that the axe is not underneath the hide, I will be open to a charge of unreliability and may incur sanctions.

4.5 The linguistic reality of the speech-act operator

Let us now turn to the specifically linguistic aspects of speech act theory. We have seen that natural language has standardized a limited number of speech act types in the form of grammatically defined sentence types. Most linguists distinguish three main grammatically defined sentence types in the languages of the world (see note 6), namely statements (assertions), questions, and commands, and we have added exhortations, rule-settings, imprecations, and vocatives, without claiming that this would complete the list. We have, moreover, decided to take the theoretical hard-line position which implies that Austin's explicit performative utterances represent statements that happen to create the state of affairs described in the proposition expressed, thus making themselves true, provided they are uttered felicitously. Their interest lies in the fact that the state of affairs that is made true by their being uttered felicitously is a piece of interpersonal, socially binding reality.

Moreover, we have, so far, taken the position that each of the five sentence types is grammatically (at SA level) defined by an overarching sentence-type operator of assertion, question, command, exhortation, or imprecation. It is shown in Section 4.5.2 that this position is too simplistic, since speech-act operators may occur some way down the SA structure. This admission may have the appearance of weakening our position. Yet the contrary is the case, since this fact by itself provides a strong argument for the linguistic reality of the speech-act operator as an expression of the socially binding force in the mental intent underlying any utterance.

4.5.1 *Ross's (1970) performative hypothesis*

The PERFORMATIVE HYPOTHESIS was put forward in Ross (1970) and further discussed and developed by other authors, notably Sadock (1975). Because of their depth and richness, both works deserve detailed discussion. For reasons of space, however, I will limit myself mainly to a discussion of Ross's study.

Ross takes as examples the sentences (4.13a–4.13c) and (4.14a, 4.14b) (Ross 1970: 222–3):

(4.13) a. Prices slumped.
b. I like you when you giggle.
c. Even Rodney's best friends won't tell him.

(4.14) a. I order you to go.
b. Go!

He then goes on to present his performative hypothesis:

> There are a number of facts which suggest that Austin's contention that sentences like [(4.14b)] contain implicit performatives is to be captured by postulating deep structures for them which are almost identical to the deep structure which has been assumed to underlie the superficially more complex [(4.14a)]. I will not discuss these arguments, for they are not central to the main thesis of this present paper. This

> thesis is that declarative sentences, such as those in [(4.13)], must also be analyzed as being implicit performatives, and must be derived from deep structures containing an explicitly represented performative main verb. (Ross 1970: 223)

Ross thus appears to take for it granted that, in principle, implicit imperatives, questions—and we may surmise also exhortations, rule-settings, imprecations, and appellations—have their explicit counterpart as their underlying SA (or, in the terminology of the day, 'deep structure'). And his thesis is that a similar assumption should be adopted for declaratives (statements, assertions). One sees, revealingly, that Ross still feels that a special case has to be made out for assertions, as against nondeclarative sentence types, which further illustrates the difficulty, repeatedly signalled above, that authors since Austin have had in recognizing the socially binding force of truth vouching. Ross's performative hypothesis is defined by the assumption that all grammatically standardized sentence types, nondeclaratives as well as declaratives, should be analysed, at SA level, as containing a speech-act operator which has the form of an explicitly represented performative main verb.

Whether by 'performative verb' Ross means a phonologically defined lexical item or a bundle of semantic features is not clear. In concrete examples, such as (4.15), he does not use a phonologically defined lexical verb for the performative verb position but a bundle of features, which, given the spirit of the day, may perhaps be taken to represent a lexical verb. Ross does not specifically comment on this aspect of his analysis. Be that as it may, he proposes (1970: 224) the following SA ('deep structure') for the declarative sentence (4.13a).

(4.15)

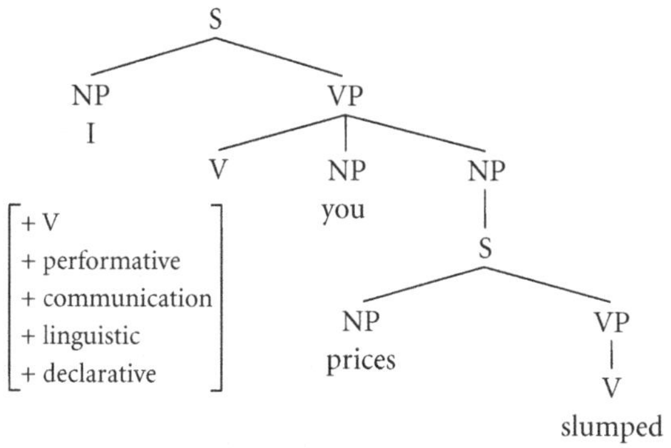

If the five features in the highest V-position are to be read as jointly defining the verb *say*, (4.15), when read out explicitly, should be taken to correspond to something like *I say to you that prices slumped.*

I disagree with Ross to the extent that he wants full lexicalizability in the SA-position of performative operators. The position I defend is (a) that all grammatically standardized sentence types should be analysed as containing a speech-act operator at SA level at the top of, or at some depth in, the SA tree, and (b) that this speech-act operator does not have the form of a lexically specifiable verb but is an abstract element that is usually deleted but may be manifest in surface structure as a particle or a morpheme (such as the question suffix in the Turkish examples (3.7a–3.7c) of Chapter 3), but never as an explicit performative verb.

Ross then proposes (1970: 249) a transformational rule of PERFORMATIVE DELETION, in a notation still widely used around 1970:

(4.16) Performative Deletion

$$\begin{array}{ccccc} \text{NP}[\text{I}] & \begin{bmatrix} +\text{V} \\ +\text{performative} \\ +\text{communication} \\ +\text{linguistic} \\ +\text{declarative} \end{bmatrix} & \text{NP}[\text{you}] & \text{VP}[\text{S}] & \\ 1 & 2 & 3 & 4 & \Rightarrow \\ 0 & 0 & 0 & 4 & \end{array}$$

Whether this rule should be taken to be obligatory or optional is left open by Ross (1970: 249). He does, however, mention the well-known *problem of infinite regress* (1970: 251–2). This problem consists in the fact that if the rule is obligatory, then a simple sentence like (4.13a) could be said to be derived not only from 'I say to you that prices slumped', but also from 'I say to you that I say to you that prices slumped', and again from 'I say to you that I say to you that I say to you that prices slumped', and so on ad infinitum. The grammar would thus assign to each declarative sentence an infinite number of underlying SA ('deep') structures, which would be unacceptable. If, on the other hand, the rule is optional, the grammar will, vacuously, allow for sentences like *I say to you that I say to you that I say to you that . . . prices slumped*, which is probably harmless, just as grammars harmlessly and vacuously allow for other kinds of unrestricted embedding, as, for example, with relative clauses.

Ross's answer is (1970: 251) 'that there is an independently necessary constraint that prohibits any verb from having a performative interpretation when it is embedded as the complement of another verb'. This, in effect, makes the rule obligatory and sensitive to the feature [+performative], while the occurrence of this feature is restricted to the position of the highest V in the SA structure. We will see in a moment that this latter condition is too strict, since speech-act operators often occur further down in the SA structure of a sentence, even within the complement of another verb. Yet the principle of restricting the position of the speech-act operator *O* in virtue of certain independent semantic principles and introducing a rule that obligatorily deletes *O* (or, exceptionally, expresses it as a particle or a bound morpheme) triggered by its being a speech-act operator, and not by any lexical form, seems sound. And it does block the vicious infinite regress signalled above.

Under this assumption, however, no evidence is left to support the view that the speech-act operator is in any sense a complement-taking lexical verb, or indeed that the speaker is lexically represented at SA level as the pronoun *I* or the hearer as the pronoun *you*. On the contrary, such a view adds a gratuitous element that should be eliminated. This has an unfortunate consequence for Ross's argumentation. For when it is no longer assumed that linguistic speech-act operators are, underlyingly, complement-taking lexical verbs with lexically realized subject and indirect-object terms, the bottom is taken out of Ross's arguments, fourteen in number, based as they are on parallels between phenomena in complementation structures and speech act phenomena.

We will now look at most of Ross's observations, leaving out those that seem too far-fetched or too weak, and show that, although they are clearly relevant and often highly interesting, they are insufficient to force the conclusion that we have to do with lexical structures fulfilling the role of speech-act operators.

Some of Ross's arguments are to do with REFLEXIVES. Consider, for example the following cases (1970: 225–30):

(4.17) a. Tom believed that the paper had been written by Ann and himself (*myself).
b. I say that the paper has been written by Ann and myself (*himself).
c. The paper has been written by Ann and myself (*himself).

(4.18) a. She told Albert that doctors like himself (*myself) are a godsend.
b. I say to you that doctors like yourself (*himself) are a godsend.
c. Doctors like yourself (*himself) are a godsend.

On the assumption that reflexivization is a matter of sentence-internal structural constraints, (4.17) would show that declaratives have an underlying first person subject, while (4.18) would show that declaratives have an underlying second person indirect object, to a performative verb of saying. However, deeper and wider probing into reflexivization phenomena shows that reflexivization has much deeper semantic and even cognitive roots. Apart from the fact that in many languages reflexive morphology—whether on verbs or on pronouns—is a great deal more complex than in English, VP-anaphora reveals the semantic and cognitive roots of reflexivization. It has often been observed that a sentence like (4.19) does not imply that Howard hurt Jack but, rather, that Howard, like Jack, hurt himself:

(4.19) Jack hurt himself, and so did Howard.

Apparently, *hurt oneself* is a predicate in its own right, with a meaning like 'self-hurt', derived from the basic predicate *hurt* by means of a productive process within the lexicon. Such a process must be taken to be at least partly based on cognitive criteria, as appears from the following example, taken from an interview with a director of the Swedish firm Electrolux, who had previously been a director of the Dutch firm Philips. At some time during the interview the man says:

(4.20) Philips usually sends Dutch management to its foreign branches, but Electrolux *does so* only in exceptional cases.

The context shows beyond doubt that what he intended to say was not that Electrolux only exceptionally sends Dutch management to the foreign branches of Philips, but that it only exceptionally sends Swedish management to its own foreign branches. The VP-anaphora *does so* picks up the relation 'send management from one's own country to one's own foreign branches', and not 'send Dutch management to the foreign branches of Philips'. This case shows that VP-anaphora (a) may function on account of world knowledge, (b) may involve forms of cognitive reflexivization that are not expressed linguistically at all. One should realize that reflexivization is an extremely complex phenomenon, whose depths we have barely started to plumb. This takes away the force of Ross's reflexivization arguments. All we know is that in any speech act speaker and listener are necessarily cognitively present. This cognitive presence is probably the basis of the use of the reflexives in (4.17c) and (4.18c).

Another, curious, argument put forward by Ross (1970: 234–6) is to do with predicates which one is not supposed to apply to oneself but only to others. Ross gives as an example (suggested by David Perlmutter) the predicate *lurk*, though he admits that many native speakers of English find *lurk* perfectly acceptable when applied by a speaker to himself or herself.[19] He also mentions (1970: 264) the verb *purport* (suggested to him by Robert Wall) as perhaps a better example. I may add the predicates shown in the examples (4.21d–4.21f). Ross observes a similar phenomenon for the phrase *according to*, which,

[19] Melissa Bowerman (personal communication) assures me that, for her, *I was lurking in a dark corner* is perfectly normal for a speaker who is a member of a society of peeping Toms.

according to him, cannot take the speaker as the referent of its object. In any case, many speakers find all or some of the following sentences awkward (or funny), for reasons that probably involve criteria of social propriety:

(4.21) a. !?I was lurking in a dark corner.
b. !?I purport to be a scholar.
c. !?According to me, food prices will skyrocket.
d. !?I am quite a character.
e. !?I am a stalwart of the early days.
f. !?I am an honest-to-God guy.

Ross's argument is that the same reaction of awkwardness is provoked by clauses with a third person singular or plural subject that is coreferential with the subject of a higher complement-taking verb attributing a mental predication to its subject referent, such as verbs of saying or believing. The examples (4.22a–4.22f) illustrate this:

(4.22) a. !?Tom believed that he was lurking in a dark corner.
b. !?Tom says that he purports to be a scholar.
c. !?Tom believes that, according to him, food prices will skyrocket.
d. !?Tom says that he is quite a character.
e. !?ITom says that he is a stalwart of the early days.
f. !?Tom says that he is an honest-to-God guy.[20]

Taking facts like (4.22a–4.22f) as basic, Ross extrapolates their structural analysis to declarative speech acts such as those in (4.21a–4.21f) for which he then assumes an underlying performative verb of saying with a lexical subject *I*. Yet the facts of (4.22a–4.22f) may not be basic at all but be regarded as resulting from a general (semantic or pragmatic) restriction on self-attribution, which manifests itself not only in (4.21a–4.21f) but also in (4.22a–4.22f). Ross's argument is typical of the period in that, in those days, it was widely, though implicitly, assumed that grammar, and syntax in particular, should bear the whole burden of explanation for all observations made. Nowadays one is more open to the view that language is but a link in a complex chain of processes mediating between the speaker's intent and the listener's adequate understanding.

[20] But note that (4.22d–4.22f) are perfectly normal with *believe* as the higher verb instead of *say*.

A further, interesting, observation regarding the cognitive presence of the speaker in any speech act is found in Ross (1970: 237–8):[21]

(4.23) Hoboken is a fine city, my (*your/*her/*his/*Bill's) love.

Again, however, there is no reason for inferring that the speaker's presence requires a lexical linguistic representation.

The same goes for the argument (Ross 1970: 238–40) that the verb *believe* can take a human anaphoric pronoun as its object term only if this pronoun 'stands in an anaphoric relationship to another NP which functions as the subject…of a verb with the feature composition [+communication, +linguistic, +declarative]', as, for example, the verb *say* in (4.24a). To maintain this generalization for cases like (4.24b), Ross proposes that the SA of (4.24b) should be taken to contain a highest clause with the subject term $_{\text{NP}}$[I] and a verb characterized by the feature complex [+communication, +linguistic, +declarative], i.e. 'precisely the claim made by postulating [(4.15)] as the deep structure of [(4.13a)]' (1970: 240):

(4.24) a. Tom said that Ann could swim, but nobody believed him (*her/*them).

b. Ann *can* swim, but if you don't believe me (*her/*them), just watch her.

Again, Ross seeks an explanation in syntactic structure, where semantic and/or cognitive conditions appear to do the work. A construction like [believe +NP] presupposes that the person referred to by the object-NP has made an assertion (has vouched for the truth of a proposition). That presupposition is satisfied by both (4.24a) and (4.24b), but not by (4.25), which should be all right according to Ross's criterion:

(4.25) !!Tom said nothing, but nobody believed him.

Ross might have added the condition that the higher verb with the feature complex [+communication, +linguistic, +declarative] should not be negated, but that would again produce heaps of counterexamples, such as:

(4.26) a. Tom didn't say THÁT, but even so nobody believed him.

b. !!Tom refused to say anything, but nobody believed him.

[21] Ross also notes (1970: 266) the obligatory occurrence of *your* in vocatives like *your worship*, *your lordship*, *your majesty*, finding this 'a total mystery'. It becomes less of a mystery when one realizes that in postmedieval Europe one was often not allowed to use the pronoun *you* when addressing highly placed persons. Instead of saying, for example, *You are surrounded by enemies*, one would have to say *Your majesty (lordship, grace*, and so on*) is surrounded by enemies*. This, as is well known, is the origin of the use of third person forms for polite address in many languages.

Owing to the contrastive accent on THÁT, (4.26a) is fine, even though the higher verb *say* is negated. And (4.26b) suffers from presupposition failure even though the higher verb *say* is not negated. In other words, a condition phrased in syntactic terms will inevitably lead to the necessity of having to plug ever newly springing leaks in ever more ad hoc ways. What is needed is a condition that is phrased in nonsyntactic but semantic and/or cognitive terms. But such a condition will no longer support Ross's argument for lexicalized speech-act operators at SA level.

A similar strategy is followed in Ross's argument (1970: 241–244) concerning the *be-damned-if* idiom, as in:

(4.27) a. I'll be damned if I'll go home.
b. Tom says that he'll be damned if he'll go home.

Ross conjectures (1970: 244) that the *be-damned-if* idiom requires a subject-NP that is identical, or coreferential, with the subject-NP of the first S up the SA tree whose head verb has the feature complex [+communication, +linguistic, +declarative]. This will explain not only (4.27b) but also (4.27a) provided it is assumed that (4.27a) contains, at SA level, a speech-act operator of the form $_{\text{NP}}$[I] + V + $_{\text{NP}}$[you], where the V-element is characterized by the feature complex [+communication, +linguistic, +declarative].

However, Ross admits (1970: 268) that sentences like (4.28a,b) are also fine:

(4.28) a. Arthur will be damned if he'll go home.
b. I heard from Ed that he'd be damned if he'd go home.

His comment is that (4.28a) is a report of a speech act, paraphrasable as:

(4.29) Arthur says that he'll be damned if he'll go home.

and he continues: 'Although I have none at present, I would hope that evidence will be forthcoming which would support the derivation of the grammatical reading of [(4.28a)] from [(4.29)]'. As regards (4.28b), his suggestion is that it should be derived from something like:

(4.30) Ed told me that he'd be damned if he'd go home.

My comment is that proposals to the effect that (4.28a) should be derived from (4.29) and (4.28b) from (4.30) are clear illustrations of the excesses of generative semantics as it was practised in those days. It is useful to see that those excesses were at least partly due to the all too narrow vision that syntax should bear most or all of the burden of explanation of observations made.

It seems to me that the *be-damned-if* idiom is best taken to be an idiomatic lexicalization of a semantic content like 'subject is strongly determined not to VP'. It does not seem necessary to assume a speech-act-report reading for sentences like (4.28a). For example, the sentence:

(4.31) The queen will be damned if she'll sack her prime minister.

does not seem to require a reading in which the queen actually said *I'll be damned if I'll sack my prime minister.* All (4.31) says, in my view, is that the queen is strongly determined not to sack her prime minister, whereby the choice of the expression *be damned if* is the speaker's responsibility.

Ross's star argument (1970: 244–5) is taken from Arabic, where, according to Michael Brame, the verb *'aquulu* ((I) say) is unique in requiring the *that*-complementizer *inna* to head its complement clause, other declarative verbs taking the complementizer *anna.* According to Ross, the sentence *The boy has left the house* can be synonymously translated into Arabic in any of the following three ways:

(4.32) a. 'Aquulu inna l walada qad taraka l bayta.
I say that the BOY_{ACC} PERF leave the $\text{house}_{\text{ACC}}$

b. ...inna l walada qad taraka l bayta
...that the BOY_{ACC} PERF leave the $\text{house}_{\text{ACC}}$

c. 'Al waladu qad taraka l bayta.[22]
the BOY_{NOM} PERF leave the $\text{house}_{\text{ACC}}$

He comments: 'I might point out that Arabic is the only language I know of where strong evidence points to a rule deleting a particular verb, instead of a proverb specified only by an abstract bundle of features.'

However, my Arabist colleague Kees Versteegh informs me that, while it is true that *inna* is unique to the verb *'aquulu* (say), it probably originates not as a complementizer but as a presentative in the form of a formulaic imperative 'look' or 'behold', usually in sentence-initial position, and with the function of presenting and emphasizing an assertion, not unlike English *behold*, as in the Biblical (Exodus 6:12):[23]

[22] Ross translates the particle *qad* as PAST, but my colleague Kees Versteegh tells me that it has a perfective-resultative meaning, so that (4.32c) should be translated as 'the boy *has* left the house'. Moreover, Versteegh informs me, the preposing of the subject *'al waladu* (the boy) forces a focus accent on that constituent. I have, accordingly, capitalized BOY and changed Ross's 'PAST' into 'PERF'.

[23] Bloch (1986: 111) also quotes instances of sentence-final *inna*, as in (i)–(iii). Note that in (i) and (ii) *inna* occurs both initially and finally:

(i) Innanī min dāka inna(h)
I am truly thus, indeed!

(ii) Inna salītan fī lhasāri inna(h)
The Salīt are truly losers, indeed!

(iii) Lā uqīmu bidāri ddulli inna.
I shall not dwell in the abode of disgrace, indeed (not)!

And Moses spake before the Lord, saying, behold, the children of Israel have not hearkened unto me.

where *behold* translates the Hebrew presentative *hēn*, a variant of *hinnēh* (still existing in modern Hebrew), which is etymologically related to Arabic *inna* (Bloch 1986) and probably originated as an imperative 'look, behold', like its French equivalents *voici* and *voilà* (Joüon 1947: 272). This would explain why *inna* takes the accusative (*walada* in (4.32a) and (4.32b)), just as French *voici* and *voilà* take accusative objects: it is the original direct object of the imperative *inna*: 'Behold the boy, he has left the house'.[24]

It would seem to me that if the rather lofty formulaic collocation of *say* immediately followed by *behold*, as in the biblical quote given above, was part of the literary standard in pre-Christian Middle-East story-telling, then all that is needed to turn *behold* into a complementizer especially reserved for the verb *say* is two processes, both well known in sociolinguistics and historical linguistics, namely a certain degree of colloquialization of the formulaic literary collocation and a subsequent grammatical reanalysis of *behold* as a standard particle to introduce the object clause after *say*. Such a double process leaves open the possibility of (the equivalent of) *behold* still standing at the beginning of an independent declarative sentence, as in Genesis 6:12:

And God looked upon the earth, and, behold, it was corrupt.

One might compare the expression *verily*, as in Matthew 5:18 or 5:26 and many other places: 'Verily, I say unto you…'. Here, too, the occurrence of *verily* (Aramaic *amēn*) is prototypically bound up with the verb *say*. One can easily imagine a variant of the form 'I say unto you, verily,…', and a subsequent reanalysis with *verily* as a newly formed complementizer, which can still stand at the beginning of an independent declarative sentence. In fact, *inna* often does have the meaning 'truly' in standard written Arabic (Bloch 1986). Sentence (4.32b) should thus mean something like 'Truly, the BOY has left the house'.

There thus does not seem to be much support for Ross's assumption that (4.32b) is derived from (4.32a) through a rule deleting the verb *'aquulu* (say). Historically, that does not seem to have been the course of events, and

The sentence-final use of *inna* is parallelled by the negative *lā* (no!), as in Bloch (1986: 111):

(iv) Laysa hāda mustaqīman lā.
This is not right, no!

[24] Bloch (1986) suggests an existential-presentative origin with the meaning 'there is/are'. That hypothesis still leaves room for the accusative case: cp. German *es gibt* (there is/are; lit.: it gives) or Modern Greek *échi* (there is/are; lit.: it has), which take the accusative.

synchronically one seems best advised to assume a sentence-initial particle meaning 'truly', distinct from a complementizer now meaning 'that' and idiosyncratically reserved for the verb *say.*

4.5.2 *Other arguments for the linguistic reality of the speech-act operator*

Ross's arguments for the performative hypothesis thus appear not to be too compelling. They fail signally with regard to the assumption of a lexical specification of the speech-act operator, but they support the view that the speech-act operator is represented at SA level as a separate structural element, which will then have to make do without a lexical specification, whether phonologically or through features. I will now add further arguments for the linguistic reality of such a speech-act operator, beyond those presented by Ross and others. Some of these arguments are purely semantic, others also have a grammatical aspect to them.

The strongest evidence comes from languages that mark speech acts formally, especially those that mark assertions, as opposed to other forms of speech act. Payne (1997: 294) quotes Tibetan as a language that marks assertions and questions, giving the following examples:

(4.33) a. Yoqöö mOOmOO sææ -p' -ree.
servant dumplings eat -PAST -ASSERTION
The servant ate the dumplings.

b. Yoqöö mOOmOO sææ -p' -repææ?
servant dumplings eat -PAST -QUESTION
Did the servant eat the dumplings?

The same is found in the Nambikwaran (Amazonian) language Sabanê, where assertion and question are marked morphologically after non-evidential tenses (Antunes de Araujo 2004: 147):

(4.34) a. Iney -i -ntal -i.
fall -VERB NONEVIDENTIAL PAST -ASSERTION
S/he fell.

b. Iney -i -ntal -a?
fall -VERB NONEVIDENTIAL PAST -QUESTION
Did s/he fall?

Such cases remind one of the Turkish examples (3.7a–3.7c) of Chapter 3, with the question suffix *-mı*, which is sensitive to the topic–comment structure of the sentence and attaches itself to the comment. They also remind one of Latin, which has the suffix *-ne* attached to the first word of the sentence, marking the fact that the sentence is a yes/no question. Latin, moreover, has sentence-initial *nonne*, marking a yes/no question that favours the answer *yes*, and the sentence-initial *num*, marking a yes/no question favouring the answer *no*.[25]

One might object that this only shows the linguistic reality of the assertion and question speech-act operators in the languages concerned and says nothing about the languages that do not mark these speech acts formally. Speech-act operators would thus be treated on a par with, for example, the category of evidentiality, which is marked in some languages and not in others. Whether or not speech acts are formally marked would thus be a question of any language-specific checklist as discussed in Section 9.4. In a way, this is true, since if a language gives formal expression to any kind of speech act, its checklist will have to contain a speech-act parameter. Yet as an objection to the language-universal nature of speech-act operators it does not hold, first because speech-act operators are semantically constitutive for any speech utterance, as is shown in the present chapter, and, secondly, because even in languages that do not mark, say, assertions formally, there are plenty of phenomena that find a natural explanation in terms of an assumed grammatically real assertive speech-act operator.

The following English sentences provide both semantic and syntactic evidence of the reality of speech-act operators (the square brackets enclose semantic material that must be mentally supplied for the sentences to make sense):

(4.35) a. Since you are here, [*I ask you*] could you tell Ann we're off?
b. Moreover, [*I ask you*] why don't you just leave?
c. Therefore, [*I advise you*] don't blame me.
d. If you're tired, [*I inform you*] I have a spare bed.
e. If you're tired, [*I advise you*] go and get some sleep.
f. If he's so clever, [*I ask you*] why isn't he rich?
g. In that case, [*I ask you*] what is your argument?
h. [*I am telling you*] quickly, ask Norbert about Abe's expenses.[26]
i. Were you at home? Since [= *I'm asking because*] the light was on.

[25] As in the famous question which, according to the Roman historian Suetonius, the emperor Vespasian asked his son Titus when holding up a piece from a chest of money collected from the tax on public conveniences in Rome that had been instituted by the father but was derided by the more refined son: 'Num olet?' (Does it stink?). Hence the saying 'Money doesn't stink'.

[26] At the 39th Annual Meeting of the Societas Linguistica Europaea, Bremen, 30 August–2 September 2006, Chris Beedham and Chris Lucas pointed out to me that in:
(i) Quickly, how much did you pay?

j. Admittedly [= *I hereby admit that*], there are some misprints in your article.

k. Before you start protesting, [*I hereby wish you*] good morning.

l. Again [= *I am repeating my assertion that*], this is too much for one trip.

m. So you did find your brother, who was in the library, wasn't he?

The semantic nature of the evidence should be obvious. Note, in particular, sentence (4.35j), where the high adverb *admittedly* implies a speech act: 'I hereby admit that', as opposed to, for example, *Reportedly, there are some misprints in your article*, where the high adverb *reportedly* stands for 'it is (has been) reported that'. The meaning 'it is (has been) admitted that' is not possible for (4.35j).

All these cases have in common that their interpretation requires an intervening speech-act operator. Sentences (4.35a, 4.35i) only make sense when *since* (note that *because* would be ungrammatical here) is read as 'I am asking because'. Sentence (4.35d) requires an assertive, and (4.35e) an imperative, speech-act operator after (below) the *if*-clause. Example (4.35m) shows that nonrestrictive relative clauses stand under their own speech-act operator. In the case at hand, the speech-act operator in question is one of suggestion, expressed by the tag question *wasn't he?* These cases show that the speech-act operator must be taken to occur, in the L-propositional or SA-structure, BELOW a higher operator or *if*–clause.

Some of the cases (4.35a–4.35m) also provide syntactic and/or lexical evidence. Sentence (4.35b), for example, would be ungrammatical with *perhaps* instead of *moreover*, the reason being that *perhaps* is not possible as a modifier of a speech-act predicate, while *moreover* is. And analogously for *therefore* in (4.35c). Example (4.35m) provides syntactic evidence, given the tag question attached to the subordinate clause—a clear indication of the presence of an underlying speech-act operator in a nonrestrictive relative clause.

Analogous examples are found in German. Consider the ambiguous sentence (4.36), with *allein* (only). This sentence is open to (at least) two mutually exclusive interpretations, namely (4.36a) and (4.36b). Example (4.36a) does not show any speech-act involvement, but (4.36b) does, in so far as the adverbial phrase *allein in Bayern* modifies the speech act operator:

the adverb *quickly* would normally be understood as a request for a quick answer. It seems to me that this might be taken to be evidence for the view that the question operator semantically includes an element of answering: 'I am soliciting you to answer (quickly)'.

(4.36) Allein in Bayern hat es fünf Fälle von Vogelgrippe gegeben.
only in Bavaria has it five cases of bird flu given
a. Only in Bavaria have there been five cases of bird flu.
b. To mention only Bavaria, five cases of bird flu have been attested there.

A further German example is (4.37a), which shows evidence for a speech-act operator, as opposed to (4.37b), which does not. Clearly, in (4.37a) the pleasure is the speaker's, while in (4.37b) the pleasure is Michael's:

(4.37) a. Du kannst gerne deine Freundin mitbringen.
you can with-pleasure your girl-friend along-bring
You are welcome to bring your girl friend along.
b. Michael bringt gerne seine Freundin mit.
Michael brings with-pleasure his girl-friend along
Michael likes to bring his girl friend along.

Consider also the difference between the following two:[27]

(4.38) a. Er wurde nicht bedient, weil er keine Kravatte trug.
He wasn't served, because he didn't wear a tie.
b. Er wurde nicht bedient, denn er trug keine Kravatte.
He wasn't served, [and I can say this because] he didn't wear a tie.

German and Dutch conditionals provide further syntactic evidence. Thus, (4.35d), when translated into Dutch or German, does not allow for the subject-auxiliary inversion that is otherwise obligatory when a subordinate clause precedes the main clause. In German, for example, (4.35d) goes like (4.39a), not like (4.39b), even though (4.39b) is the normal form for sentences with a fronted subordinate clause. Sentence (4.39b) conveys the implausible message that the speaker has a spare bed in case the addressee is tired (and probably not if the addressee is not tired):

(4.39) a. Wenn du müde bist, ich habe ein Gästebett.
b. Wenn du müde bist, habe ich ein Gästebett.

Cases like (4.35a–4.35m), (4.36), (4.37a, 4.37b), (4.38a, 4.38b), or (4.39a, 4.39b) should suffice to convince one that any adequate semantic representation of

[27] It was pointed out to me, at the meeting mentioned in note 26, that when German *weil* takes main-clause word order instead of the regular verb-final subordinate-clause word order, which it often does, it is ambiguous between regular *weil*, as in (4.38a), and *denn*, as in (4.38b).

the sentences concerned must be taken to contain a structural element of some sort symbolizing the appropriate speech-act operator.

They should, moreover, suffice to convince one that, contrary to what is maintained by Ross (1970), speech-act operators do not necessarily occur at the very top of the SA structure. It is perfectly possible to qualify them by means of even higher operators. This in itself is a powerful argument in favour of the reality of speech-act operators as elements in semantic representations. The possibility of qualifying speech-act operators by means of even higher operators is not limited to implicit speech acts. The same goes for explicit speech acts, as becomes clear from (4.40a–4.40c):

(4.40) a. It is with great satisfaction that I hereby inform you that you have lost the elections.
b. You should all be aware of the fact that I hereby admit that I lied.
c. I am very sorry to have to hereby ask you to leave.

A further argument for the linguistic reality of speech-act operators may be derived from the use of so-called EPITHETS in language (see note 8). An epithet is a definite noun phrase whose noun is a predicate that is satisfied not so much by the ontological properties of the object referred to as by an evaluation, usually negative, on the part of the speaker. Epithets are normally reckoned to have the status of 'pronoun with an edge'. To fulfil their epithetic function they have to be unaccented. Examples are (the epithets are in italics):

(4.41) a. I hope *the runt* won't be re-elected.
b. Where's *my little darling*?
c. *The fool* didn't even ask for my name.
d. I don't want *that creep* on the committee.
e. Does *that creature* have to be present while we talk?

Truth-conditionally, these sentences are equivalent to their counterparts with an ordinary definite pronoun in lieu of the epithet. Semantically, however, they are not. How to account for the semantic difference is a question that has not, so far, received much attention in the literature. In Seuren (1985: 27, 348) it is proposed to insert a speech-act operator of the kind 'I hereby call *x* such-and-such'. Sentence (4.41a) would thus be represented, at SA level, as something like 'I hope [not [the *x*—and I hereby call *x* a runt—will be re-elected]]'. And similarly for (4.41b–4.41e). Further research is obviously required.

4.6 The logical connectives from a speech-act point of view

To close this chapter, let us have a look at the logical connectives from the point of view of speech act theory. The point is that *not*, *and*, and *or* are often used in ways that escape description in logical terms. Leaving *not* aside for the moment, we concentrate on *and* and *or*. Consider the following two sentences:

(4.42) a. Please do as I say and you will be a rich man.
b. Don't laugh or I'll shoot.

One notes first that *and* and *or* in these sentences are not symmetrical. Inversion of the conjuncts or disjuncts results either in a very different message or in nonsense (marked by '!!'):

(4.43) a. You will be a rich man and please do as I say.
b. !!I'll shoot or don't laugh.

Sentence (4.43a) is a double speech act consisting of a promise to the addressee that he will be a rich man, followed by a request to do as the speaker says, without any connection between the two. By contrast, (4.42a) is a request to do as the speaker says with the added promise that if the addressee complies with the request he will be a rich man. In (4.42a), the second conjunct's becoming true depends on compliance with the request expressed in the first conjunct. No such link is present in (4.43a). Then, (4.42b) is not a 'disjunctive' speech act in the sense that either or both are to be interpreted as having force, in the given force field. Such 'disjunctive' speech acts appear to be impossible, as one sees from (4.43b), which fails to make sense.

How to describe speech acts like those in (4.42) is a question to which no adequate answer is available in the existing literature. It would seem that *or* in (4.42b) expresses a speech act whereby the speaker leaves the addressee the choice between (a) not laughing with no consequences specified and (b) laughing with the consequence that the speaker will shoot. The speech-act operator would then translate as something like 'I hereby leave you the choice between not laughing and being shot.' Perhaps *and* in (4.42a) does the same, but with different (a) and (b): 'I hereby leave you the choice between (a) doing as I say and being promised great wealth and (b) not doing as I say and not being promised great wealth.' Does this mean that there is a further linguistically fixed category of speech-act operator in the form of 'I hereby leave you the choice ...'? It would be premature to try to answer this question here and now.

A pragmatic answer appears to be insufficient. There are too many specific structural constraints on the interpretations at hand for it to be plausible that they are generated as extrapolations of the logical meanings of the connectives concerned based on pragmatic factors like common sense reasoning or principles of proper conversational interaction. A more systematic analysis along the lines pursued here will hopefully lead to an adequate theory of the logical connectives functioning at speech-act level. Cases like these are somehow, perhaps closely, related to conditionals of the type 'if P then Q', but so far it has not been possible to lay bare the hidden links. Problems like these are, of course, of the greatest interest and it is high time that semanticists look at them more closely.

That same admonition applies to cases like the following, where the second conjunct or disjunct refers back to the first:

(4.44) a. Stop drinking now, and I mean it.
b. Is she divorced? Or am I not supposed to ask that question?
c. Let's go home, or do you have a better idea?

In these cases there is a metalinguistic element, which in itself should not cause surprise. It is an obvious corollary of every utterance that sounds, words, and larger expressions are used and thus come into being as possible reference objects—a fact that both speakers and listeners are fully aware of. That being so, there is no reason why these sounds, words, and larger expressions could not be the situationally given antecedent of referential uptake in the form of pronominal anaphora or a lexically filled definite description.

But apart from sounds, words, and larger expressions, speakers and listeners, apparently, also have sufficient awareness of speech acts and the propositions embedded in them for these to be antecedents of referential uptake. In (4.44a) and (4.44b), for example, the referring expressions *it* and *that question*, respectively, refer back to the immediately preceding speech act. And in (4.44c), the second disjunct implies a reference ('a better idea than us going home') to the proposition 'us going home' embedded in the immediately preceding speech act (compare the examples (4.1a, 4.1b) in Section 4.1 above).

Metalinguistic reference to or correction of a lexical selection just made is a perfectly normal phenomenon as well, especially by means of the connective *or*:

(4.45) a. That's awful, or should I say 'dreadful'?
b. That's awful, if that is the right word.
c. That's awful, and it is the right word.
d. That's an awful, or rather, despicable thing to do.

The use of *and*, as in (4.45c), is perhaps less problematic, since (4.45c) can be taken to consist of two assertive speech acts, in the second of which reference is made to a word used in the first. But the use of *or* or *if* in (4.45a, 4.45b, 4.45d) is harder to interpret. For one thing, *or* and *and* in (4.45a, 4.45c, 4.45d) cannot be expanded to *either ... or* or *both ... and*, respectively. But *if* in (4.45b) can be expanded to *if ... then*: 'If *awful* is the right word, then that's awful'.

As regards negation, Horn (1985) showed that the negation *not* is often used metalinguistically to deny the appropriateness of linguistic material, as in the following examples:

(4.46) a. Not LIZZY, you twit, but the QUEEN is wearing a funny hat.
 b. From now on, we'll no longer have a 'šédjul' but a 'skédj'l'.
 c. Miss Debenham is not a WOMAN, she is a LADY!

Sentence (4.46a) is naturally taken as an admonition to the addressee not to use the expression *Lizzy* but rather *the Queen* for the person referred to, besides, of course, the possible interpretation which has the speaker say that the one who is wearing a funny hat is not Lizzy but the Queen. Sentence (4.46b) is natural in a situation in which a speaker of British English wants to accommodate to an American-speaking environment, where the word *schedule* is pronounced [skédj'l] rather than the British [šédjul]. And (4.46c)—with thanks to Agatha Christie—does not deny that Miss Debenham is a woman, but rather says that *lady* is a more appropriate expression, in the case at hand, than *woman*.

As regards the negation operator, there is the further fact that emphatic negation, realized as a separate word (and not, for example, as a bound morpheme), in construction with the finite verb (at least in English), and not commanding a negative polarity item, can be used to deny a presupposition of the argument L-proposition. Horn (1985) classifies this use of the negation along with metalinguistic cases like those in (4.46a–4.46c). It was shown in Seuren (1988, 2001: 333–83) that although the emphatic presupposition-cancelling negation is clearly metalinguistic in nature, it must be taken to form a separate class, distinct from phonological or lexical correction cases like those presented in (4.46a–4.46c).

In connection with negation, one might also look at uses of the reply or exclamation words *yes* and *no*, especially the latter. Consider the following cases:

(4.47) a. Just before a crash: 'NO!!!' (I don't want this to happen)
 b. Instructor: 'No' (this is not the right way to do it)

c. Surgeon: (nurse handing instrument) 'No' (this is not the right one)

d. How about going for a walk? 'No' (I reject)/'Yes' (I agree)

Here again, there is a variety of uses or meanings, all to do with rejection or acceptance or agreement, but without—as far as I can see at the moment—there being a clear overall meaning that can be caught under one single precise formula.

There is also the curious fact that in many languages, such as Arabic, the equivalent of *yes* means 'I agree with the speech act you have just expressed', while the equivalent of *no* means 'I don't agree with the speech act you have just expressed'. Thus one finds the following contrast between, say, English and Arabic, where English echoes the negation in the proposition expressed, while in Arabic one agrees or disagrees with the speech act just expressed:

(4.48)	(a)	So you're not coming, are you?	English: No (I'm not coming) Arabic: Yes (I'm not coming)
	(b)	That's not right!	English: No (it isn't) Arabic: Yes (it isn't)

Some languages have a special variant of 'yes' to express the rejection of a negation in the proposition just expressed by the other speaker. French, for example, has *si* instead of *oui* for such cases and German has *doch* for the otherwise normal *ja*:

(4.49) a. Il ne veut pas partir, n'est-ce pas? Non/Si!
(He doesn't want to leave, does he? No/Yés he does!)

b. Das war jedenfalls nicht richtig. Nein/Doch!
(That was anyway not right. No/Yés it was!)

All this serves to show that there is much more to the logical connectives *and*, *or*, *if*, and *not* (and the expressions *yes* and *no*) than meets the eye. Owing to the almost compulsive focusing, in semantics, on the logical aspects of meaning, at the expense of the wider and more general semantic aspects of language and language use, facts such as those shown in (4.42)–(4.49), although well known to linguists generally, have hardly been paid attention to in the semantic literature.

The pragmatic literature fares much better in this respect, but pragmatics, unfortunately, tends to leave the holy shrine of logic untouched, but for minor adjustments and refinements. For facts that escape the powers of the standard logical machinery, pragmatics seeks, on the whole, an explanation in conditions of use and in considerations of common sense. The position defended in this book is that the pragmatic approach is too weak, mainly because many of the phenomena concerned appear to be subject to precise, formally statable, structural restrictions. If the phenomena at issue were a question of general common sense or of conversational principles, such structural restrictions ought not to exist. And if the phenomena at issue are a question of conventionalized meanings, there hardly seems any reason not to call them semantic.

What is needed is a semantic account of the expressions concerned in terms that are general enough to cover all their uses and precise enough to exclude all their non-uses. The purely logical aspects of the meanings of the connectives usually called 'logical' would then presumably be seen to follow from their overall description, and the logic of language would indeed be seen to be a specific consequence of the more general semantic description of the logical constants involved. This, however, is an ideal that I cannot hope to achieve in the present study.

5

Intensionalization

5.1 The problem stated

The question of the interaction of extensional and intensional predicates and reference objects constitutes an area of investigation that was neglected under the old regime, but demands exploration now that reference to virtual objects has been given recognition. Given the novelty and the complexity of the issues, the answers provided in the present book can only be partial and provisional, and further research is needed to provide more adequate analyses. But we can make a start.

One, widespread, phenomenon in natural language, and one that causes serious complications for any semantic theory, is what is called here INTENSIONALIZATION, due to the mind's power to form representations, thereby creating virtual worlds and objects. Intensionalization is the phenomenon that a predicate F, whose extension is strictly actual in virtue of its satisfaction conditions, is sometimes legitimately interpreted as having an extension that optionally includes virtual objects to which the satisfaction of the conditions of F in some virtual world has been ascribed. In such a case we say that F is intensionalized. The question that presents itself is: under what conditions are predicates intensionalized?

To see what is at issue we must revert to Section 3.5.1, where a distinction is made between EXTENSIONAL and INTENSIONAL term positions. One recalls that an extensional term position under a predicate F requires an actually existing object (or a reification that is reducible to actual objects) for truth to arise under F. For the predicate *hit*, for example, both the subject term position and the object term position are extensional, since for it to be true that Tim hit Tom, both Tim and Tom must be, or have been, actually existing persons. Not so, however, for it to be true that Tim hates Santa Claus, since although for truth to arise under the predicate *hate* actual existence is required for anything referred to by the subject term, the object term is free to refer to virtual entities, such as Santa Claus. We say, therefore, that the predicate *hate* is intensional with regard to its object term position. In general, we say that a predicate F is intensional with regard to a term position t_i when F allows for the reference value $\rho(t_i)$ of t_i to be an intensional object.

As is argued in Sections 3.5.1 and 6.2.2, existence predicates, such as *exist* or *imaginary*, are intensional with regard to their subject term, allowing but not requiring virtual entities as ρ-values. *Exist* then has the satisfaction condition that this ρ-value be instantiated by a corresponding actual object, while *imaginary* requires that the ρ-value of its subject term not be instantiated.

This analysis looks straightforward enough, even if in many ways it does not conform to accepted doctrine. Yet, apart from accepted doctrine, there is the problem of intensionalization, signalled by a small number of authors, in particular Linsky (1967), Parsons (1980), Zalta (1983, 1988), Castañeda (1985/86: 58), Lejewski (1985/86: 232). As was shown in Section 2.1, the problem was recognized in Montague-type possible-world semantics, but the solution proposed there has proved to be inadequate on grounds directly connected with the very foundations of possible-world semantics, in particular the 'programme of extensionalization of intensions'. Since our strategy for dealing with intensionalization phenomena rests on totally different premisses, we have no choice but to make a totally new start.

As far as can be judged, given the present, primitive state of the enquiry, it seems that two forms of intensionalization should be distinguished. First, there are cases like (5.1a, 5.1b), where one naturally takes (5.1a) to be true in a sense in which (5.1b) is naturally taken to be false:

(5.1) a. Sherlock Holmes is an Englishman.
b. Sherlock Holmes is a Frenchman.

Here the intensionalized predicates are *Englishman* and *Frenchman*, respectively. Both have a strictly extensional extension, since to be an Englishman or a Frenchman one has to be able to claim actual existence. Both are therefore strictly extensional with regard to their definite subject terms, which means that the entity referred to must be an actually existing entity for truth to arise. Yet even though Sherlock Holmes is not an actually existing entity but a fictitious one, speakers detect truth in (5.1a), as opposed to (5.1b) which is considered false. The sentences (5.1a) and (5.1b) are, apparently, treated as though Sherlock Holmes were an actually existing person, and an Englishman at that, despite the fact that he has so far lacked any incarnation as an actual person. It also appears that this has been made possible by the fact that he figures in a widely known set of stories, together forming a socially powerful intensional context. This form of intensionalization, which is discussed more fully in Section 5.3, we call IMPLICIT INTENSIONALIZATION.

The second form of intensionalization, called VIRTUAL OBJECT ATTRACTION or VOA, involves quantification under an intensional matrix predicate, as in:

(5.2) a. A cat is worshipped in that shrine.
b. Some Englishmen are imaginary.

Here the matrix predicates *be worshipped* and *imaginary* are intensional with regard to their subject term, in the sense that the set of things that are worshipped or imaginary may or must contain virtual entities. This causes a problem for sentences (5.2a) and (5.2b). For in standard modern predicate calculus the existential quantifier requires a non-null intersection of the matrix set and the restrictor set for truth to arise. Now the set of cats is strictly extensional and thus contains only actually existing cats, and analogously for the set of Englishmen. This would mean that only an actually existing cat or Englishman can make (5.2a) or (5.2b), respectively, true. But (5.2a) is also clearly true when an imaginary cat is worshipped in the shrine in question. And (5.2b) is made true by notorious instances of fictitious Englishmen, such as Conan Doyle's detective Sherlock Holmes or Shaw's professor Henry Higgins.[1] This problem is discussed in Section 5.4. A more formal account is given in Section 8.1.1 in Volume II.

5.2 'Ceci n'est pas une pipe': confusion between object and representation

Before we can embark upon a discussion of intensionalization phenomena of the kinds illustrated above, a few things must be said on the widespread confusion, not only in ordinary speech but also in professional linguistic discourse, between actual or virtual objects on the one hand and their representations on the other. A striking example, in the form of a pun, is the text on Magritte's famous painting of a pipe:[2]

(5.3) Ceci n'est pas une pipe.
(This is not a pipe.)

At first sight, those who look at the painting with this text on it are puzzled, because what is represented is very obviously a pipe—perhaps not an actual

[1] Higgins appears to be, at least in part, a blend of the historically actual English phoneticians Henry Sweet and Daniel Jones (see Collins and Mees 1999: 97–103).

[2] The painting, which dates from 1928–29, is aptly called 'The treachery of images'. Magritte repeated the theme in 1964, when he made a painting of an apple with the text 'Ceci n'est pas une pomme'.

pipe but in any case a virtual one. The naïve viewer wonders why there should be such an obviously false statement on the painting. But then, driven by the OSTA principle set out in Section 3.4.2, which makes one search for sense, truth, and actuality, the viewer will detect the truth of the statement by letting the word *ceci* refer not to the object represented, as would be normal, but to the representation itself—a legitimate, though perhaps not a primary, interpretation. This makes the statement true, since a representation of a pipe, whether actual or virtual, is not a pipe. Owing to OSTA, the viewer thus decides that this is, apparently, what must have been meant by the statement (5.3) made by Magritte on his painting. Yet (5.4) would also have been true, if made in the same circumstances. Its truth would result from the normal interpretation, in which *ceci* refers not to the representation but to the object represented, which is either an actually existing or a virtual pipe:

(5.4) Ceci est une pipe.
(This is a pipe.)

Both (5.3) and (5.4) may, therefore, be regarded as ambiguous. The word *ceci* is open to (at least) dual use, with the result that (5.3) and (5.4) can be taken to be both true and false at the same time, though not of the same object, because, again, a representation of a pipe is not a pipe. It seems that one has to say that (5.3) and (5.4) are *differently keyed* in each of the two possible readings, since in (5.3) *ceci* is keyed to the representation, whereas in (5.4) it is keyed to the, perhaps virtual, object.

It is clear that this type of ambiguity, general and widespread as it is, is best regarded as being part of the machinery of human cognition and human language. Sentences (5.5a, 5.5b), for example, are perfectly normal expressions for what would be more pedantically formulated as (5.6a, 5.6b), respectively:

(5.5) a. Ann is in the middle of the picture.
b. Am I in the story?

(5.6) a. Ann's picture is in the middle of the picture.
b. Is there mention of me in the story?

A crass example is (5.7a), found in a professional linguistic publication. A more appropriate formulation would be (5.7b):

(5.7) a. No language has a morphological class of expensive nouns.
b. No language has a morphological class of nouns reserved for reference to expensive objects.

In (5.7a), the expression *nouns* stands for the representational element but is used, incorrectly one would say, for the objects that form their extension. In this respect, (5.7a) differs from (5.5a, 5.5b), where the expressions *Ann* and *I*, which are normally used to refer to the real thing, are used for their mental or linguistic representations—a use that strikes one as being legitimate.

Since representations create intensional objects but are not identical with them, the phenomena exemplified in (5.3)–(5.7), ill-understood as they still are, have to be mentioned and set apart before we can start the discussion of intensionalization phenomena in language.

5.3 The truth about Sherlock Holmes and Cy: tacit intensional operators

5.3.1 *Implicit intensionalization*

One feels a great deal of sympathy with Linsky when he says:

> It is all right to say that the sentences of a novel do not have truth-values, but there are problems here. One wants to say that 'The home of Ulysses was in Italy' is just false, his home was in Greece. When Homer says that the home of Ulysses is in Greece, what he says is neither true nor false. But when *we* say that it was in Italy what we say is false. (Linsky 1967: 28)

The sympathy does not extend to Linsky's view, clearly influenced by Strawson's theory of presupposition, that the sentences of a novel do not have truth-values, since, as is argued in Section 3.1.5, interpretation involves keying and therefore necessarily comes with a truth-value, even if that value falls outside the standard, strictly bivalent, system of truth-values. In actual fact, most sentences in a fictional story that is otherwise fully coherent and interpretable will, strictly speaking, have the third value of radical falsity in virtue of presupposition failure. But one will readily agree with Linsky's view that the truth-value of a sentence in a fictional story may well differ according to whether the story is being unfolded by its originator—supposedly the poet Homer, himself half-mythical, in Linsky's example—or whether it is repeated by a later speaker or writer who merely wants to relate the contents of the original story. In the former case, the sentences of the story are for the most part radically false (Linsky's 'neither true nor false'), while in the latter case the sentences specify what the story is like and are thus naturally taken to be simply true or simply (minimally) false, according to whether the account they give of the original story is or is not correct. Linsky thus has a valid and important point when he says that (5.8a) is false, and (5.8b) true, when said by us, because *in the story* Ulysses' home is in the Greek island of Ithaca:

(5.8) a. The home of Ulysses was in Italy.
b. The home of Ulysses was in Ithaca.

When interpreted as said by us, the predicates *be in Italy* and *be in Ithaca* have been intensionalized in the sense indicated above. Although both predicates have an extension that is strictly limited to *actual* objects since for anything to be in Italy or in Ithaca it has to have actual existence, in (5.8a, 5.8b) they are legitimately interpreted as having an extension that includes *virtual* objects owing to the fact that, in our culture, talk about Ulysses necessarily takes one back to Homer's Iliad and Odyssey, which place Ulysses' home in Ithaca and not in Italy.[3]

A similar point, although from a slightly different angle, is made by Fauconnier (1985: 21), who gives examples such as the following:

(5.9) The girl with blue eyes has brown eyes.

Besides the obvious interpretation which makes it inconsistent, (5.9) has at least two perfectly consistent interpretations, both involving a tacit intensional operator such as 'in the picture' or 'in the story' besides a reality operator 'in reality'. Sentence (5.9) can thus be read as either (5.10a) or as (5.10b):

(5.10) a. *In the picture*, the girl who *in reality* has blue eyes, has brown eyes.
b. *In reality*, the girl who *in the picture* has blue eyes, has brown eyes.

One notes that (5.10a) is strictly speaking to be read as 'The girl who in reality has blue eyes, is represented in the picture as having brown eyes', while in (5.10b) the relative clause *who in the picture has blue eyes* should, strictly speaking, be read as 'who in the picture is represented as having blue eyes', as was explained in the previous section.

Philosophically minded critics might feel the urge to sneer at this analysis, since, they might say, this would make any false or exotic sentence true: all one would have to do to make it true is posit the existence of some tacit intensional operator under which the state of affairs reported in the sentence obtains in a virtual way. This would then take away all force from the basic distinction between truth and falsity. But such a criticism is too hasty. For it is not enough merely to *posit* the existence of some tacit intensional operator to make any exotic or false sentence true. The intensional context to which such an operator refers has to be *actually there and known*, explicitly spelled out or visible in all necessary detail and constraining the cognitive keying process.

[3] The examples are found not only in classical Antiquity or in Conan Doyle's novels. Recently, a participant in a German TV-quiz made a considerable amount of money by giving the correct answer, namely 'a Peugeot 403 convertible', to the question 'What car does Inspector Colombo drive?'.

This is so because truth and falsity require *keying*, and without there being an *actual and known* painting, movie, story, fairy tale, or what not, specifying the objects and the situations, there are no virtual objects and there is thus nothing to key on to.

Now consider sentence (5.11a), where the name *Hans Brinker* is taken to refer to the boy who, in Mary Mapes Dodge's fictional story, saved the city of Haarlem from being flooded by putting his finger in the dike where it was leaking. Some readers will be unfamiliar with the story. For them, as has been explained in the preceding paragraph and also in Section 3.1.5, (5.11a) will be uninterpretable for lack of anchoring and keying and hence lack a truth-value (unless they happen to know a Hans Brinker in some other context). Some readers, however, will recognize the name as occurring in that story, which provides them with a referent for the name. For them, (5.11a) is properly keyed so that it will have a truth-value to be determined by what the story says about the colour of Hans Brinker's hair:

(5.11) a. Hans Brinker had dark hair.

b. Hans Brinker did not have dark hair.

If Mary Mapes Dodge, in the process of inventing the story, had produced (5.11a), it would have been well-anchored contextually, and thus also well-keyed, but radically false owing to presupposition failure, since the Hans Brinker of the story never existed in the actual world, which makes it impossible for him to have had hair of any colour. Then also, the story would have specified that Hans Brinker's hair was dark, so that (5.11a), if taken as a report of the story, would be true. But in fact the story says that Hans Brinker was 'sunny-haired'—that is, blond—which makes (5.11a) minimally false (false without presupposition failure) and (5.11b) true.

But now suppose the story had failed to specify the colour of the boy's hair. Does this mean that (5.11a) would be false and that therefore (5.11b), its negated version, true? It seems that the answer has to be affirmative. The problem is that (5.11b), when interpreted purely extensionally, invites the pragmatic inference that Hans Brinker's hair was of a different colour, given that persons are normally endowed with hair and hair has a colour. But in intensional contexts, such inferences are unwarranted, since intensional contexts are always indeterminate as regards those properties of the objects created in them that are left unspecified. If the colour of Hans's hair is unspecified in the story, then a listener who knows the story well is likely to interpret (5.11b) as 'In the story it is not specified whether Hans Brinker had dark hair'. The implicitly understood intensional operator 'in the story' eliminates all kinds of pragmatic inferences based on world knowledge, precisely because virtual objects are by definition incompletely defined in any given intensional context. They can, as a matter of principle, always be fitted out with more properties, which would make them less incompletely defined than they are (irrelevant though that may be for the story at hand), but they can, again as a matter of principle, never be defined for all appropriate parameters the way actual objects are. The complications that this observation brings about for the semantics of negation in implicit intensional contexts have not been investigated so far.

The overall conclusion is that any proposition whose interpretation can involve embedding in a known intensional context creates an ambiguity between a (nondefault) extensional reading and a (default) intensional reading. The latter is available only to those speakers and listeners who are familiar with the story, myth, legend, poem, or visual representation constituting the intensional context in question, and also to the author or originator of the intensional context who looks back upon his or her creation in hindsight. The extensional reading for such propositions is highly artificial—and hence nondefault—since, as was pointed out by Frege, what counts in intensional contexts is interpretation, not truth-values:

> When we listen, for example, to an epic poem it is, besides the euphony of the language, only the meanings of the sentences and the representations and feelings brought about by them that captivate us. But when we ask about the truth we leave the area of esthetic pleasure and turn to a scientific mode of consideration. (Frege 1892: 33)

Given that Conan Doyle's stories about Sherlock Holmes leave no doubt about Holmes' nationality, (5.12a) is true in its intensional reading but (radically) false in its extensional reading, whereas (5.12b) is true in its extensional reading (with the negation taken in the radical sense of denying a presupposition) but false in its intensional reading:

(5.12) a. Sherlock Holmes is an Englishman.
b. Sherlock Holmes is not an Englishman.

This analysis is not at all philosophically disastrous, as Quine would have us believe. It is, on the contrary, philosophically relevant in at least two ways. First, if the truth of an assertive utterance is taken to depend on the correspondence between what is said or thought on the one hand and the state of affairs keyed on to on the other, we now see that, in cases when the proposition in question is embedded in an intensional context known to speaker and listener, the state of affairs keyed on to need not be an *actual* state of affairs but may be a *virtual* state of affairs described in the intensional context in question. The second reason why this analysis is philosophically relevant lies in the role of cognition in defining the notion of truth. In Section 3.4.1, the

Principle of Cognitive Independence (PCI) was rejected for the semantics of natural language on account of the fact that, at least for occasion sentences, the mind is an indispensable factor in the truth relation. This conclusion is now reinforced, since virtual objects and states of affairs are now seen to intrude upon traditional extensionalist territory, and virtual objects and states of affairs are products of the thinking mind only, as there is no other source for their being.

Zalta (1988) uses the terms EXEMPLIFICATION and ENCODING for what we call extensional and intensional readings, respectively. In the extensional reading of (5.12a), for example, an actually existing person Sherlock Holmes should *exemplify* the predicate *Englishman* for truth to arise, but in the intensional reading all that is required for truth is that the property of being an Englishman has been *encoded* for the virtual object Sherlock Holmes. This terminology is happily chosen, for the term *exemplify* naturally evokes actual existence of anything that 'exemplifies' (the term occurs casually, in precisely the meaning intended by Zalta, in the opening paragraph of Parsons 1980). Moreover, the term *encode* drives home the fact that it is through ascriptions by the creative mind, and not through the causal forces of the outside world, that virtual objects acquire their properties—though it should be remembered that 'encoding' is required for all forms of predication, since some form of mental registration is required, as a matter of principle, for both the production and the comprehension of all uttered sentences. This is what *anchoring* in discourse domains is all about. One recalls from Chapter 3 that anchoring is a prerequisite to production and interpretation, and thus also to any truth-value assignment. Since the mental process of anchoring requires a memory system in the form of a *discourse domain*, it follows that some form of discourse-driven or cognition-driven encoding is necessary for anchoring and hence for truth-value assignments—also in cases of exemplification.

5.3.2 *Substitutivity, identity, and the value-assigning predicate* Be_V

As is observed by Parsons (1980: 34), *substitution of co-referring terms salva veritate*, also known as the *Principle of Substitutivity* (see Section 6.1 for full commentary), applies in full force to sentences whose main predicate has been implicitly intensionalized. Whatever is said of Zeus is said of Hera's husband, since, in Greek mythology, Zeus and Hera's husband are the same mythical figure. This is, of course, in full agreement with the Aristotelian notion of truth as correspondence, which in itself does not exclude reference to virtual objects. As was said in Section 2.1, the fleshing out of the concept of truth as correspondence will vary with the kind of ontology one embraces. If the ontology comprises virtual objects, then the correspondence that makes for truth applies to these the way it does to actual objects.

We now understand why the problem of nonsubstitutivity in intensional contexts, which was diagnosed by Frege in his famous article of 1892 and gave rise to much semantic theorizing during the twentieth century, is restricted to cases where reference is made in an intensional context to objects that have actual existence outside the intensional context—Parsons' 'immigrant' objects, as opposed to 'native' objects which are restricted to the intensional context and lack actual existence (Parsons 1980: 51). In intensional contexts an 'immigrant' object may receive distinct representations (addresses) and may thus be encoded for different properties, thereby instantiating distinct virtual objects.[4]

In order to make substitutivity work, we must define the IDENTITY PREDICATE in terms of the theory as it has been developed so far. We define the identity predicate as an intensional binary predicate that merges two object representations (addresses) into one, so that one single new virtual object comes about, to be instantiated by one single actual object, if any.[5] It would be more appropriate, therefore, not to speak of the 'identity' but of the 'identification' predicate. We will, however, stick to the established term.

The identity predicate is intensional with regard to both its terms, because it is possible to utter a true identity statement not only with regard to two actually existing objects, as in (5.13a), but also with regard to two virtual objects, as in (5.13b). One notes that both are genuine identity statements, as one can say, for example, *Mount Everest and Chomolungma are the same mountain*, or *Zeus and Jupiter are the same god.*

[4] An amusing example is the following, hit upon by chance (Kiprian 2002: 28–9):

A certain archimandrite Antonius Karjavine, who was an inspector of the Ecclesiastical Academy of Moscow and subsequently archbishop of Tver, and who died on March 16th, 1914 . . . , was famous for having discovered, in his master's thesis, two philosophers, Descartes and Cartesius. In their systems, for that matter, he had found close convergencies but also significant differences.

[5] An early reference to the 'merging' character of the identity predicate is found in Strawson (1974). Speaking of a discourse domain as a 'knowledge-map' with dots representing objects and lines representing relations, he writes:

We might improve this model, perhaps. Instead of thinking of the man as operating on his knowledge-map, when his knowledge-state is changed, we may think simply of the knowledge-map as becoming changed. When he learns something from an ordinary predication, new lines inscribe themselves on his map, attached to the appropriate dot or joining two different dots. When he learns from an identity-statement, the two appropriate dots approach each other and merge, or coalesce, into one, any duplicating lines merging or coalescing at the same time. (Strawson 1974: 55)

(5.13) a. Mount Everest is Chomolungma.

b. Zeus is Jupiter.

c. The ghost that appears in your dreams is your father.

A difficulty arises when one term of the identity predicate refers to an actually existing object while the other term refers to a virtual object, as in (5.13c). If one accepts that one can say *The ghost that appears in your dreams and your father are the same person*, one implicitly accepts the possibility of simultaneous being at two levels, the actual level of existence and virtual being. If one feels that one cannot meaningfully say that, one will have to specify in what sense the identity is meant to hold. One may say, for example, that the ghost 'represents' the addressee's father, as in *The man in the middle of the photograph is your father.*

In the process, the property encodings of the two old addresses are pooled under one new address. The semantic description of the identity predicate '=' is thus as in (5.14) (one remembers the method of defining predicate extensions shown in Section 3.3.2). (The asterisk following a term variable signals the optional virtuality of any object satisfying the predicate in question and thus the intensional character of the corresponding term position; '**o***' stands for an actual object if there is one and otherwise for a virtual object):

(5.14) $[\![=]\!] = \{ <x^*,y^*>|\ x^* \text{ and } y^* \text{ are one and the same object } \mathbf{o}^* \}$

In terms of discourse semantics, this means that a pair of object representations in the truth domain, or in some intensional subdomain, are henceforth merged so that they represent one single virtual object which, if instantiated, must be instantiated by one single actual object for truth to arise. For example, let there be a discourse domain **D** containing an address a_1 for the object known as Mount Everest and another address a_2 for the object known as Chomolungma (one remembers that the two names denote one single, huge, actual object). When sentence (5.13a) is interpreted, the addresses a_1 and a_2, representing the virtual objects *MOUNT EVEREST* and *CHOMOLUNGMA*, respectively, merge into a new address a_3, which is characterized by the union of all property encodings of a_1 and a_2. Clearly, when two virtual objects are encoded for incompatible properties, their identification is blocked, as the discourse would become incoherent. For example, if Harry believes that Mount Everest is on the Nepalese border and that Chomolungma is in Kenya while he knows that the one property excludes the other, he cannot coherently identify the two by means of the identity predicate.

The situation is more complex when Harry used to think that Mount Everest and Chomolungma are two distinct objects, just like his friend

Thomas, while Harry knows that Thomas thinks that Chomolungma is in Kenya but that Mount Everest is on the Nepalese border. Now a_1, the Everest address, is encoded, in Harry's mind, for the property P(be believed by Thomas to be on the Nepalese border) while a_2, the Chomolungma address, is encoded for the property P(be believed by Thomas to be in Kenya). Now suppose Harry discovers, unbeknownst to Thomas, that (5.13a) is true and he asserts (5.13a), thereby creating a new address a_3 for Mount Everest also known as Chomolungma. The question is what happens with the property encodings P(be believed by Thomas to be on the Nepalese border) in the old Everest address and P(be believed by Thomas to be in Kenya) in the old Chomolungma address. Clearly Harry's new address a_3, which unites the old addresses a_1 and a_2, cannot be encoded for these two 'emigrant' properties (to vary on Parsons' terminology), not only because that would make a_3 inconsistent but also because Thomas has no beliefs about whatever confirms a_3. The answer would seem to be that in such a case the old addresses a_1 and a_2, though merging into a_3 in Harry's truth domain, remain intact in the intensional subdomain encoded for the property P(be believed by Thomas)—that is, in the belief-domain ascribed to Thomas.

It is important to observe, in this connection, that the identity predicate, lexicalized as *be* in English, must be sharply distinguished from the VALUE-ASSIGNING PREDICATE Be_v, as in (5.15a–5.15f), which can be regarded as answers to the corresponding questions given within brackets:

(5.15)	a.	The number of planets is nine.	(What is the number of planets?)
	b.	His name is 'George'.	(What is his name?)
	c.	She has two children[6]	(How many children does she have?)
	d.	Her age is twenty-two.	(What is her age?)
	e.	The ship's captain was Drake.	(Who was the ship's captain?)
	f.	What she did was cry out.	(What did she do?)
	g.	The winner was Nixon.[7]	(Who was the winner?)

[6] This example shows that pragmatic accounts of the fact that *two* often means 'exactly two' are unnecessary: *two* means 'exactly two' when it is the value of a parameter of quantity and does not represent an existential quantifier (see Seuren 1993; Scharten 1997). Numerals can also indicate a maximum, as in *The cottage can sleep five* or *I can lend you ten quid.* Here, too, semantics rather than pragmatics seems to provide the answer, as it looks very much as if it is the semantics of *can* that is responsible for the 'maximum' reading.

[7] This example is in reference to Kripke (1980):

Suppose that someone said, pointing to Nixon, 'That's the guy who might have lost'. Someone else says 'Oh no, if you describe him as "Nixon", then he might have lost; but, of course, describing him as the winner, then it is not true that he might have lost'. Now which one is being the philosopher, here, the unintuitive man? It seems to me obviously the second. The second man has a philosophical theory. The first man would say, and with great conviction, 'Well, of course the winner of the election might

In (5.15a–5.15g) the grammatical subject expresses the topic of the sentence (see Section 3.2) and refers to an abstract object that has the nature of a parameter (number, name, age, and so on) requiring a value. The grammatical predicate nominal expresses the comment and specifies the value in question. The difference with the identity predicate stands out clearly when one considers the utter impossibility of an intensional context, a story for example, with two virtual beings, one referred to by the expression *the number of planets* and the other by the expression *nine*, and a correction of that story by the utterance of (5.15a), as a result of which there now is one virtual object, alternatively called *the number of planets* and *nine*.

The semantic description of the value-assigning predicate Be_v is given in (5.16), which, of course, is totally different from the identity predicate defined in (5.14):

(5.16) $[\![Be_V]\!] = \{<P_{o*},v> \mid$ v is the value on the parameter P of an object o*$\}$[8]

One notes that Be_v is intensional with regard to the term referring to the object **o**, because not only actually existing, but also virtual objects may have values or parameters. For example, the virtual object Sherlock Holmes has the actual name 'Sherlock Holmes', the word *name* being an expression denoting a parameter, and 'Sherlock Holmes' being its value for the object at hand. By contrast, *P* and *v*, standing for parameters and values, respectively, range over objects that actually subsist in the sense of Section 2.2 (Meinong's 'higher-order objects').

For some inexplicable reason, no logical or semantic analysis has ever been provided of sentences in which a value is assigned to a parameter. Quine incorrectly treats the value-assigning predicate Be_v as the identity predicate, falsely assuming (1953: 143–4) that (5.15a) is an identity statement of the logico-semantic form 'The number of planets = 9'. This false assumption fatally flaws his argument, widely accepted in the philosophical world, that modalities create intensional contexts and thus block SSV.[9] According to Quine,

have been someone else'. The actual winner, had the course of the campaign been different, might have been the loser, and someone else the winner; or there might have been no election at all. So, such terms as 'the winner' and 'the loser' don't designate the same objects in all possible worlds. (Kripke 1980: 41)

Kripke resorts to possible worlds to account for the notion of parameter, obviously unaware of the difference between the identity *be* and the value-assigning Be_v. In the first man's sentence *The winner of the election might have been someone else*, the NP *the winner of the election* sets the parameter and *someone else* denotes the value. No appeal to possible worlds is necessary to account for the fact that a parameter may have different values, or has different values at different times.

[8] The syntactic structure corresponding to an L-proposition of the form $Be_V(P_{o*},v)$ is, generally speaking: *The P of o* is v.* Sometimes there is a parameter-denoting copula verb, such as *weigh*, or *be called* instead of *is*.

[9] In fact, Quine's argument has been so influential that formal semanticists are easily misled into associating Frege's discovery of the problem of nonsubstitutivity in intensional contexts with

the incorrectness of the following argument is due to illegitimate SSV in an intensional modal context:

(5.17) The number of planets is nine.
Nine is necessarily greater than seven.
Ergo: !The number of planets is necessarily greater than seven.

But (5.17) is invalid for at least two reasons, neither of which is related to intensional contexts or modalities. One reason, pointed out in Seuren (1985: 391), specifically concerns the ambiguity (well known in medieval philosophy) of the word *number*, which denotes either a unit of arithmetic (as in 'the number 9'), or the parameter 'cardinality of a set' (as in 'the number of planets'), or the set itself (as in 'a number of people'). A more general reason is that no rule of philosophy or linguistics allows for the unrestricted substitution *salva veritate* of a parameter name for an expression specifying its value, or vice versa.[10] Mutual substitution of a parameter name for an expression specifying its value is generally possible within the restrictions of a specific formal calculus, but not in language. Sentence (5.18) is as invalid as (5.17), even though no modality or other kind of intensional context is involved, and (5.19) even leads to ungrammaticality in the conclusion:

(5.18) The number of stars dwarfs the number of planets.
The number of planets is nine.
Ergo: !The number of stars dwarfs nine.

(5.19) This package weighs twenty-two pounds.
Her age is twenty-two.
Ergo: *!This package weighs her age pounds.

modalities, not with belief-contexts. For example, Dowty *et al.* write (1981: 142): 'Frege noticed the problem with modal operators', falsely attributing to him the example *Necessarily the Morning Star is the Morning Star* versus *Necessarily the Morning Star is the Evening Star.* See also Section 7.2.1.3 in Volume II. In fact, Frege wouldn't touch modalities with a barge pole.

[10] In August 2006, it was decided at the Prague Convention of the International Astronomical Union to rule out Pluto as a planet, its size being too small. Given this fact, one may say in truth:

(i) The number planets has been reduced from nine to eight.

According to Quine, this would have to be equivalent to:

(ii) Eight [or, for that matter, nine] has been reduced from nine to eight.

It is not hard to see that this stretches the boundaries not only of common sense but also of good method in linguistics beyond tolerance.

In language, substitution is sometimes, but not always, possible when the parameter name stands under a predicate different from Be_v, as in (5.20a–5.20d) ('$\equiv_{mk}$': 'equivalent modulo key'):

(5.20)	a	I wrote his name.	$\equiv_{mk}$	I wrote 'George'.	cp. (5.15b)
	b.	The ship's captain wore a hat.	$\equiv_{mk}$	Drake wore a hat.	cp. (5.15e)
	c.	What she did was brave.	$\equiv_{mk}$	Crying out was brave.	cp. (5.15f)
	d.	The winner was exuberant.	$\equiv_{mk}$	Nixon was exuberant.	cp. (5.15g)

Obviously, a great deal of research still has to be done on sentences assigning values to parameters and on the grammar and semantics of parameter expressions in general.

5.4 Unicorns and ritual cats: virtual-object attraction (VOA)

Virtual-object attraction, or VOA, is a corollary of the fact that if a definite term t has an actual or virtual referent $\rho(t)$, it follows that $\rho(t)$ has being, or that there is an object of any kind answering the description given in t: if it is true that the moon circles around the earth, then there is a moon and there is an earth. Quine, for whom being equalled existence, spoke of EXISTENTIAL GENERALIZATION, but this term is, for obvious reasons, inappropriate in the present context. We prefer to speak of the ENTAILMENT OF BEING.

In itself, the entailment of being induced by definite terms is not problematic, provided we read the existential quantifier as licensing an entailment of being, not of existence. It begins to be problematic when one considers that a sentence like (5.21a) entails sentence (5.21b):

(5.21) a. The unicorn thought by John to have rampaged through his garden is worshipped on that hilltop.

b. A unicorn is worshipped on that hilltop.

Let the predicate *worship* be defined as follows (the asterisk indicates intensionality of the term position in question):

(5.22) $[\![\text{Worship}]\!] = \{ <x,y^*> \mid$ x shows reverence, in a ritual manner, for y^*, which is considered a higher being $\}$

That the predicate *worship* is intensional with regard to its direct-object term accounts for the truth of (5.21a) if indeed John thinks a unicorn has rampaged through his garden and if indeed that unicorn is worshipped on the hilltop in question. That being so, it follows that (5.21b) must also be true, owing to the entailment of being.

But now there is a problem, because, as is shown in Section 2.3.5.2 in Volume II, the existential quantifier yields truth just in case there is a non-null intersection of the MATRIX SET, in this case the set of things worshipped on that hilltop, and the RESTRICTOR SET, here the set of unicorns. But the set of unicorns is strictly extensional given the satisfaction conditions of the predicate *unicorn*, which are defined in physical terms only. In this respect, unicorns do not differ from cats: to be a unicorn, or a cat, one has to have actual existence; mere virtual being does not suffice. We know, of course, that, in this world, the set of unicorns is null, which excludes any non-null intersection with any other set. This would make (5.21b) unavoidably false in our actual world. But it is not, because, as has been shown, if John does indeed think that a unicorn has rampaged through his garden and that this unicorn is worshipped on the hilltop in question—which is quite possible in the fairground of surprises that our world is—then not only (5.21a) but also (5.21b) is true. In fact, the entailment from (5.21a) to (5.21b) remains intact when *unicorn* is replaced with *cat*, and it still does not follow that an actually existing cat is worshipped on that hilltop.

To solve this anomaly we invoke VIRTUAL-OBJECT ATTRACTION, or VOA, as a mandatory principle intensionalizing the restrictor predicate in cases where the matrix predicate is intensional with regard to a term t_i that is quantified over. VOA says:

> VIRTUAL-OBJECT ATTRACTION (VOA):
> When the main (or matrix) predicate is intensional with regard to a term t_i and t_i is quantified, then the predicate defining the corresponding restrictor set **R** is intensionalized.

That is, when **R** is strictly extensional on account of the satisfaction conditions of the predicate F defining **R**, then, as a result of VOA, **R** will also include all virtual objects encoded for the property expressed by F.

VOA appears to apply not only to existential but also to universal quantification, though there is a question as regards the use of universal *any* for *all*—a question that will be left undiscussed here. It seems, anyway, that both (5.23a) and (5.23b) may be true even when John has no children at all. All it takes for (5.23a, 5.23b) to be true is for Ursula to be so out of touch with reality that she has spun a myth around imaginary children of John's, some or all of whom, she thinks, need her prayers:

(5.23) a. Ursula prays for some of John's children.
b. Ursula prays for all John's children.

The same goes for (5.24a, 5.24b), where the phrase *all heroes* applies to historically real as well as to merely fictitious heroes:

(5.24) a. In ancient Greece, all heroes had a shrine dedicated to them.
b. In Homer, all heroes have superhuman powers.

It may be observed, in this context, that higher operators of time and place, such as 'in ancient Greece' in (5.24a), apply to extensional and intensional contexts alike. This principle is further illustrated by sentence (5.25), which is subject to implicit intensionalization:

(5.25) In America, Santa Clause rides in a sleigh, but in Holland he rides a horse.

But let us concentrate, for the moment, on existential quantification. Consider the following two sentences:

(5.26) a. A cat is sitting on the window sill.
b. A cat is worshipped in that shrine.

Sentence (5.26a) requires for truth that the intersection of the strictly extensional set of real cats and the equally extensional set of things sitting on the window sill be non-null. By contrast, (5.26b) is considered true even when the cat in question is not a real cat but a fictitious one, perhaps part of a mythology. It is clear that the reason for this anomaly lies in the fact that the matrix predicate *worship* is intensional with regard to its object term, since virtual objects can be worshipped as easily as actual objects can be.

The restrictor sets of (5.26a) and (5.26b) thus differ, owing to the fact that the predicate *worship* is intensional with regard to its object term. In order to account for the undoubted truth of (5.26b) in cases where a virtual cat is worshipped in the place indicated, it is necessary to apply VOA, so that the restrictor set of (5.26b) is made to include virtual cats besides actual cats. One notes that the device of intensionalization by means of VOA leaves the semantics of the existential quantifier as standardly defined for extensional cases intact. But it weakens existential import.

Sentence (5.26b) remains interpretable even when no intensional context is given for the cat in question. In such a case, the intensional context creating the cat worshipped in the shrine is constituted by sentence (5.26b) itself, which establishes an intensional subdomain of things worshipped in the place mentioned and introduces a virtual cat in that subdomain on account of the fact that existential quantification is the standard means of introducing new addresses in (sub)domains.

VOA is not affected by negation. If the negation is the bivalent negation of standard logic, the truth of, say, (5.26b) implies the falsity of its negation, *No cat is worshipped in that shrine*, and vice versa, so that a virtual cat can be a counterexample to the negation of (5.26b). Nor is VOA affected when the negation is taken to be the minimal or presupposition-preserving negation, since, as is shown in Chapter 2 in Volume II, the extensional or intensional character of a predicate with regard to a given term position is treated as a presuppositional phenomenon which is, therefore, not affected by negation.

One notes the differences between implicit intensionalization and VOA. In cases of implicit intensionalization one has to do with the default intensionalization of the normally extensional main predicate, caused by the fact that a term referent is virtual and occurs in a known intensional context. VOA differs from implicit intensionalization in that (a) the intensionalized predicate is not the main (matrix) predicate but the predicate defining the restrictor set, (b) it is caused by the main predicate which is intensional with regard to a term t_i, and (c) implicit intensionalization is cognition-driven and works by default, whereas VOA is language-driven and subject to a mandatory principle.

Now consider (5.2b), repeated here as (5.27):

(5.27) Some Englishmen are imaginary.

Since *imaginary* is defined as being intensional with regard to its subject term (see (6.6b) in Section 6.2.2), the restrictor set of Englishmen is intensionalized so as to include virtual Englishmen as well. Now all that is needed for the truth of (5.27) is a non-null intersection of, on the one hand, those virtual objects that are denied instantiation in the shape of an actual object—that is, all imaginary objects—and the intensionalized set of Englishmen on the other. Clearly, the famous Sherlock Holmes can be produced as an element in that intersection, making the sentence true. For this to work, however, it is necessary that (5.1a), *Sherlock Holmes is an Englishman*, should be considered true. This shows that implicit intensionalization is a necessary prerequisite for VOA to work.

5.5 Nonspecific readings and prelexical syntax

A particular, and highly interesting, complication arises with predicates like *look for*, *look like*, *need* and a few others. Consider the sentence pairs:

(5.28) a. John was looking for the unicorn he thought had ruined his garden.
b. John was looking for a unicorn.

(5.29) a. This chair looks like the tower of Pisa.
b. This chair looks like a tower.

(5.30) a. John needed the drink that had been put before him.
b. John needed a drink.

Clearly, the (a)-sentences entail the (b)-sentences, in virtue of the entailment of being. But the (b)-sentences are ambiguous, and they are entailed only in one reading, which we call the *specific reading*. Sentence (5.28b) is entailed in the specific reading 'there was a specific unicorn John was looking for', not in the equally natural, or perhaps even more natural, nonspecific reading 'John was looking for anything that would qualify as a unicorn'. Likewise for (5.29) and (5.30): (5.29b) is entailed only in the specific reading 'there is a tower that this chair looks like', and (5.30b) is entailed only in the specific reading 'there was a drink John needed', not in the nonspecific reading 'John needed a drink, no matter what'. This is remarkable, since (5.21b), which is likewise entailed by (5.21a) in virtue of the entailment of being, is not that way ambiguous.

Now (5.28b), in the entailed reading, involves VOA, since the unicorn that serves as its truth-maker is a virtual object, but (5.29b) and (5.30b) do not, as their truth-makers are actually existing objects. Yet they could have involved VOA, because the predicates *look like* and *need* are, like *look for*, intensional with regard to their object term. Instead of (5.29a, 5.29b) we might have had (5.31a, 5.31b), and (5.30a, 5.30b) may be replaced with (5.32a, 5.32b), without any difference as regards the ambiguity of the (b)-sentences and their being entailed in their specific readings only:

(5.31) a. This chair looks like the throne of Zeus.
b. This chair looks like a throne.

(5.32) a. John needed the drink that had been promised him.
b. John needed a drink.

Thanks to VOA, the specific readings of the (b)-sentences are thus unproblematic. It is their nonspecific readings that pose a problem. And the problem is: how are these readings accounted for? It seems that an answer can be provided in terms of the theory of prelexical syntax, presented by McCawley during the late 1960s (McCawley 1973) and discussed in detail in Section 8.6. This theory posits that at least some predicates are to be semantically described as incorporating a bit of lexicalized syntactic structure. McCawley's analysis of the verb *kill* as incoporating the syntactic structure schema CAUSE(x, [BECOME DEAD(y)]) has become famous (see Section 8.6.1). In similar manner we may describe the verb *look for* as incorporating the syntactic structure schema

TRY(x, [FIND (x,y)]), where the higher predicate TRY presupposes an intensional context that may contain both 'native' and 'immigrant' objects—to use Parsons' terminology again (see Section 5.3.2).

Thus, instead of defining *look for* as in (5.33), we use the definition (5.34):[11]

(5.33) $[\![\text{Look for}]\!] = \{ \langle x,y^* \rangle \mid x \text{ tries to find } y^* \}$

(5.34) $[\![\text{Look for}]\!] = \{ \langle x,y^* \rangle \mid S_1 \qquad \}$

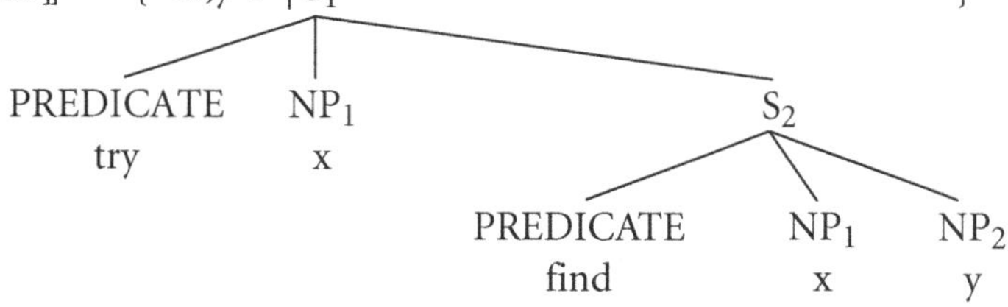

Look for is thus defined as intensional with regard to its object term: it expresses a relation between an actual person and a possibly virtual object, whose location is sought. The predicate that actually induces VOA, in this analysis, is *try*, which allows for unsuccessful attempts.

The advantage of (5.34) over (5.33) is that (5.34) involves a syntactic structure, which makes it possible to insert quantification in two places, as shown in (5.35a) and (5.35b). Sentence (5.35a) has the specific reading 'there is a (possibly virtual) unicorn that x tries to find' and (5.35b) the nonspecific reading 'x tries to find anything qualifying as a unicorn'.

(5.35) (a)

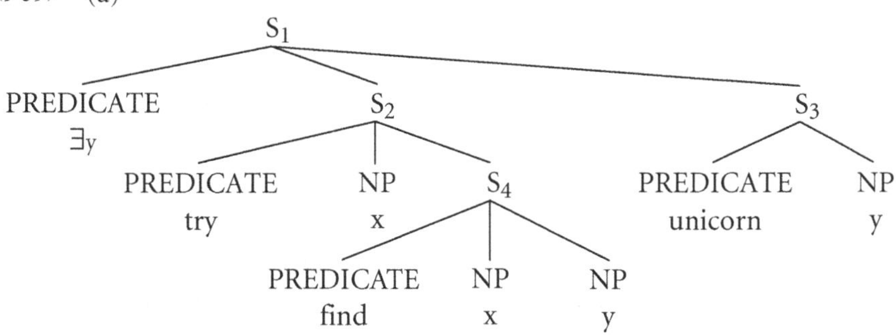

[11] Aficionados of syntax will be pleased to know that by prelexical cyclic DELETION of the lower NP_1 and by prelexical cyclic RAISING of the lower predicate, both induced by *try*, the complex predicate TRY (to) FIND comes about, with the terms $_{NP}$[x] and $_{NP}$[y]. It is also important to remember that the NPs in the tree structure of (5.33) are *definite* NPs, such as *John* and *the unicorn*, which, therefore, require addresses in any current discourse domain.

(b)

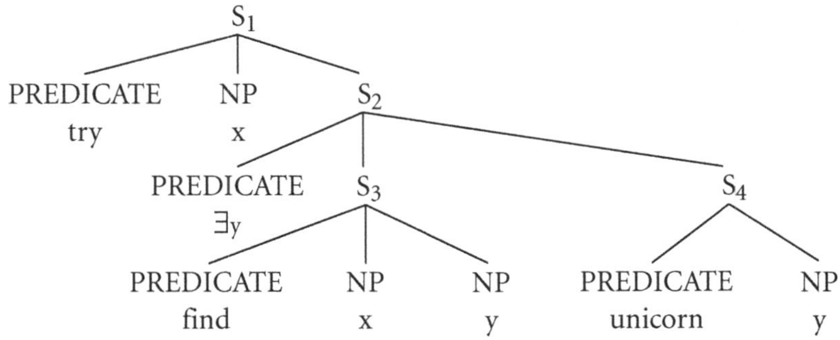

In similar fashion we may analyse the predicate *look like* as incorporating the schematic prelexical syntactic structure Appear, ([Similar(x,y)]) and *need* as incorporating the structure Necessary, ([Have(x,y)])—both with a layered structure that allows for quantifier insertion at two levels.

5.6 There are winged horses but not everything is water

In Section 3.5 it was concluded that, contrary to established views in logic, the truth of an affirmative existential statement does not entail the *actual existence* of the elements in the non-null intersection concerned. What is entailed is mere *virtual being*, actual existence being a possible corollary induced by the satisfaction conditions of the matrix predicate of the sentence at hand. The absence of existential import in the semantics of the existential quantifier is emphsized by Parsons, who gives the following example (Parsons 1980: 7):

(5.36) There are winged horses—Pegasus, for example.

In the theory presented here, the sentence *There are winged horses* is analysed as follows, in terms of generalized quantification:

(5.37) $\exists x(\text{Obj}(x), \text{Winged horse}(x))$

The universe of objects **OBJ** is denoted by the abstract predicate *Obj* extending over all objects, actual and virtual, and expressed in English as the matrix predicate *be there*. The predicate *Obj* is defined as follows:

(5.38) $[\![\text{Obj}]\!] = \{\, x^* \mid x \in \mathbf{OBJ}\}$

Sentence (5.37) thus says that there is a non-null intersection between **OBJ** and the set of winged horses. Since *Obj*, the matrix predicate, is intensional with regard to its subject term, the set of winged horses is intensionalized in virtue of VOA, making it possible for the virtual object Pegasus, the well-known winged horse of Greek mythology, to serve as a truth-maker for the sentence *There are winged horses.* In general, ontological generalizations like (5.36), where it is not obvious that the quantifier expresses a relation between two sets, are taken to involve the set **OBJ** comprising all objects in the universe, including any virtual objects, and denoted by the predicate *Obj.*

In (5.37), *Obj(x)* is the matrix predicate and *Winged horse(x)* is the predicate defining the restrictor set. By contrast, (5.39a), which reflects the ontology of the pre-Socratic philosopher Thales of Miletus, is analysed as (5.39b), with *Obj(x)* as the predicate defining the restrictor set and *Water(x)* as the matrix predicate:

(5.39) a. Everything is water.
b. ∀x(Water(x), Obj(x))

No VOA is possible here, since the main predicate, *Water(x)*, is strictly extensional. By contrast, the predicate *Obj(x)* extends over both actual and virtual beings. Since, by definition, a set containing virtual objects cannot be included in a strictly extensional set, (5.39a) must be considered necessarily false, at least in an ontology that includes virtual being. If the predicate *Obj(x)* is taken to be strictly extensional—which it may be in a philosophical construction—(5.39a) is contingently false, or else we must be fundamentally mistaken about the physical nature of the universe.

Not so with Quine's facetiously meant ontological statement (5.40a), analysed as (5.40b) (Quine 1953: 1):

(5.40) a. Everything is.
b. ∀ x(Obj(x), Obj(x))

Here *Obj(x)* is both the matrix predicate and the predicate defining the restrictor set, which makes VOA redundant. It also makes the statement tautologous, no matter whether *Obj(x)* is taken to be strictly extensional or intensional, and thus philosophically vacuous, since by definition any set is included in itself (at the level of constructed cognition—see Chapter 3 in Volume II).

The nontautological statement (5.41a) is analysed as (5.41b)—read as 'some things that are there actually exist'—where the intensional matrix predicate *Exist(x)* (see Section 6.2.2) allows for virtual objects as truth-makers:

(5.41) a. Some things exist.
b. ∃x(Exist(x), Obj(x))

We thus see that the conclusion of Section 3.5 to the effect that, in natural language as opposed to standard logic, the existential quantifier lacks existential import, though it does have the entailment of being, and that existential import is induced by the satisfaction conditions of the matrix predicate of the sentence at hand, remains applicable even in extreme cases of constructed ontological generalizations such as (5.36), (5.39a), (5.40a), or (5.41a).

6

Extensions and intensions: language, mind, and world

6.1 Frege's analysis and the standard system

In his famous article, Frege (1892) extrapolated the traditional medieval distinction between the EXTENSION and the INTENSION of a predicate (a) to (nominal or sentential) argument terms of predicates and (b) to full sentences, or rather the L-propositions that underlie them. Frege's terms for what is now normally called extension and intension were *Bedeutung* (reference) and *Sinn* (sense), respectively. The reason why Frege felt this distinction had to be extended from predicates to argument terms and nonembedded and embedded L-propositions lies in a particularly nasty quandary for the Aristotelian theory of truth. The quandary that Frege discovered[1] consists in the fact that Leibniz's principle of SUBSTITUTION OF CO-REFERRING TERMS SALVA VERITATE, or the principle of SSV, also called the principle of SUBSTITUTIVITY, fails to apply in so-called intensional contexts. This calls for some comment.

The principle of substitutivity, formulated repeatedly by the German philosopher Leibniz at the end of the seventeenth century, runs as follows:[2]

> PRINCIPLE OF SUBSTITUTIVITY (SSV)
> Terms that refer to the same object can always be substituted for each other *salva veritate*—that is, without affecting the truth-value of the sentence in question.

The principle follows directly from the Aristotelian notion of truth as correspondence, since all that matters for that theory is that the property assigned

[1] Or rather, rediscovered, since it is precisely one of the Greek philosopher Eubulides's famous paradoxes, the Electra paradox. See Seuren (1998: 381, 2005) for details.

[2] In actual fact, Leibniz looked at SSV from an inverted perspective. His problem was the question of ontological identity, and his formulation, accordingly, ran as follows (with minor variations from one locus to another): 'Eadem sunt quae mutuo substitui possunt salva veritate', best interpreted as: 'Things are identical when the terms that refer to them can always be substituted salva veritate'. Leibniz thus derived identity from truth, not the other way round—a fascinating perspective in the light of Kant's epistemological argument discussed in Section 2.2.2, as truth is seen as a Cartesian mental experience that cannot be denied.

to an object in the mental act of a proposition really adheres to that object, regardless of what name is used to refer to the object in question.

Frege discovered, presumably to his horror, that this principle does not always apply, which might shake one's faith in the Aristotelian notion of truth as correspondence. The Principle of SSV fails to apply, notably (but not exclusively), in L-propositions embedded as a term under a predicate that attributes, one way or another, a PROPOSITION (a 'thought' for Frege) to a thinking subject, as in:

(6.1) a. Henry believes that the morning star is inhabited.
 b. Henry believes that the evening star is inhabited.

Clearly, it is possible for (6.1a) to be true and for (6.1b) to be false at the same time and of the same person, or vice versa, even though the terms *morning star* and *evening star* actually refer to the same object, the planet Venus. If Henry is so uninformed as to think that the morning star and the evening star are different planets, then he may very well also think that the one, but not the other, is inhabited by an extraterrestrial race. In cases such as (6.1a, 6.1b), therefore, the Principle of Substitutivity appears to break down. This would have calamitous consequences for the age-old Aristotelian theory of truth and it would drag the whole of logic along in its fall. Since this was abhorrent to Frege, he felt he had to save Aristotle's theory of truth.

He did so by taking two theoretical steps. First, he posited a distinction between the *extension* (Bedeutung) of an argument term (whether nominal or sentential) and its *intension* (Sinn). The extension of a definite (nonquantified) nominal term is the object it refers to. What should be taken as the intension of a definite term is, however, not made unambiguously clear by Frege. He speaks (1892: 26) of 'the mode of presentation' (die Art des Gegebenseins) of the reference object, but what exactly this should be taken to mean is not obvious. Dummett (1981: 44–6) discusses this question extensively and does not come to a definite conclusion. One of the points Dummett makes is that if the intension of a definite nominal term is its 'mode of presentation' of the object(s) referred to, then there can be no intension when there is no object, since in that case there is nothing to present (1981: 45). Yet definite terms are often used without there being a corresponding actually existing reference object (virtual objects were not part of Frege's ontology). It is proposed in Seuren (1998: 373–4) that the intension of a definite term is best described as the linguistic 'search procedure' or 'route' employed to get at the reference object that constitutes its extension. Dummett (1981: 45) considers the same answer and quotes Frege (in Patzig 1969: 95) as speaking of 'Weg' (road) in this connection. Now the problem of failed reference does not arise, since a search procedure may well fail to yield a result, just as a road may lead nowhere.

But let us take an example where reference is successful. In order to refer to the Eternal City, I may use the expression *Rome* but also the expression *the capital of Italy*, or *the city where Byron died*, and so on. We have thus at least three different expressions, each referring to the same object, namely the city of Rome. In the 'search procedure' interpretation of the Fregean notion of intension, the actual city of Rome is the extension of the terms in question, and the different ways in which each of the three expressions 'leads to' the city of Rome are the different intensions of the terms in question. Clearly, when I say *Rome is situated on the river Tiber*, then its truth-value does not change when I put any of the other terms in the place of the term *Rome*, since what matters for the truth-value of the statement is that that specific city be in possession of the property of being situated on the river Tiber, regardless of the linguistic means I use to refer to or 'get at' that city. This is one way of explaining the Principle of Substitutivity, but it leaves open the question of what such a 'search procedure' actually consists in.

As regards sentential terms, Frege considered the extension of a sentence, or rather L-propositional token, to be its truth-value. Its intension, for him, was the underlying 'thought', in our terminology the underlying proposition—that is, the mental act by which a property is assigned to one or more (actual or virtual) objects. In principle, the same extension–intension distinction should also apply to L-propositions that are embedded as a term under a higher predicate, let us say *S-terms*, as opposed to *nominal terms*. Consider, for example, (6.2a, 6.2b), where the phrase *the morning/evening star to be invisible* represents an embedded object-L-proposition, or object-S:

(6.2) a. The clouds caused the morning star to be invisible.
b. The clouds caused the evening star to be invisible.

In this case the Principle of Substitutivity applies in full force: the S-terms *the morning star be invisible* and *the evening star be invisible* are fully interchangeable *salva veritate*, as well as the nominal terms *morning star* and *evening star* contained in them.

Not so, however, in sentences (6.1a) and (6.1b) with the predicate *believe*, and indeed in the vast majority of cases where a predicate takes an S-term as argument. It is to put that right that Frege took his second step, which consisted in stipulating that for some predicates that take an S-term as argument the extension of the embedded L-proposition is not, in any given token occurrence, its normal extension—in his perspective, its truth-value—but its intension, which, one presumes, thus suddenly finds itself in the position

	Extension	Intension
Definite term	Object	Search procedure
Predicate	Set of objects	Concept (satisfaction conditions)
L-proposition (Satz)	Truth-value	Thought (proposition)
L-proposition in intensional term position		

FIGURE 6.1 Frege's position with regard to extensions and intensions

of being both extension and intension. The predicates that do this to their S-terms are normally called INTENSIONAL PREDICATES, and the S-term position in which an L-proposition's extension is its normal intension is standardly called an INTENSIONAL CONTEXT. Thus, the predicate *believe*, as in (6.1a, 6.1b), is intensional with regard to its sentential object term and the object term position that comes with it constitutes an intensional context. As has been said, the majority of predicates that take an S-term as one of their arguments are intensional with regard to the embedded L-proposition. In all these cases, the Fregean extension of the L-proposition that fills the term position in question is taken to coincide with its Fregean intension. Frege's position can be represented diagrammatically as in Figure 6.1 (see also Section 10.1.3 in Volume II).

Frege's decision to let the intension of a main L-proposition P become its extension when P is embedded under an intensional predicate is not a conjuring trick or a sleight-of-hand, but the expression of a nontrivial and probably correct theoretical insight. The insight is that the object of one's belief, hope, wish, knowledge, or whatever other mental attitude, is always a particular proposition, or, in Frege's terminology, a particular thought. A belief, for example, is a particular mental attitude with regard to a proposition *p*, the attitude being one of considering *p* to be true (as a result of an act of the will, not necessarily on grounds of fully determinate information). Frege implicitly assumed, in addition, that the intension of a sentence—its underlying thought—is built up from the intensions of its constituent parts. Therefore, the intension of *The morning star is inhabited* differs from the intension of *The evening star is inhabited*, since the terms *the morning star* and *the evening star*, though coreferential, have different intensions. This, then, is what we take to be Frege's solution:

> The fact that the intensions of the L-propositions *The morning star is inhabited* and *The evening star is inhabited* are different means that they have different extensions when embedded as an S-term under an intensional predicate, so that the question of substitutivity *salva veritate* of either the whole L-propositions in question or of any co-referring terms occurring in them does not arise.

So far, we have no quarrel with Frege. But he did more. As has been said, he took the extension of an L-proposition to be its truth-value. And here we have serious reservations, shared by many philosophers. The point is of considerable importance, since the practice of taking an L-proposition's truth-value in any token occurrence as its extension is universally followed in modern logic, model theory, and formal semantics, and has led to a predicament that has so far not found a solution in terms that are compatible with these disciplines.

Let us have a closer look. Taking a well-anchored and well-keyed L-proposition's truth-value as its extension makes it necessary to accept Truth and Falsity as some kind of objects present in the real and any other possible world. This is a necessary consequence of this theory, since, for Frege as for others, extensions are the things that populate worlds. Now it may be *convenient* for propositional logic to have truth-values as extensions for the propositional terms of the truth functions, defined as functions that take truth-values and deliver truth-values, recursively. But it is also extremely *odd*. Not only do Truth and Falsity stick out like two sore thumbs in the strictly nominalist and extensional worlds favoured by logicians and logically minded philosophers, there is also the strange anomaly that, apparently, any world must have Truth and Falsity as fixed denizens, even those worlds that contain no cognizing individuals and therefore have to get along without any propositions and hence without any truth or falsity. This would confer upon Truth and Falsity some august, yet spurious, metaphysical status, independent of any world, actual or virtual.

The anomaly can be softened, but not removed, if it is assumed that truth and falsity come about in any world as an automatic result of a proposition being conceived, so that we should speak not of Truth and Falsity as such, but rather of 'the truth (or falsity) of *p*', where *p* is any given proposition. Then, however, we are forced to conclude that it is the truth (or falsity) of *p* that makes *p*, and any L-proposition expressing *p*, true (or false), since, for Frege as for others, it is extensions that produce truth or falsity. This would be like saying that it is John's tallness that makes John tall, or, to speak with Molière's Argan in *Le malade imaginaire*, that it it is the soporific power ('virtus dormitiva') of opium that makes it soporific—not very enlightening when one is interested in the causes of John's tallness or of the soporific effects of opium.

In a sufficiently flexible ontology it may make perfect sense to regard truth and falsity as *properties* of propositions and their symbolic expressions, and also to treat the truth or falsity of a proposition *p* as an abstract, reified object of a kind that can be said to 'subsist' in the actual world, to use Meinong's terminology. But taking the truth (or falsity) of *p* as the *extension* of *p* or its symbolic expression as an L-proposition leads to circularity, as long as the principle is upheld that it is extensions that make truth or falsity. In simple terms, Frege's decision to take truth-values as extensions of tokens of assertive utterances (L-propositions) leads straight into a quagmire of hopeless philosophies. Or, in the words of Michael Dummett (1981: 42): 'This conception leaves us weltering in philosophical perplexity.'

It must be said immediately, at this point, that this is not meant as a debunking of Frege's work in the philosophy of language. On the contrary, if Frege erred, he did so only because he was among the first to broach questions of this nature. We are even prepared to defend him to a point, as it must have been quite a discovery for Frege to see that the Boolean functions over 1 and 0 actually compute the truth functions of propositional calculus (a fact that gave rise to the widespread convention of using '1' for truth and '0' for falsity). Wishing to put this discovery to good use, he will have thought that the obvious thing to do was to take truth-values as extensions of (tokens of) L-propositions. The philosophical complications arising out of that decision could hardly have been clear to him at that stage (or if they were he did not want to be bothered: Frege often insists, in his writings, that he is not a philosopher but a mathematician). But the philosophical complications have become so clear now that they can no longer be ignored.

Moreover, and this is perhaps even more serious than the philosophical complications, closer inspection of the facts reveals that, from an empirical point of view, the situation as regards SSV is a great deal more complex than the oversimplified picture emerging from Frege's analysis. Thus, as shown in Section 6.2.3.1, the predicate *cause* and its synonyms allow for free SSV of nominal terms in the clausal object term. This would suggest that *cause* is extensional with regard to its object term, which would then take its truth-value as extension. But then a sentence like *A short-circuit caused the house to burn down* should be freely exchangeable *salva veritate* with any sentence *A short-circuit caused*... followed by a clausal object term with the same truth-value as *the house burned down*. But this is absurd, since it may well be true that a short-circuit caused the house to burn down, while it is false that a short-circuit caused the sun to rise at 7.35, although *The house burned down* and *The sun rose at 7.35* (said of the same day) may both be true. More examples of empirical complications are presented in the remainder of this chapter. Yet the relevant literature, mostly written by philosophers and logicians, concentrates mainly on more or less formal constructions intended to account for *believe*-sentences and shows little awareness of the empirical complexity of the situation in this regard.

By contrast, the philosophical complications have not passed unnoticed in the literature. Besides the extensive exegetic and critical writings by Michael Dummett on Frege (Dummett 1973, 1981), where this aspect of Frege's analysis is dissected and criticized at regular intervals, one finds other authors also occasionally expressing worries about what is called 'Frege's Platonism'. But nothing is ever done about it.

The problem of lack of substitutivity *salva veritate* in intensional contexts, (re)discovered by Frege, was, in fact, the starting point of twentieth-century formal or model-theoretic semantics introduced by Richard Montague during the 1960s, in the wake of work done by Alfred Tarski, Donald Davidson and other philosophers and logicians. In logical semantics, Frege's solution was taken over in principle, but not entirely. While Frege's view that the extension of an L-proposition is its truth-value was accepted, as well as Frege's decision to let the extension of an L-proposition in an intensional context coincide with its intension, it was not accepted that the intension of an L-proposition is its underlying thought, thoughts being considered too unmathematical and too hard to handle by formal means. Instead, the intension of an L-proposition was defined as the set of possible worlds in which the L-proposition is true, and the belief relation was seen as a relation between a believer and a set of possible worlds. Accordingly, a proposition became a set of possible worlds, strange though that may sound.

The immediate intuitive resistance against such an exotic notion of proposition was soon reinforced by a real and insurmountable problem. The problem is that truth-conditionally (and hence logically) equivalent L-propositions are by definition true in exactly the same set of possible worlds and thus take identical intensions. Therefore, any two truth-conditionally equivalent L-propositions should be mutually substitutable *salva veritate* in intensional contexts. And this is obviously not the case, as appears most clearly with either necessarily true or necessarily false L-propositions. Any necessary truth is truth-conditionally equivalent with any other necessary truth, and likewise for any necessary falsity. By definition, any necessary truth takes the full set of all possible worlds as its Montagovian intension and any necessary falsity takes the null set as its Montagovian intension. This means that the sentences

(6.3a) and (6.3b), or (6.4a) and (6.4b), must always have the same truth-value: they must always be either both true or both false at the same time of the same Henry:

(6.3) a. Henry believes that $2 + 6 = 13$.
b. Henry believes that he is both dead and alive.

(6.4) a. Henry believes that $2 + 6 = 8$.
b. Henry believes that the sum of the angles of a triangle equals 180°.

However, it is clear that Henry may very well believe that $2 + 6 = 13$, yet be convinced that he is not both dead and alive. Likewise, Henry may very well believe that $2 + 6 = 8$, yet be totally ignorant of the geometry of triangles. This problem is known as THE PROBLEM OF PROPOSITIONAL ATTITUDE SENTENCES.

Montagovian model-theoretic semantics is tripped up by the fact that humans are, on the whole, not logically and mathematically omniscient. The philosopher David Lewis was among the first to call attention to this problem:

> But one part of the tradition about propositions must be given up: propositions understood as sets of worlds cannot serve as the meanings of sentences that express them, since there are sentences—for instance, all the logical truths—that express the same proposition but do not, in any ordinary sense, have the same meaning. (Lewis 1970: 46)

As was pointed out in Section 2.1, Dowty, Wall, and Peters, in their excellent *Introduction to Montague Semantics*, are also frank about the matter:

> [T]he problem of propositional attitude sentences is a fundamental one for possible world semantics, and, for all we know, could eventually turn out to be a reason for rejecting or drastically modifying the whole possible world framework. (Dowty *et al.* 1981: 175)

Meanwhile no solution has been offered that would save possible world semantics. Nor does it look as if any will be found. It is to be feared, therefore, that this version of formal semantics is fundamentally flawed and must indeed be rejected, as Dowty, Wall, and Peters were afraid might have to happen. Its main failure lies precisely in its incapacity to deal with cognitive phenomena. The rejection of the Fregean notion of the underlying thought as the intension of L-propositions and the replacement of this notion with the idea that the intension of an L-proposition is the set of possible worlds in which it is true appear to have sealed its fate.

There is some irony in the fact that possible world semantics rejected what was promising in Frege's analysis—the thought factor—and took over the element that contributed to its own insoluble predicament, namely, truth as extension. That logically (truth-conditionally) equivalent L-propositions are not substitutable *salva veritate* in intensional contexts should, in principle, not be a problem for Frege, since logically equivalent thoughts (propositions) may well differ in content, so that logically equivalent L-propositions may well have different Fregean intensions. The problem with Frege's solution was, rather, that he failed to provide an explicit analysis of the notion of content, his 'Sinn' or intension (besides, of course, his decision to take the truth-value of a token utterance of an L-proposition as its extension).

A further flaw in Frege's solution is what may be seen as its overkill: his solution can be upheld in principle without extrapolating the extension–intension distinction from predicates to terms and L-propositions. In order to solve the problem of why SSV is blocked in intensional contexts, it suffices to define the nature of the extension of nominal and S-terms according to the restrictions placed upon them by the predicate in question. For most predicates it is stipulated that their nominal argument terms require as extensions actually existing world objects (often categorially restricted to some subset of **OBJ**). For S-terms, predicates normally require the proposition expressed by them as their extension. Occasionally, however, something other than the default value is required. In such cases the extension of the nominal term or the S-term must consist in something else with which the term in question is somehow uniquely connected in the overall system of language, mind, and world.

The Fregean notion of intension, moreover, lacks unity, as it is not clear what unites the intensions of nominal terms, predicates and L-propositions as defined by him. These notions appear to have little more in common than the label 'intension' (Sinn) given them by Frege. Predicate intensions, for example, are type-level, lexically fixed bundles of satisfaction conditions, whereas term intensions and intensions of L-propositions are, in Frege's analysis, token occurrences. In this respect, model-theoretic semantics does better, taking all intensions to be functions from possible worlds to the corresponding extensions (Dowty *et al.* 1981: 149). Unfortunately, however, this does not work, as we have seen.

All that has to be shown, in order to solve the problem of nonsubstitutivity in intensional contexts, is that a particular value, essential for truth-value assignment and required for S-terms in intensional term positions, may change when nominal terms that are in actual fact co-referential are mutually substituted. That will block the free substitution of the S-terms in question *salva veritate* and hence of the co-referring nominal terms occurring in them.

This perspective is elaborated in the following section, where an alternative solution to Frege's problem is presented.

6.2 An alternative solution to Frege's problem

Unlike Frege we do not extrapolate the distinction between extension and intension from predicates to nominal and sentential terms and to main L-propositions. We consider the distinction useful for predicates only, just as the tradition before Frege had it. The EXTENSION OF A PREDICATE F is, for each world state or situation (or event), the set of objects that are truthfully characterized by F, and its INTENSION is, in our perspective, the lexically fixed SATISFACTION CONDITIONS to be fulfilled by any object or n-tuple of objects for them to produce truth under F. As regards main L-propositions and nominal and S-terms, they do not seem to be usefully characterized as having extensions or intensions. We say that a well-anchored and well-keyed L-proposition, if true, is made true by an ACTUAL FACT. If false, it is made false by the lack of an actual fact to make it true. It then only has a virtual fact corresponding to it in natural ontology. A main L-proposition, moreover, EXPRESSES A PROPOSITION and may be seen as having a VALUATION SPACE associated with it (in a sense that would have to be shown to be psychologically plausible; see Section 2.3.3 of Volume II).

The primary property of definite nominal terms and embedded S-terms, is that they have a reference or ρ-value, which is the value of a so far largely unformalizable reference function, part of the keying process in the case of actual ρ-values, but an automatic result of the mental coming about of the term in question in the case of virtual ρ-values. But before we can elaborate this any further, something must be said about actual and virtual objects.

6.2.1 *Actual and virtual objects*

Being objects of one kind or another, ρ-values come in different varieties. We stipulate that each predicate specifies the variety of ρ-value needed for its terms so that truth results. As regards nominal terms, the predicate usually requires actually existing objects for truth to arise. For example, for anything to be *blue* or to *weigh five pounds*, it has to be an actually existing object. These predicates require an actual ρ-value for their subject term if truth is to result (they generate a presupposition of actual existence). We say that they are extensional with regard to their subject term. Since most term positions under predicates are extensional, we consider that to be the default case. By contrast, for anything to be *talked* or *laughed about*, or to be *worshipped*, or to *instil fear*, it does not have to belong to the world of actually existing things. These predicates allow their (subject) terms to take virtual objects as a ρ-value, thought up by the infinitely creative thinking mind and defined by the properties assigned to them in the course of the thinking process. We say that these predicates are intensional with respect to the argument term in question, which means that the argument term in question may take not only an actual but also a virtual ρ-value, defined by the properties assigned to it by the thinking mind.

Twentieth-century Anglo Saxon philosophy has, unfortunately, banned virtual objects from its ontology—other than as objects in an ill-fated ontology of possible worlds. As was shown in Sections 2.2 and 3.1.3, they were the object of much philosophical debate during the nineteenth century, but owing to a variety of factors discussed in Chapter 2—such as the mathematization of logic that broke through at the beginning of the twentieth century and the general trend permeating that century to write anything to do with the mind out of the script of science—philosophers, logicians, and now also linguists in their wake, show a tendency to become a little nervous when virtual objects are brought up, especially since Quine's rhetorical blast (1953: 4), quoted and commented upon in Section 2.2.2. Yet Quine's diatribe is hardly supported by sound argument. Quine applies criteria of actual existence, such as a well-defined identity and being defined for all physical, mental and historical details, to virtual objects, which are only defined to the extent that their mental representation is, and then concludes, unsurprisingly, that virtual objects do not satisfy the criteria of actual objects.

Moreover, Quine and the philosophical tradition following his lead have overlooked the monumental fact that natural language refers to and quantifies over virtual objects in exactly the same way as it does with regard to actual objects. Linguistically, there is nothing odd about a sentence like (6.5a), which functions in precisely the way (6.5b) does. Yet, as is argued at length in Chapter 2, twentieth-century analytical philosophy and semantics have no satisfactory analysis for (6.5a), precisely because it involves virtual entities:

(6.5) a. Two gods were worshipped in that temple.
b. Two statues were found in that temple.

The reader will have become accustomed by now to the fact that the entire approach developed and advocated in this book is crucially dependent on the assumption of the reality though not the actual existence of virtual objects and virtual facts created by the thinking mind and projected onto what we take to be the actual world as something like templates that may or may not fit the sense data that support the language user's theory of what constitutes actual reality. Consider the case of a man who intends to write an article but has not yet written it. Yet he already lists this intended article in the bibliography of a grant application (although he is not supposed to do so). Should we now say that he has made no reference because the article does not actually exist, or should we say that he has made an inappropriate reference? It seems that we should say the latter, not the former. The assumption of virtual objects is made necessary by the fact that we can interpret rules like 'Do not refer to publications not yet written'. We will, therefore, treat the verb *refer* as being nonextensional with respect to its direct object position, so that one can say in truth that a reference is made to an article not yet written. And we reject the notion, standardly accepted in the philosophical literature, that one can only refer to actually existing objects.

Rejecting the assumption of the reality-without-existence of virtual objects means a continuation of the failure to account for the cognitive dimension in language and language use, stifling a proper insight into the nature of language and thinking. The formal semanticists' failure to come up with a proper answer to the dramatic collapse of possible world semantics illustrated in the preceding section is a telling detail in this connection. The assumption of virtual objects and virtual facts appears to be a necessary but perfectly affordable price to pay for an adequate theory, not just a theory of truth and reference in language, but of cognition in general.

On this assumption, we say that an actual object **o** instantiates the corresponding virtual object ***o*** projected onto the world by the thinking and interpreting mind. We argue, in addition, that the projection of virtual objects and facts is a necessary prerequisite for the psychological process of interpretation of sense data: *no interpretation without mental representation and hence prior virtual reality.* Interpretation of sense data is an act, less open to voluntary decision than most other acts, but still an act. That this act is open to wilful decision at least to a certain extent, is demonstrated by the well-known gestalt experiments of the 1930s, which show that certain figures can be seen this way or that way, depending on how one sets one's mind (think of the vase versus two opposing faces, and so on).

The mental representation [**o**] of an object is by itself not sufficient for interpretation, because a representation without the virtual object ***o*** created by it is not a representation but just a configuration of materials, perhaps brain cells: what makes a representation a representation is the very fact that it creates a virtual reality. This is borne out by natural linguistic usage. Not even Quine (1953: 2) could deny the fact that when we speak about Pegasus, we do not speak about the mental representation (Quine's 'idea') of Pegasus. What we do speak about is, in the view defended by his imaginary philosopher

Wyman but in fact attributable to Meinong and supported by the present author, a virtual object, and not nothing, as Quine would have it.

It will be remembered from Chapters 2 and 3 that for the mind, and therefore also for language, the class of things that are treated as objects far exceeds the class of things that are construed as forming individual entities in the world. As was pointed out in Section 3.1.3, the human mind is strongly *object-oriented*, in the sense current in computer science. It has a knack of taking all sorts of elements from the world together and bundling them up as one 'abstract' object. Such abstract objects are usually called 'reifications'. Humans know, of course, that what they denote by means of reifying nominal expressions are not individual entities. Nobody will require one to actually produce the average Boston cab driver, because the average Boston cab driver is a product of reification, not a person. In Meinong's terms, the average Boston cab driver does not *exist*, but he *subsists*, in that he owes his being to a number of really existing entities. Reifications are useful to the extent that they allow for generalizations to be made over a population or a system. They should be well defined, although in practice they often are not, which quickly leads to sloppy thinking. Like mental representations of individual entities, reifications may be actual or virtual. When they are actual, they are based on complexes of entities construed as having actual existence. When they are virtual, they are at least partially based on complexes of entities construed as being merely thought-up.

6.2.2 *Actual and virtual ρ-values for nominal terms*

The machinery set up so far implies that a predicate F produces truth just in case the ρ-values of its terms satisfy the conditions associated with F. Or, as was stipulated in (3.22) of Section 3.3.2, a properly anchored and keyed L-proposition $F(a_1,\ldots,a_n)$ is true just in case $< \rho(a_1),\ldots,\rho(a_n) > \in [\![F]\!]$. It often occurs that no *actual* ρ-value is available for a term under a given predicate—the all-pervasive problem known by the misnomer 'reference failure'. In such cases there always is a *virtual* ρ-value, in virtue of the fact that a proposition has been conceived. Virtual ρ-values come into being as an automatic corollary of the very fact that a definite nominal term is used in an L-proposition expressing a proposition. The fact that a proposition *p* occurs in a given mind *M* implies that *M* has a representation of the object or objects to which the property at issue in *p* is assigned. This representation may or may not represent an actually existing object, just as a representation of a dog may or may not represent an actually existing dog. When it does not, it represents a virtual dog.

Accordingly, we posit that every assignment of an actual ρ-value is necessarily and automatically preceded by the assignment of a virtual ρ-value which is, in the case of an actual ρ-value, INSTANTIATED by the corresponding actual object on grounds of relevant sense data and/or other cognitive clues in terms of the available world construal. A virtual ρ-value is not necessarily matched by an actual ρ-value, since the search procedure for instantiation may fail to yield a positive result. Reference failure, in the accepted sense, is thus seen to be no more than the lack of instantiation of a ρ-value under a predicate that requires an instantiated ρ-value for truth. The term *reference failure*, however, should be avoided, since it is normal for virtual objects to be referred to. Such forms of reference occur both frequently and naturally in ordinary speech. It is more appropriate to speak of the 'lack of instantiation' of a ρ-value in cases where the ρ-value in question needs to be instantiated for truth to arise, but has been found not to be.

Each term in an L-proposition needs a ρ-value for the L-proposition to function as a token and to have a truth-value. What kind of ρ-value is required or allowed for truth to arise depends on the kind of term and on the semantics of the predicate in question. For definite nominal terms the default case is that the predicate F requires an actually existing object as ρ-value for truth. Term positions that fall under this default are called extensional term positions. But F may also be lexically specified as being intensional with regard to a specific term position. In that case, F is allowed, but not obliged, to produce truth with a virtual object as the ρ-value of a term in that position. For example, the binary predicate *worship* is specified for actual ρ-values with regard to its subject position since only actually existing beings can worship, but the term in object position optionally allows for virtual ρ-values, since what is worshipped does not have to exist in the actual world and may be a merely thought-up, virtual being. Or, to quote Linsky (1967: 18), 'Referring, unlike hanging, can have a non-existent object', the reason being that *refer* is intensional, while *hang* is extensional, with regard to its object term.

The predicates *exist* and *imaginary* are intensional with regard to their subject term, which thus may refer to either an actually existing or a merely thought-up entity. The reference hierarchy of Section 3.5.2 makes actual ρ-values take precedence over virtual (intensional) ρ-values, which can function as ρ-values only when no actual ρ-value is available. One remembers from Section 3.5.2, that the ρ-value of a definite term is fixed in a stepwise procedure, guided by the so far unformalizable process of intentional keying. Every interpretable definite term has a virtual ρ-value, or else it cannot be used in an interpretable (well-keyed) L-proposition. Some interpretable definite terms have an actual ρ-value instantiating their virtual ρ-value. When an actual

ρ-value is available, the interpretation machinery settles for that as the final ρ-value, but when no actual ρ-value is available, the machinery must be content with the virtual ρ-value which is always there.[3]

The predicates *exist* and *imaginary* can be defined as follows (*o* stands for the virtual object underlying any actual object **o** if there is one, and otherwise for a mere virtual object):

(6.6) a. [[Exist]] = { o* |*o* is instantiated as an actual object **o**}
(the extension of *exist* is the set of actual objects **o** or virtual objects *o*, such that *o* is instantiated as an actual object **o** in the actual world)

b. [[Imaginary]] = { o* | *o* is not instantiated as an actual object **o** }
(the extension of *imaginary* is the set of actual objects **o** or virtual objects *o*, such that *o* lacks an instantiation as an actual object **o** in the actual world)

In terms of discourse semantics this means, for *exist*, that a virtual object represented in some intensional subdomain (possibly the truth domain itself), is given a representation in the truth domain to which the speaker is committed (if there wasn't already one there), and, for *imaginary*, that such a representation is withheld.

This analysis avoids the problems that present themselves in strictly extensionalist ontologies, which do not accept virtual objects as ρ-values for nominal terms and restrict being to actual objects. Such ontologies get into trouble with true negative statements of existence like *The king of France does not exist*, since the subject term now suffers from what these theories consider to be reference failure, which would exclude the assignment of a truth-value at all. Normally, Russell's Theory of Descriptions (Russell 1905*a*) is called upon, which remodels definite terms as a form of existential quantification, turning the sentence just quoted more or less into 'it is not true that there exists an x such that x is a king of France and no one else is king of France', but, as was shown in Chapter 2, this analysis, though still widely accepted in the

[3] 'Established' presupposition theory, which is based largely on Strawson's contributions to the field, is still unable to explain why a sentence like (i) does not presuppose that Pegasus actually exists, which would make the sentence semantically inconsistent:

(i) Pegasus does not exist.

Some say, following Strawson (1964), that this is because *Pegasus* is, though subject, not the topic of the sentence. For them, the reference of an NP is presupposed only if that NP is topic. That answer can be dismissed right away, first, because topic status can be determined only in context and (i) is given without any context; secondly because if we make *Pegasus* the topic in any accepted sense of the term, as in (ii), there is still no presupposition of actual existence:

(ii) As regards Pegasus, he does not exist.

philosophy of language and semantics, is thoroughly flawed. This becomes apparent when one considers that, for example, the statement *The king of France does not exist* would be true if there were two actually existing kings of France, and a sentence like *The god Apollo was worshipped in the island of Delos* would entail the actual existence of Apollo. The recognition of virtual objects together with an analysis of the predicate *Exist* as given in (6.6a) makes it possible to take definite terms for what they are and to avoid their unwanted remodelling in terms of existential quantification.

The analysis given in (6.6a) explains why SSV is allowed under the predicate *Exist*. Consider the sentences:

(6.7) a. The morning star exists.
b. The evening star exists.

Given that the DAs for the terms *the morning star* and *the evening star* are intentionally keyed to what is in fact one and the same actual object, the planet Venus, it follows that in all cases where the virtual object ***the morning star*** is instantiated in the actual world, the virtual object ***the evening star*** is likewise instantiated in the actual world, and by the same actual object. This suffices for the terms *the morning star* and *the evening star* in (6.7a, 6.7b) to be freely substitutable *salva veritate* in nonintensional contexts.

A third kind, besides actual and virtual objects, of possible ρ-values for definite nominal terms is provided by their TOKEN-REFLEXIVE USE consisting of a mention of linguistic material, as in *'Rome' consists of four letters*, or *You just said 'Oops'.* In medieval philosophy this token-reflexive use of a definite NP was called *suppositio materialis*, as opposed to *suppositio formalis*, which is the normal, non-token-reflexive form of reference to objects. In modern literature, the term *hypostasis* is often used for token-reflexive use.[4]

There can be no question of a general ban on token-reflexive use on the grounds that object language (the language we use to say things about the world) and metalanguage (the language we use to say things about the

[4] As was noted in Seuren (1985: 366), anaphoric pronouns can be used token-reflexively when the antecedent is used referentially, as in:

(i) ✓He has lived in Dnepropetrovsk for 20 years, but he still can't pronounce *it* properly.

But when the antecedent is a token-reflexive noun phrase, subsequent referential anaphora seems to be ruled out in some cases, but not in others. Thus (ii) seems to be unacceptable, but (iii) appears to be all right:

(ii) ! *Esperanto* means 'one who hopes' in *it*.
(iii) ✓It is called *punch* because it has *it*.

It cannot be our task to solve this problem here.

language) should be kept strictly apart. In standard modern logic such a strict separation is considered mandatory on pain of falling victim to some variety of the Liar Paradox: when I say *This very utterance is false*, then, in case the utterance is true it is false, and when it is false, the fact that it is false makes it true. To avoid this trap, logicians, following Russell and Tarski, introduced the principle 'Never mix object language with metalanguage in the same sentence'. The sentence *This very utterance is false* would then constitute a violation of this principle, since the predicate *false* then applies both to the utterance itself, which belongs to the metalanguage, and to the utterance referred to in the subject term, which would belong to the object language. But, as Kneale and Kneale rightly observe (1962: 228), such a principle would exclude all sorts of harmless forms of self-reference, such as *This very sentence is a sentence of English.* In fact, one may add, natural language use is replete with perfectly natural and harmless mixings of object language and metalanguage, which, of course, nobody will be able to stop.

Token-reflexive use must be permitted, by the theory as well as in practice, since what we say at any time constitutes by itself an object in the world, and it would be absurd to allow one to speak about any object at all, actual or virtual, reified or not, except about the objects we produce when speaking or writing. The Liar Paradox is avoided not by a prohibition imposed by logicians but by the very mechanism of linguistic interpretation (see Seuren 2001: 119–35 and Section 3.1.1 for a detailed discussion). In a sentence such as *This very sentence is false*, the term *this very sentence* must refer to a proposition expressed as an L-proposition, since the predicate *false*, like the predicate *true*, requires propositions expressed as L-propositions to apply to (see (6.16) below). We thus need a proposition to serve as the ρ-value to which the predicate *false* is mentally assigned. But all there is to be found is the very same attempted but noncompleted proposition, which is then again said to be false, and so on ad infinitum. There is thus no proposition of which it can be said that it fails to satisfy the satisfaction conditions of the predicate *false.* The sentence *This very sentence is false* is thus by definition uninterpretable. By contrast, a sentence like *This very sentence is a sentence of English* is fully interpretable, since the predicate *be a sentence of English* does not require a proposition as the ρ-value of its subject term, but merely a well-formed sentence, which is obviously available.

It is not just virtual objects, with or without actual objects instantiating them, that are associated with definite nominal terms. Each definite nominal term in a token occurrence of an L-proposition has associated with it a SEARCH PROCEDURE triggered by the definite determiner and relating it to its actual or virtual ρ-value. This search procedure necessarily involves the identification of a mental representation or DISCOURSE ADDRESS (DA) of the object or objects referred to, as part of a discourse domain **D** and keyed to a virtual and often also an actual object. We thus associate with each definite nominal term under a predicate in a token L-proposition:

(a) a virtual object ***o***, automatically created by the DA, and, in case there is one, an actual object **o** that is the instantiation of ***o*** in terms of the world construal of the interpreting subject;
(b) a mental object representation in the form of a discourse address (DA) that is part of a larger **D**;
(c) a referential search procedure identifying first the DA with its correlate ***o***, and then, if any, an actually existing object **o** instantiating ***o***.

Virtual and actual objects, mentioned under (a), can serve as ρ-values for nominal terms under a predicate, but DAs and search procedures, mentioned under (b) and (c), respectively, are, apparently, not the kind of object that predicates can be lexically defined as applying to.

6.2.3 *Possible ρ-values for S-terms and substitutivity*

For embedded L-propositions, or S-terms, there are three possible kinds of ρ-value:

(a) the proposition expressed in the S-term;
(b) the L-proposition underlying the S-term;
(c) the corresponding actual or virtual fact.

Which of these three is to be the required type of ρ-value depends on the predicate standing over the S-term in question. In normal or default cases, the ρ-value of an embedded S-term is the *proposition* expressed by it—its *propositional* ρ-value. Truth predicates, in particular, seem to be best described as requiring *L-propositional* ρ-values. Other predicates appear to require *facts* as the ρ-value of the embedded S-term concerned—the S-term's *factual* ρ-value. Since it is well-known that 'fact' is a slippery notion, let us see what we can do to make it manageable without getting embroiled in a hopeless ontological tangle.

6.2.3.1 *Actual and virtual facts as ρ-values* As with objects, ACTUAL FACTS are distinguished from VIRTUAL FACTS. We start with the notion of actual fact, saying that in all cases where a properly anchored and keyed token occurrence of an L-proposition $F(a_1, \ldots, a_n)$ is true, there is an ACTUAL FACT

$$\mathbf{Fact}[P(F)(\rho(a_1), \ldots, \rho(a_n))]$$

that makes it true. Actual facts are truth makers for propositions and their expressions as L-propositions. Let the notion 'actual fact' be defined as follows:

> ACTUAL FACT:
> If a well-anchored and well-keyed L-proposition P of the form $F(a_1, \ldots, a_n)$ expressing the proposition p, is true, both P and p are made true by the actual fact **Fact**$[P(F)(\rho(a_1), \ldots, \rho(a_n))]$—that is, by a state of affairs (key) in which the n-tuple of ρ-values of the terms $a_1, \ldots, a_n$ is an element in $[\![F]\!]$ defined by the property P(F), or $<\rho(a_1), \ldots, \rho(a_n)> \in [\![F]\!]$.

Thus, when a properly anchored and keyed token occurrence of the L-proposition underlying the sentence *The policeman caught the thief* is true, the corresponding actual fact that makes it true consists in a state of affairs (or event) that the L-proposition in question is keyed to and in which the actual policeman and the actual thief were indeed related to each other the way a catcher stands to a catchee, regardless of whatever else was or was not the case in the rest of the world (the situation). When, on the other hand, the policeman and the thief in question failed to satisfy the *catch*-condition in the state of affairs (event) concerned, or when there was no policeman and/or no thief, there is nothing in the actual world to make the L-proposition (or the proposition expressed by it) true. In that case, the L-proposition is false and lacks a corresponding actual fact.

Next to actual facts we posit VIRTUAL FACTS. These are mentally constructed, thought-up states of affairs, represented in the mind as propositions. Obviously, virtual facts may contain actually existing objects, just as they may contain merely thought-up, virtual objects. What makes a fact virtual is the assignment of a property to one or more (actual or virtual) entities. Virtual facts, however, are not themselves propositions, but are created and projected as virtual states of affairs in the world by propositions that will be true when the virtual facts created by them turn out to be instantiated as actual facts in terms of the world construal held by the interpreting subject. We use the notation:

> ***Fact***$[P(F)(\rho(a_1), \ldots, \rho(a_n))]$

for virtual facts.

Any L-proposition $F(a_1, \ldots, a_n)$ expressing a corresponding proposition has a corresponding virtual fact ***Fact***$[P(F)(\rho(a_1), \ldots, \rho(a_n))]$, but only true token occurrences correspond to an actual fact **Fact**$[P(F)(\rho(a_1), \ldots, \rho(a_n))]$ instantiating it. The analogy between propositions and terms with their DAs, which always have a corresponding virtual object created by them but are instantiated by an actual object only when there is one in the interpreting subject's construal of the actual world, should be obvious.

Sometimes, a predicate that takes an S-term requires an actual fact as ρ-value for the S-term, for truth to come about. Obvious cases are the so-called FACTIVE PREDICATES, which require the truth of their embedded S-term (in fact, they *presuppose* the truth of the embedded S-term—an aspect that is not taken into account here but is further elaborated in Chapter 10 in Volume II). One such predicate is *realize*, as in:

(6.8) John realizes that the morning star is uninhabited.

When one realizes something, one realizes an actual fact, not a proposition: instead of (6.8) one may say *John realizes the fact that the morning star is uninhabited.* If it were false that the morning star is uninhabited, sentence (6.8) would be false ('radically false' in a presuppositional perspective), when said of the intended referent John at the time referred to by the present tense. Note, however, that SSV is blocked under factive predicates: sentence (6.9) may well be true while (6.8) is false at the same time, and vice versa:

(6.9) John realizes that the evening star is uninhabited.

The blocking of SSV follows from the semantic description of the predicate *realize* (as it does in the case of most other factive predicates), which involves the condition that the referent of the subject term—John in the case at hand—knows (his world construal entails) that the proposition expressed by the S-term in question is made true by the corresponding actual fact. Since, in all cases where the terms *the morning star* and *the evening star* denote different discourse addresses and hence correspond to different virtual objects, the proposition expressed by *The morning star is uninhabited* is different from the proposition expressed by *The evening star is uninhabited*, and since John's world construal may, as yet, not entail that the two addresses should be amalgamated for an adequate world view to arise, it follows that SSV is blocked in the S-terms of (6.8) and (6.9), and generally within S-terms under factive predicates.[5]

Other predicates require virtual facts as ρ-values for their S-terms. The modal predicates of futuricity in different languages, such as English *will* or the abstract predicates underlying future-tense morphemes, seem to be a case in

[5] In general, factive predicates block SSV in their factive clauses, except when the factive clause stands in subject position, as in (i), where the term *the butler* is freely substitutable *salva veritate* with any co-referring term. Note the difference with (ii), which does not allow for free SSV for *the butler*:

(i) That the butler was upstairs proves that he is innocent.
(ii) John proved that the butler was innocent.

point. By definition, any future fact is virtual, correctly or incorrectly predicted in virtue of one's predictive mental powers. The question of what makes (L-)propositions under a futuricity predicate true is a famous ancient riddle, first formulated by Aristotle in his *On Interpretation*, where he depicts a situation where seamen and soldiers are busily preparing a fleet for battle and a spectator says 'There will be a sea battle tomorrow'. The question is now whether or not this utterance has a truth-value at the moment of utterance. If one assumes it does, one is committed to determinism as regards courses of events in the actual world, since, on this view, all future events must be predestined. If, more reasonably, one assumes that the utterance in question does not have a truth-value at the moment of speaking but will acquire one only when the predicted moment of truth has arrived, then one must accept that there are meaningful assertive utterances without a truth-value—a conclusion Aristotle appears to be resigned to accept. The antiquity of this riddle, together with the fact that it still gives rise to a lively literature, would suggest that an easy solution is unlikely to be within immediate reach. All we can do here is present a new perspective.

The predicate *cause* and its synonyms take actual facts as their object-term extension. The semantics of *cause* may be (trivially) described as follows (**o** ranges over actual individual or reified objects, **f** ranges over actual facts:

(6.10) $[\![\text{Cause}]\!] = \{ <\mathbf{o},\mathbf{f}> \mid \mathbf{o} \text{ brings about } \mathbf{f} \}$

That is, when the clausal object term of *cause* refers to an actual fact **Fact**[P(F)($\rho(a_1),\ldots, \rho(a_n)$)] brought about by some **o**, then '**o** caused **Fact**[P(F)($\rho(a_1),\ldots, \rho(a_n)$)]' is true. But when the object term **f** gets no further than a virtual fact (following the reference hierarchy of Section 3.5.2), there being no actual fact to refer to, then '**o** caused **Fact**[P(F)($\rho(a_1),\ldots, \rho(a_n)$)]' can only be false. In fact, it will then be *radically* false as the precondition of actuality for **f** has not been satisfied. A sentence like *The rain caused the roads to be flooded* is, and is felt to be, radically false when the roads were not flooded at all.

One notes the parallel with *exist* as defined in (6.6a) above. One also notes that SSV is freely allowed under *cause*. This follows from (6.10), which implies that any term a_i in an S-term $F(a_1,\ldots, a_i, \ldots, a_n)$ under *cause* may be freely replaced, *salva veritate*, with any other term b_i that happens to have the same ρ-value, so that any L-proposition $F(a_1,\ldots, a_i,\ldots, a_n)$ may be replaced *salva veritate* with $F(a_1,\ldots, b_i,\ldots, a_n)$. For what matters is that the n-tuple of ρ-values be an element in $[\![F]\!]$, regardless of what linguistic means are used to 'get at' them.

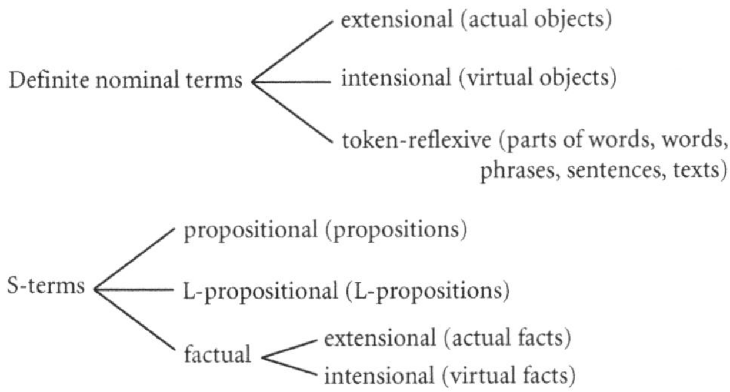

FIGURE 6.2 Different kinds of ρ-values for nominal terms and S-terms

6.2.3.2 *Propositional and L-propositional ρ-values and substitutivity* As was said above, the default ρ-value of an embedded S-term is the proposition expressed by it. However, some predicates, in particular the truth predicates, including the operators of propositional logic, are best seen as taking L-propositions as ρ-values of their terms. And some remaining predicates, such as *cause*, take factual ρ-values. This completes the picture of possible ρ-values for argument terms under predicates. There are now three kinds of possible ρ-value for definite nominal terms, namely extensional (actual), intensional (virtual), and token-reflexive, and likewise three kinds of possible ρ-value for S-terms, namely propositional, L-propositional, and factual, where the class of factual ρ-values is again subdivided into extensional for actual facts and intensional for virtual facts. Figure 6.2 shows these distinctions and, in addition, the kinds of object, in the world, in the mind, or in language, that correspond to the various sorts of ρ-value.

The question now is: what explains the blocking of SSV for some kinds of propositional ρ-value for S-terms? Why is SSV, in most cases, blocked for propositional ρ-values, as in the case of (6.1a, 6.1b) given above, with the superordinate predicate *believe*? Let us concentrate, for the moment on the predicate *believe*. Under the uncontroversial assumption that the S-term in direct object position under *believe* takes a propositional ρ-value, it must be shown that the substitution of one co-referential term for another may yield different propositions. Consider the following approximate semantic description of the predicate *believe* (*a* ranges over cognizing beings; *p* ranges over propositions):

(6.11) ⟦Believe⟧ = { <a,*p*>| a has, by an act of the will, decided to accept *p*, and any L-proposition P expressing *p*, as true }

We have defined a proposition as the mental act of assigning a property to one or more virtual objects, represented as DAs. Each DA creates a virtual object, which may or may not turn out to be incarnated or instantiated as an actual object. Thus, given an L-proposition underlying the sentence *The morning star is inhabited*, the proposition expressed by it assigns the property expressed by the predicate *be inhabited* to the virtual object *the morning star* by adding this predicate to its DA. It is possible, of course, that the believer in question is well informed and knows that the morning star is identical with the evening star. In that case the believer will have one single virtual object with one single DA (and one key) for both terms, and the property expressed by *be inhabited* will be added to that address. Now SSV goes through unimpeded. But it is also possible that the believer thinks that the morning star and the evening star are two distinct heavenly bodies. In that case he will have two distinct DAs and two distinct corresponding virtual objects, ***the morning star*** and ***the evening star***, even though, unbeknownst to the believer, the same actual object instantiates the two virtual objects. Now the L-proposition underlying the sentence *The morning star is inhabited* expresses a different proposition from the L-proposition underlying the sentence *The evening star is inhabited*, because in the former, the property expressed by *be inhabited* is assigned to ***the morning star***, whereas in the latter it is assigned to ***the evening star***. This blocks SSV, even though in actual fact the two terms refer to the same object, the planet Venus.

This, then, is our solution to the problem of nonsubstitutivity of co-referring terms in intensional contexts. (One notes that the Montagovian problem of improper substitutivity of logically equivalent L-propositions has vanished, since logically equivalent L-propositions will, in most cases, correspond to different underlying propositions. Only when they do not, should there be substitutivity *salva veritate*.)

It is now also clear why, as is shown in Section 3.2, different topic–comment structures cannot be substituted *salva veritate* in emotive intensional contexts. Consider again the examples (3.11a, 3.11b) of Section 3.2, repeated here as (6.12a, 6.12b):

(6.12) a. Ann was angry that JOHN had sold the car.
       b. Ann was angry that John had sold the CAR.

The reason why it is possible for (6.12a) to be true while (6.12b) is false, and vice versa, now appears to be the following. As shown in Section 3.2, the surface structure *JOHN has sold the car* expresses a different proposition from *John has sold THE CAR*. The difference is reflected in L-propositional form, where the predicate is *Be* JOHN in the former, and *Be the* CAR in the latter sentence:

(6.13) a. Be_v JOHN(the x[x sold the car])

b. Be_v the CAR(the x[John sold x])

The two propositions are made true by different actual facts, as follows from the definition of 'actual fact' given above. Sentence (6.13a) is made true by the fact that the value on the parameter 'who sold the car' was John; (6.13b) is made true by the fact that the value on the parameter 'what did John sell' was the car. These two facts are co-extensive in the sense that whenever the one is a fact so is the other. And since our innate truth theory 'knows' that co-extensive facts are substitutable *salva veritate*, *cognitive* intensional predicates such as *believe* are insensitive to the difference as that between (6.12a) and (6.12b): the proposition underlying the one is freely substitutable with the proposition underlying the other. But this does not hold for *emotional* intensional predicates such as *be surprised* or *be angry*. Consequently, emotional intensional predicates are to be defined as extending over pairs of *cognizing subjects* on the one hand and *actual facts* on the other. Now the guarantee of free substitutability is gone, because there is no guarantee that an emotional attitude with regard to one of a pair of co-extensive facts remains the same with regard to the other member of the pair.

This leaves the question of why for some predicates L-propositional ρ-values do not block SSV but freely allow for substitution *salva veritate* of co-referring terms. We are thinking in particular of the abstract operator-predicates of propositional logic discussed in Section 2.3.4 in Volume II.

To answer this question we must first have a renewed look at the notion of *valuation space* (VS) of an L-proposition, which we can now derive from the notion of virtual fact. As we have seen, every proposition *p*, and every L-proposition P of the form $F(a_1, \ldots, a_n)$ expressing *p*, automatically create a virtual fact ***Fact***$[P(F)(\rho(a_1), \ldots, \rho(a_n))]$. This virtual fact can now be seen as defining the set of possible situations $/F(a_1, \ldots, a_n)/$—that is, those possible situations that have the virtual fact ***Fact***$[P(F)(\rho(a_1), \ldots, \rho(a_n))]$ in common, or those possible situations that are defined for the condition that $<\rho(a_1), \ldots, \rho(a_n)> \in [\![F]\!]$, though, apart from that, the situations in question may differ in innumerable ways. It is this set $/F(a_1, \ldots, a_n)/$ of possible situations that we have called the valuation space of the L-proposition $F(a_1, \ldots, a_n)$.[6]

One should keep in mind that the valuation space of an L-proposition is defined MODULO KEY (that is, under a referential key K_i, as defined in Section 2.3.3 in Volume II). From this it follows that valuation spaces obey

[6] Zalta's notion of situation strongly resembles ours (1988: 61): '[A] situation ... is defined by the propositions that are true in that situation'. Unfortunately, he still finds it necessary to index situations for possible worlds.

the Principle of Substitutivity: /F(a)/ = /F(b)/ when $\rho(a) = \rho(b)$. Given the two sentences:

(6.14) a. The morning star is inhabited.
b. The evening star is inhabited.

and given the fact that the two nominal terms *the morning star* and *the evening star* refer to the same actual object—the planet Venus—it follows that /(6.14a)/ = /(6.14b)/, since there is only one set of possible situations characterized by the condition that the planet Venus is inhabited.

We posit, finally, that the operators of propositional logic, *not*, *and*, and *or*, are predicates which take L-propositional ρ-values for their terms: they are taken to say something about L-propositions, yet in such a way that SSV is saved. Still taking predicates as functions from ρ-values to truth-values, we can now say that the predicates of propositional logic are functions from the L-propositions expressing their underlying propositions to truth-values. At this point we simply present the definitions of negation ($\neg$), conjunction ($\wedge$) and disjunction ($\vee$), which are repeated and further discussed in Section 2.3.4 in Volume II ('P^+' stands for any *n*-tuple of L-propositions):

(6.15) a. $[\![\neg]\!] = \{ P \mid s_{act} \in /\overline{P}/\}$
(the extension of $\neg$ is the set of all L-propositions P such that the actual situation s_{act} is a member of the complement of /P/)

b. $[\![\wedge]\!] = \{ P^+ \mid s_{act} \in (\bigcap_{/P/^+})\}$
(the extension of $\wedge$ is the set of all sets of two or more L-propositions P such that s_{act} is a member of the intersection of all $/P/^+$)

c. $[\![\vee]\!] = \{ P^+ \mid s_{act} \in (\bigcup_{/P/^+})\}$
(the extension of $\vee$ is the set of all sets of two or more L-propositions P such that s_{act} is a member of the union of all $/P/^+$)

One notes that although the *propositions* underlying (6.14a) and (6.14b) above may be different, depending on whether the terms *the morning star* and *the evening star* are represented by the same address or by different addresses, their *valuation spaces* are identical, since their subject terms have a common key, the planet Venus, and valuation spaces are defined modulo key, as stated above.

Given the definition of negation in (6.15a), it follows that SSV applies in negated sentences the same way it does in non-negated sentences. This is so because the satisfaction condition of $\neg$ is not sensitive to differences in propositions but only to differences in corresponding valuation spaces, and valuation spaces remain constant under substitution of co-referring terms, owing to the modulo-key condition. Therefore, since /(6.14a)/ = /(6.14b)/, it follows that $\overline{/(6.14a)/} = \overline{/(6.14b)/}$. Now the condition that sit_{act} is an element in $\overline{/(6.14a)/}$ differs in no way from the condition that sit_{act} is an element in $\overline{/(6.14b)/}$. Analogously for the operators of conjunction and disjunction.

This explains why SSV is saved in logical compositions. One notes that in these cases SSV is saved not by the nature of the ρ-values concerned but by the fact that the satisfaction conditions associated with the logical operators are defined in terms of valuation spaces, which are invariant under substitution of co-referring terms.

SSV is likewise freely allowed under predicates such as *true*, *false*, *probable.* Consider the semantic definition of *true*, when applied to L-propositions (as in (3.18) in Chapter 3, repeated here as (6.16)):

(6.16) $[\![\text{True}]\!] = \{ \text{F}\ (a_1,\ldots,a_n) | < \rho(a_1),\ldots, \rho(a_n) > \in [\![\text{F}]\!] \}$
(The extension of the predicate *true* is the set of well-anchored and well-keyed token occurrences of any L-proposition F $(a_1, \ldots, a_n)$ such that the *n*-tuple of objects referred to by $a_1, \ldots, a_n$ is an element in the extension of the predicate F.)

The predicate *true* also occurs as a predicate over a nominal subject term, as in *The sentence 'John is ill' is true*, or *'John is ill' is true*, where *'John is ill'* stands for the underlying L-proposition, in which case definition (3.18) in Chapter 3 applies. But what we wish to consider here is sentences of the form *That John is ill is true*, or *It is true that John is ill*—that is, sentences with a genuine embedded S-term, not an S-term that is encapsulated in an NP-frame. In these cases, SSV applies freely within the S-term, since what matters is that the *n*-tuple of ρ-values $< \rho(a_1), \ldots, \rho(a_n) >$ be an element in $[\![\text{F}]\!]$, regardless of what linguistic means are used to 'get at' them. That being so, when $\rho(a_i) = \rho(b_i)$ under the predicate F, the satisfaction condition of *true* is automatically satisfied by $\text{F}(a_1, \ldots, b_i, \ldots, a_n)$ whenever it is satisfied by $\text{F}(a_1, \ldots, a_i, \ldots a_n)$ and vice versa. This makes both the S-terms as a whole and any co-referring terms within them freely substitutable *salva veritate.*

7

The ecology and machinery of language

7.1 The ecology of mind, language, and world

7.1.1 *De Saussure's speech circuit*

The functioning of language in speech has often been the subject of analysis during the twentieth century. There is, first, De Saussure's famous *speech circuit* (circuit de la parole), which sees the use of language as a circular flow from speaker to hearer and back, sounds being produced and received on both sides, passed through the mind, where they are processed and associated with thoughts and concepts, which leads to new sound productions, and so on. Figures 7.1a and 7.1b reproduce De Saussure's own renderings of the speech circuit (De Saussure 1916: 27–8).

What De Saussure failed to take into account is the fact that all use of language is REFERENTIAL or INTENTIONAL: speech is only speech if it is ABOUT something, which is either in the actual world as it exists today, complete with its history, or in a virtual world or situation that has been thought up, one way or another, by the individuals that produce and understand speech. Figure 7.2 is like Figure 7.1a but it incorporates the 'aboutness' or referential aspect of speech, without which any analysis of the speech process is basically flawed.

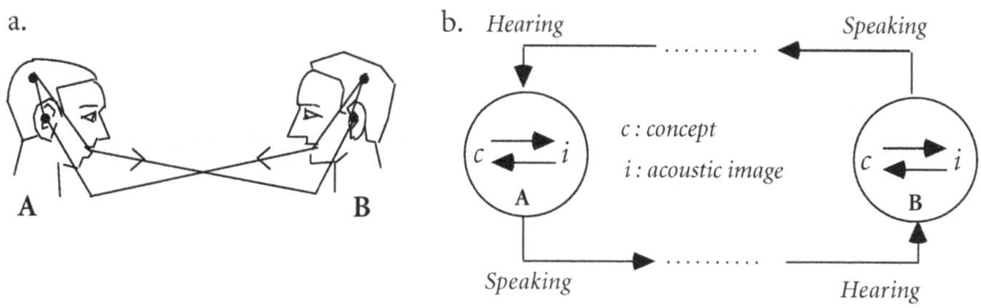

FIGURE 7.1 De Saussure's speech circuit without 'world'
Source: De Saussure 1916: 27–8.

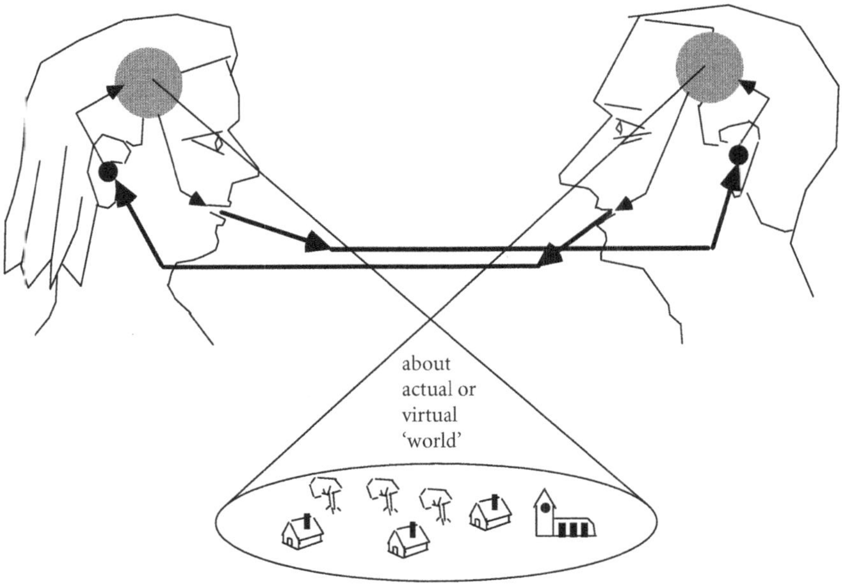

FIGURE 7.2 The speech circuit with language, mind and 'world'

It shows how the speaker and listener's minds are intentionally focused on an actual or virtual situation or set of situations, which forms the reference domain of their utterances.

7.1.2 *Ogden and Richards' semiotic triangle*

The speech circuit depicted in Figure 7.2 has a parallel in Ogden and Richard's famous semiotic triangle, reproduced in Figure 7.3 (Ogden and Richards 1923: 11). Here, speaker and listener have been collapsed into what these authors call 'thought or reference' but is preferably called 'mind' nowadays. Their term 'symbol' stands for the linguistic utterance, and 'referent' stands for what we call 'world' or 'situation' spoken about. The symbol symbolizes a given thought, either top-down, in the production of an utterance, or bottom-up, in its comprehension. Either way, the process is considered *causal*, in that the mind causes an utterance to come about and the utterance causally produces, among other things, a mental representation of what the utterance describes. This symbolization is subject to criteria of CORRECTNESS, or, in modern terminology, grammaticality on a reading.

The 'thought' is intentionally related to whatever it is about, called the 'referent' by Ogden and Richards. This relation is again considered causal, even though no causal account of the aboutness relation had been presented in

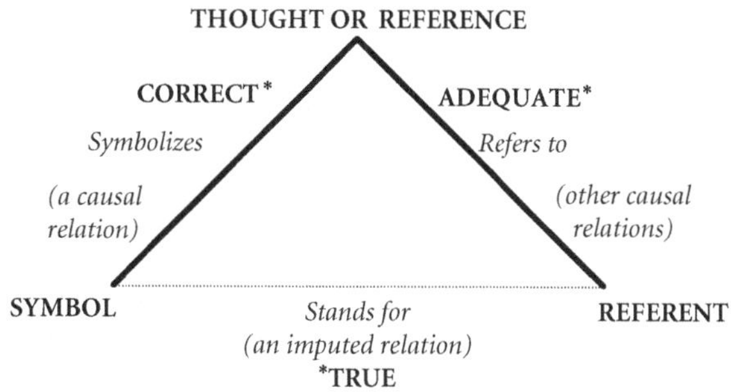

FIGURE 7.3 Ogden and Richards' semiotic triangle

their day—a state of affairs that has not changed since. The authors regard the relation between a thought and the object or situation it is about as subject to a criterion of ADEQUACY. This is, however, nothing but a terminological expedient designed to avoid having to speak of truth and falsity, which Ogden and Richards consider to be properties of speech utterances, not of thoughts. Truth and falsity are called upon by these authors to do duty on the base line of the triangle, where one finds the relation between utterance and 'world' (not a word about virtual situations). This truth relation is not treated as being causal, but is called 'imputed', by which the authors must have meant 'mediated by the thought component'. The fact that any relation between utterance and 'world' is mediated by the mind is also apparent from Figure 7.2, where no direct connection is shown between the speech sounds and any possible situation, and where any relation with a (possible) situation is represented as holding between it and the mind.

In our view, it is more appropriate to speak of TRUTH, and not of ADEQUACY, as a criterion for the relation between thought and (possible) situations, and to disregard the indirect or 'imputed' relation between utterance and 'world' altogether. The question that presents itself immediately when one sees Ogden and Richards' 'imputed' truth relation is: imputed by whom? The only empirically correct answer, and surely the answer Ogden and Richards had in mind, must be that the imputing is done by and in the speakers' and listeners' minds. But if that is so, the base line had better be scrapped, since the 'imputing' is already expressed in the analysis by the relations between mind and utterance on the one hand and between mind and 'world' on the other.

A further, compelling, reason for scrapping Ogden and Richards' 'imputed' truth relation between utterance and world is the following. The human mind is not at all like a truth-value-assigning algorithm, a rubber-stamp machine that looks at the utterance and at the world and then says 'true' or 'false'. On the contrary, the processing of linguistic structures by and in the mind requires all kinds of 'external' cognitive input, not contained in the linguistic structures themselves. To eliminate the nonlinguistic compartments of the mind from the truth relation between language and world is, therefore, empirically inadequate. The 'imputed' relation may be useful in mathematical model theory, which analyses the relation between a formally defined language and a formally defined 'world'. But in an analysis of the ecological environment and the internal structure of the mental machinery that processes linguistic structures, the mind as a comprehensive cognitive system cannot be dispensed with.

7.1.3 *The semiotic architecture in a modern perspective*

Both De Saussure's speech circuit and Ogden and Richards' semiotic triangle lack detail. They are almost entirely promissory in that they bank on future developments that will hopefully reveal the details of the mechanisms they anticipate. Meanwhile, the best part of a century has passed, and it has not passed without showing results in many areas, while, of course, many uncertainties still remain.

In 1989 Levelt's influential book *Speaking* appeared, which deals with the many questions related to the speaking process. Levelt shares the methodological premises of the present author, in particular the emphasis on natural cognition, the position of cognitive realism as non-hardware realism, while hardware (neurophysiological) realism should be pressed for wherever possible—in short, the position of weak reductionism (see Section 1.3). He also subscribes to the ecological point of view upheld throughout the present book:

> Many aspects of a speaker's information processing cannot be correctly evaluated if we lose sight of the canonical ecological context of talking: the speaker's participation in conversation. (Levelt 1989: 2)

Levelt's book is of particular importance because it incorporates what were, in 1989, the latest insights into the processes involved in the activity of speaking. On page 9 he presents what he calls 'a blueprint for the speaker', though it encompasses also a schematic 'blueprint' for the listener for the purpose of self-monitoring (Levelt 1989: 13). Levelt's model is reproduced here as Figure 7.4.

There are, of course, some differences in terminology and in degree of detail. For example, Levelt's 'preverbal message' corresponds to our 'intent' consisting of a commitment operator and a proposition (Levelt 1989: 73). But the attentive reader will easily detect the basic similarity of Levelt's blueprint

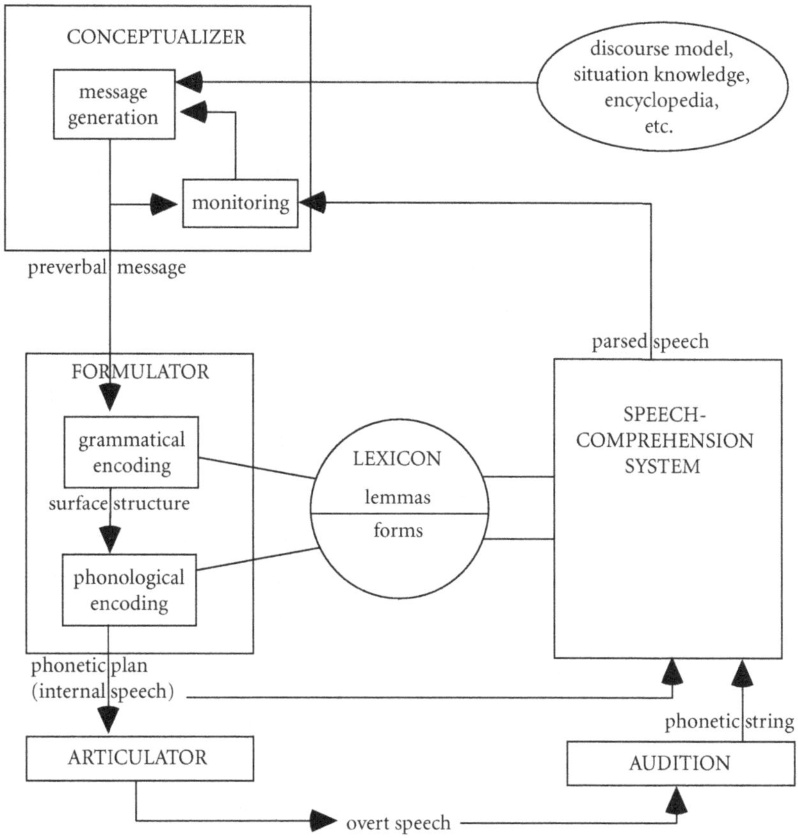

FIGURE 7.4 Levelt's (1989) 'Blueprint for the Speaker'

Note: Boxes represent processing components; circle and ellipse represent knowledge stores.

with my attempt, shown in Figures 7.5 and 7.6, at setting up a flowchart for the production and interpretation of linguistic utterances in terms of a 'semiotic architecture', in the light of up-to-date theoretical considerations and, as far as possible, experimental results.

So as to avoid symbol crowding, I have set up separate models for production and comprehension/interpretation, which has the additional advantage of showing more clearly the differences between the processes involved in the comprehension and interpretation of linguistic utterances on the one hand and the machinery involved in production on the other (Levelt 1989, 1999). Production is strongly deterministic in that the form of an actual utterance is to a large extent determined by a given intent as input and the specific language, with its lexicon, grammar, and phonology, in which the intent is to be

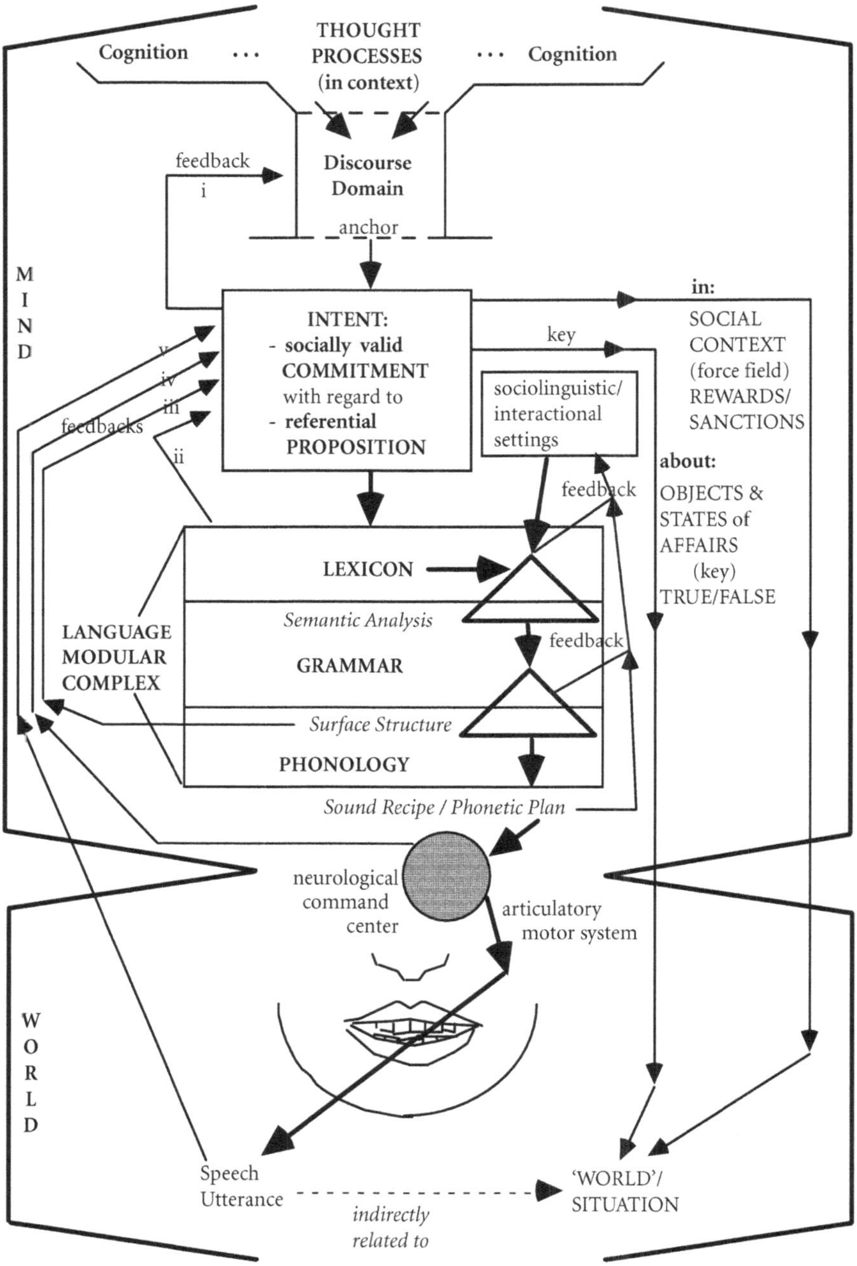

FIGURE 7.5 The semiotic architecture: production model

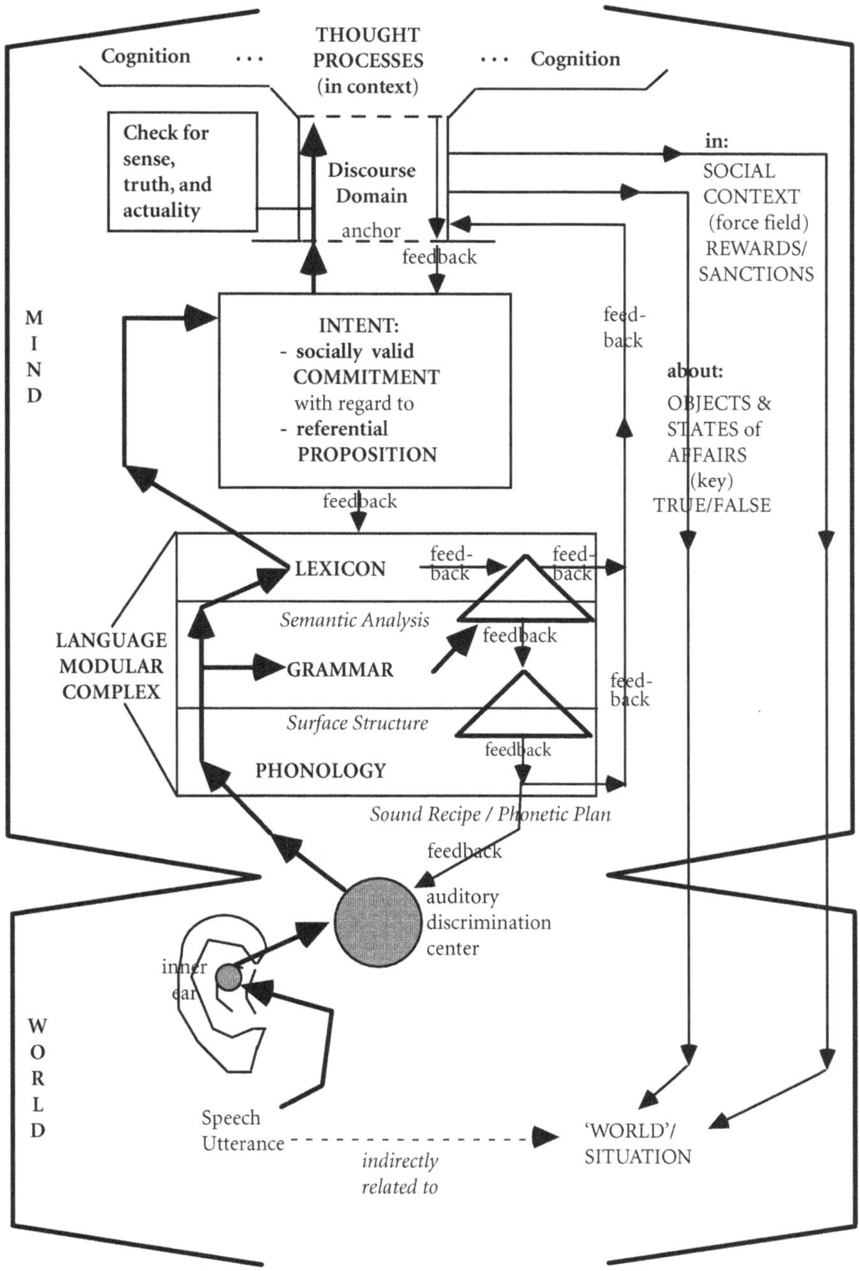

FIGURE 7.6 The semiotic architecture: tentative comprehension/interpretation model

expressed. Moreover, given the fact that speakers always calibrate their lexical choices, their grammar, and their phonology with regard to the sociolinguistic and interactional setting in which the utterance-to-be is intended to land (Seuren 1982), it must be assumed that there is a system of sociolinguistic and interactional 'settings' guiding lexical choices, grammatical constructions, and phonetic realizations.

In Section 7.2.1 it is argued that once the main lexical items have been selected, the further processing by the grammar is MODULAR in the sense that the processes involved are domain-specific, rapid, automatic, and inaccessible to any form of consciousness or awareness. Once the lexical items have been selected, the grammar module is set into motion yielding a surface structure. Kempen and Hoenkamp (1987) argue that this process allows for an 'incremental' implementation in the sense of parallel, stepwise processing of subsequent sentences realized as token utterances. This must be correct, if only because everyone is familiar with the phenomenon of speakers not properly finishing a running sentence and starting a new sentence before the previous sentence has been brought to an end. This is only possible if the new sentence is already being planned at higher levels before the running sentence is brought to an end at the lower levels of actual realization.

There are, however, a number of interfaces where feedback control is possible (see Section 7.2.1). One important typical feature of feedback-control mechanisms is the fact that it is relatively easy to raise them to some level of consciousness or awareness—an achievement that is, in principle, impossible with regard to modular systems. When and to the extent that awareness is raised, the time span involved in the production process becomes longer.

It is assumed that the first stage at which feedback is possible is where the intent has been put together: speakers can take their time to ensure that what they are going to say is precisely what they want to say. The next possible feedback stage is after the selection of lexical items, since any lexical selection is open to immediate, possibly conscious, control, which includes the given sociolinguistic and interactional settings. It would seem that the speaker can monitor again at surface-structure output, via feedback control, to see whether the morphosyntactic plan for further realization corresponds to the intended cognitive input. The subsequent phonology is a further module leading to a plan for phonetic realization, which is, again, subject to possible feedback control so as to ensure that the original intent will indeed be adequately expressed when the phonetic plan is realized. Meanwhile all outputs concerned may be checked by feedback routes for sociolinguistic and/or interactional appropriateness. Under normal conditions, the final physically realized acoustic output is once again fed back and controlled for semantic adequacy and sociolinguistic and/or interactional appropriateness. At that final stage, however, any correction, or, in current terminology, any repair, has become a public affair.

In Figure 7.5, the heavy lines from intent to speech utterance indicate the direct path along which an intent is realized as an utterance if all lexical selections are made and all the rules of the grammar and the phonology are followed without any feedback controls. More often than not, however, actual practice is different, quite apart from any feedback routes. It appears that competent speakers tend to make heavy use of ready-made phrases and other complex expressions which are, apparently, kept in store, either in the lexicon or in a less permanent compartment kept especially for the purposes of a given discourse, and which are called upon for the expression of often larger than lexical units, including idioms.

The upward pointing thin lines on the left represent the feedback controls assumed to be operative from intents back to the discourse domain and from the lexicon, the surface structure, the phonetic plan, and the final acoustic output again back to the intent. (They have been numbered from i to v according to the five interfaces listed at the end of Section 7.2.1.) The upward pointing thin lines on the right stand for feedback relations with regard to the sociolinguistic and/or interactional settings for the utterance in question. The downward pointing thin lines on the right show the double intentional relation between 'mind' and 'world': (a) the speech-act relation of social commitment in a force field, with its sanctions and rewards, and (b) the referential propositional relation in virtue of which the mind focuses on a given situation and given objects in a relation of truth or falsity.

By contrast, the processing involved in the comprehension of utterances appears to be far less deterministic and much more dependent on feedback processes driven by hypotheses regarding the likely intent underlying the speaker's utterance. There can hardly be any doubt that this largely inductive process is supported not only by the strict boundaries set by the grammar and lexicon of the language concerned, but also by general cognition as well as by the discourse domain that has been built up in interaction between speaker and listener and is thus their common property. The discourse-driven topic–comment structure of utterances, with its highly structured question–answer game (see Section 3.2 and Chapter 11 of Volume II), must take a heavy burden of explanation.

An overarching problem in the study of linguistic comprehension is posed by the fact that comprehension is, in most cases, extremely fast, not accessed by introspection and immediately correct, whereas the multiple feedbacks and inductive guesses that most comprehension models, including Figure 7.6, assume to be involved would make one expect comprehension to be slow, insecure, and much more open to introspection. It is true that sometimes comprehension is indeed slow. It may take a listener from full seconds to full minutes to hit upon the intended meaning, during which time the comprehension machinery is in top gear. But these cases are rare in comparison with what normally occurs, when comprehension is close to immediate.

In Figure 7.6, we assume, first, a direct and largely deterministic path, indicated by heavy lines, from physical sound via auditory and cerebral discrimination to the phonology and the lexicon, whereby a fair amount of grammatical, especially morphological (see below), structure may also be taken to be reconstructed, perhaps supported by statistical frequency measures (Cutler and Clifton 1999). In most cases, however, this is too little to make for interpretative success. On grounds of a provisional determination of the main lexical items, a grammatical structure is reconstructed and fed into the phonology for comparison with the original input and thus for confirmation or disconfirmation. At the same time, both the phonology and the lexicon establish a direct feedback with the discourse domain at hand to see if expectations are confirmed or disconfirmed. Given a tentative interpretation, the discourse domain automatically activates the production machine, again to see if the result matches the original acoustic input and thus if the hypothetical interpretation can be confirmed or must be disconfirmed.

It would seem, however, that the rich machinery which listeners have at their disposal need not always be put to full use. We may assume that the parsing and comprehension problem is facilitated in a number of ways. For one thing, there is the important fact (unrecognized in the parsing and the comprehension literature) that, *in cases of a question–answer-driven topic–comment structure, all that is needed for comprehension and interpretation is an identification of the comment—the topic being already available.* In the vast majority of cases, therefore, the main task of the interpretation machinery consists in the identification of the comment, usually one single word or constituent, taken from a range of possibilities that is already strongly limited by contextual factors. In this perspective, the fact that comments are always strongly accented, while topics stay under a weak intonational pattern, makes good functional sense. This is why the topic–comment structure of utterances is said to bear a large part of the burden of explanation.

Moreover, some authors (Sanford and Garrod 1998; Sanford and Sturt 2002; Ferreira and Patson 2007) defend a 'quick-and-dirty' approach to comprehension. According to these authors, full and detailed reconstruction of the incoming message is not necessary for adequate comprehension, since a good contextual fit ('scenario mapping') will suffice, even if some grammatical details are left unprocessed or are ignored. This point of view seems to me to be entirely plausible.

A strongly related factor, which, in addition, constitutes an argument for an analysis-by-synthesis theory of utterance comprehension of the kind proposed here, consists in the fact that listeners often *anticipate* or *predict* the way an utterance started by the speaker is going to end. This is not only an everyday experience,[1] it has also been confirmed experimentally. Using magnetoencephalography (MEG), Helenius *et al.* (1998) found strong activation in the left superior temporal cortex when subjects were reading sentences with a semantically inappropriate final word. Similar findings are reported in Wicha *et al.* (2004) on the basis of event-related potential (ERP) data (see also Federmeier 2007). Van Berkum *et al.* (2005) took sentences whose final word or phrase was positively expected. They showed that unexpected utterance endings systematically provoke an ERP effect the moment flectional morphology begins to exclude the expected utterance ending. Similarly, in self-paced reading experiments, unexpected sentence endings slow down the reading process—again an effect that sets in the moment morphology begins to be inconsistent with the expected sequel. The expectations entertained by the listener may be based on anything at all:

> [W]e suggest that prediction about how an utterance will unfold can draw upon information from any relevant interpretive domain—the prior discourse, a scene, a much earlier conversation, general world knowledge, cospeech gestures, inferred characteristics of the speaker, and so forth—as long as this information is made relevant or recruited by locally unfolding constraints. . . . That is, any context made relevant by the currently unfolding sentence will do. (Van Berkum *et al.* 2005: 462)

Pickering and Garrod (2007) agree that comprehension is to a large extent a matter of analysis-by-synthesis (see also Townsend and Bever 2001, discussed at the end of Section 7.2.3). The model they describe shows a remarkable similarity to our Figure 7.6. They write:

> We present the case that language comprehension involves making simultaneous predictions at different linguistic levels and that these predictions are generated by the language production system. Recent research suggests that ease of comprehending predictable elements is due to prediction rather than facilitated integration, and that comprehension is accompanied by covert imitation. We argue that comprehenders

[1] It is a common experience, in ordinary conversation, to see that speaker and listener finish an utterance simultaneously. Consider the case of a man accused of stealing money from an elderly lady. During his interrogation at the police station, his and the police officer's speech may naturally overlap in something like the following way:

> Suspect: That may be true, but I never stole anything from her.
>
> Officer: . . . you never stole anything from her. I know, I know . . .

use prediction and imitation to construct an 'emulator', using the production system, and combine predictions with the input dynamically. Such a process helps to explain the rapidity of comprehension and the robust interpretation of ambiguous or noisy input. This framework is in line with a general trend in cognitive science to incorporate action systems into perceptual systems and has broad implications for understanding the links between language production and comprehension. (Pickering and Garrod 2007: 105)

(For further comment on the 'analysis-by-synthesis' hypothesis and the 'imitation' component of comprehension, see Section 7.2.3.)

A further simplification of, in particular, the parsing problem is provided by the fact that nominal and verbal flectional morphology makes for automatic and fully determined parsing. In fact, this may well be a general *raison d'être* for morphology in natural language. Consider the highly morphological language Turkish, where one finds sentences like:

(7.1)	Bu	kitab - ı	yaz -	dır -	a -	ma -	malı -	y -	dı.
	this	book - ACC	write -	CAUS -	POSS -	NEG -	NECESS -	COP -	PAST

(It should not have been possible to get someone to write this book.)

Here, the left-branching verb form *yaz-dır-a-ma-malı-y-dı*, where the operator scopes descend from right to left, each operator taking its left-hand side neighbour as its scope, immediately reduces to a chunk of SA tree structure where all elements unambiguously find their place in virtue of a relatively simple algorithm (Seuren 1996: 350; Pembeci *et al.* 1998). Since it has been found (Aksu-Koç and Slobin 1985; Topbaş *et al.* 1997) that, in spite of the highly complex nominal and verbal morphological processes in their language, Turkish children acquire these processes rapidly and easily, it appears that native acquisition benefits from the modularity of the processes involved—a most gratifying agreement of theory, data, and computational modelling. I am not aware of any data on morphological processing by adult native speakers of Turkish, but one would expect an analogous ease and rapidity.

Be all this as it may, once all the input checking and feedback controls have been gone through, the listener settles for a final interpretation in the form of a reconstructed speaker's intent, which is integrated into the listener's existing discourse domain and classified as 'speaker's intent'. Yet even after this point, intensive double-checking may go on for quite a while—for so long even that the process opens up to awareness.[2]

[2] Neuroimaging techniques have recently yielded interesting results with regard to the brain processes corresponding to the hearer's reconstruction of the speaker's intent. Hagoort *et al.* (2004) conducted experiments at the F. C. Donders Centre for Cognitive Neuroimaging at Nijmegen, Netherlands, which showed that whenever there is, without discourse-conditioned preparation, a

This is as much as can be said in the present context. It is clear, however, that much further research will have to be done, not only theoretically but also experimentally, in particular in the area of neuroimaging, for greater clarity and certainty to be attained.

7.2 Is there a self-contained modular complex for language?

7.2.1 *Claim: speech production is modular*

Given the deterministic character of utterance production, we take it that production is channelled through a MODULAR COMPLEX consisting of at least (we shall be more specific below) a GRAMMAR MODULE, fed by general cognition (in a propositional format) and by the lexicon, and a PHONOLOGY MODULE, fed by the grammar module.

Let us first consider the notion of module, which was introduced into the literature by Jerry Fodor in his influential 1983 book *The Modularity of Mind.* Yet we do not take over Fodor's notion lock, stock, and barrel. Instead, we try to reduce the notion of module to its essentials, taking into account, as much as possible, the latest data on the genesis and neurophysiology of the human brain. Given these restrictions, we say that modularity implies (a) that the neural structures involved are domain-specific, and (b) that the machinery is informationally encapsulated.

Domain-specificity is taken to imply that the neural structures in question are specifically fit for the tasks they are meant to perform. In Levelt's words (Levelt 1989: 14–15): 'a processing component is a *specialist*'. Although this

semantico-pragmatic mismatch, as in *Dutch trains are sour*, or an obvious falsity, as in *Dutch trains are white*, (everyone in the Netherlands knows that Dutch trains are predominantly yellow), the brain responds in a basically comparable way: both interpretative problems elicit the so-called N400 effect in brain potentials (although the effect is somewhat stronger for semantic mismatches) and both engage the left inferior frontal gyrus (LIFG) of the human brain. In a later paper, Hagoort and Van Berkum conclude:

> ... the findings consistently point to a one-step model of language interpretation. Not only core linguistic information about phonology, syntax and semantics of single words and sentences, but also discourse information, world knowledge and non-linguistic context information immediately conspire in determining the interpretation of compound expressions. Language input seems to be mapped onto a discourse model that takes communicative acts, including eye gaze, iconic gestures, smiles and pointing, into consideration (Clark 1996). (Hagoort and Van Berkum 2007: 808)

The results show that the N400 responses in question occur the very moment a mismatch of the sorts described has actually occurred. As Hagoort and Van Berkum emphasize, this rules out any strictly two-step model in which sentence comprehension temporally precedes any further contextual and/or situational integration. In terms of the semiotic model for comprehension/interpretation shown in Figure 7.6, which is not a two-step model, this would mean that N400 response is closely connected with the OSTA Principle (optimization of sense, truth, and actuality) introduced in Section 3.4.2, repeatedly referred to in this and other chapters. Right now, however, it is too early to say anything more definite about these and related matters.

must have hardware implications, it seems that domain-specific functions can be taken over by other brain areas or brain paths in cases of severe damage or otherwise inflicted severe dysfunction. Training may thus, up to a point, compensate for incurred losses.[3]

Informational encapsulation implies that the modular machinery is screened off not only from introspection but also from any kind of external interference. A module has an input, cast in a specific format, and an output, again in a specific format, and between these two there is an automatic, autonomous, algorithmic processing machinery producing the output from the input. The module, so to speak, 'prints out' any well-formed input, and it does so at a speed that approximates zero time.

That the brain must be taken to be endowed with a probably large number of sometimes quite small *cognitive modules* (for the term see Fodor 1983: 37) seems obvious: 'the brain in many ways appears to carry hallmarks of specialization' (Marcus 2006: 444). Yet there is a question as to the hardware physical substrate of cognitive modules, in particular the relation between genes and their 'expression' in high-level cognitive modules. The concept of module, as defined above, seems clear enough. Yet the reduction of theoretically postulated modules to actual brain structures turns out to be problematic. According to Fodor (1983: 98–9), each cognitive module must be taken to be a unique piece of hardware brain machinery, genetically predisposed to develop into a module for its predefined specific purpose, such as the use of language, given sufficient external stimuli. Thus defined, a module must have a precisely defined location in the brain. It is now clear, however, that such a view of modularity is in need of some further comment.

First, modularity does not necessarily imply one separate and well-defined local area in the brain specifically reserved for the modular function involved. One may imagine a modular structure that is dispersed over one or more related brain areas that serve other purposes as well. In such a case one may see the modular structure as a special facilitation of the function involved, more or less in the way cycling is facilitated by the existence of special bicycle paths, either alongside main roads or independently. These paths are specifically

[3] Marcus (2004) confirms this general point, but fails to provide hardware specifics:

To an earlier generation of scholars, the evidence for innateness and the evidence for flexibility seemed almost irreconcilable. Most scholars simply focused their attention on the stream of evidence that they were more impressed with. Nativists gathered examples of what a child could do without the benefit of experience; empiricists gathered examples of how much the basic structure of the brain could change in response to challenges from the environment. Both sides have their point. The brain is capable of awesome feats of self-organization—and equally impressive feats of experience-driven reorganization. But the seeming tension between the two is more apparent than real: Self-organization and reorganization are two sides of the same coin, each the product of the staggering power of coordinated suites of autonomous yet highly communicative genes. (Marcus 2004: 147–8)

reserved for cyclists and may, in fact, be built in such a way that no other traffic can enter them and that any cyclist must reach an outlet to get out of the bicycle-path system. In that sense, bicycle paths can be termed 'modular'. Yet such a 'module' for cyclists need not be restricted to one specific part of the countryside where no other roads would be allowed. On the contrary, it may co-exist with any other road system and any particular location may be reachable by a variety of road-types. But be all this as it may, the truth is that we know far too little, as yet, about these matters to be in a position to put our foot down and make definite claims. Any claim we do make must be absolutely minimal.

Moreover, it seems that ancestral structures are often reduplicated in phylogenetic development, whereby the function of the new copy is modified for some special range of purposes (Marcus 2004: 112–24; 2006: 450–1). This would help to explain the well-known fact that distinct modules often show remarkable analogies and that a genetic defect may be expressed in distinct modules in analogous ways.[4]

Then, one must assume that a given module may need input from one or more other modules to be able to work. We may thus have a network of modules working together to produce an output. In such a case we speak of a *modular complex*. When a modular complex is serial in that the output of one module forms the input of the next module (as is probably the case with grammar and its subsequent modules), we may speak of a *modular cascade*.

It is also possible, however, for a piece of machinery to need input provided by general cognition over and above input from other modules. For example, as argued in Section 3.4, the proper interpretation of linguistic utterances appears to depend crucially on nonlinguistic information drawn from general cognition, such as information on probable truth. I argue that, while speech *production* is, in principle, a modular cascade involving the grammar and subsequent modules on the way to actual sound or writing, speech *interpretation* is a more complex process requiring recourse not only to the grammar and subsequent modules but also, at many points, to central cognition. In such a situation, it is no longer appropriate to speak of a modular complex *tout court*, precisely because general cognition may, and usually does, enter the modular complex at almost any stage. For such cases, we reserve the term *weak modular complex*, as opposed to *strong modular complex*, which involves only modules, without any intermediate assistance from general cognition.

[4] Marcus claims (2006: 450): 'Although modules are, by definition . . . computationally distinct, they need not be genetically unrelated.' Yet the link between abstract computational modules and neural hardware is in no way clarified.

Taking these considerations into account, we assume that, in principle, speech production takes place through a top-down strong modular complex—that is, it takes its input from central cognition and the lexicon, and produces an output in the form of a sound (or writing) specification and hence in actual sound or writing. By contrast, we surmise that speech interpretation occurs in terms of a weak modular complex, as it requires input from general cognition—that is, from encyclopedic and contextual knowledge—throughout. More is said about this in Section 7.2.3.

The function of the lexicon is, though obviously of extreme importance, less clearly delimitable, given the present state of our knowledge. What is clear is that the lexicon involves extremely rapid search, retrieval, and evaluation procedures, which can only be modular since they are domain-specific, automatic, rapid, and totally closed to introspection or any other form of awareness. (It is still unknown, for example, according to what system the masses of lexical items that speakers have at their disposal—some tens of thousands on average—are stored and accessed in their memories. The alphabet is an unlikely storing principle, but what is?) On the one hand, therefore, introspection is kept at bay in processes involving the lexicon. On the other, however, any lexical selection may be scrutinized intensively for appropriateness, along a variery of parameters, in the given context.

In utterance production, all introspection can do, as far as lexical selection is concerned, is mentally 'hold up' a possible lexical choice and evaluate its expected effect on the listener/reader according to semantic, stylistic, and pragmatic parameters. The item (or idiom) selected may then be accepted or rejected. In the latter case a more appropriate item (or idiom) is searched for. In utterance interpretation, lexical items are identified and again evaluated for semantic, stylistic, and pragmatic appropriateness. When the item appears to make no sense, the OSTA Principle (see Section 3.4.2) forces the listener to try alternatives that still seem to fit the phonetic input. When no proper fit occurs, the interpretation process runs into perplexity. In order to understand how these procedures work, it is probably necessary to postulate a weak modular complex, involving a spread of input from general cognition and various modules. But there is, on the whole, little room for greater specificity.[5]

[5] Compare Bierwisch and Schreuder (1992: 24):

Whatever the modular nature of the pertinent mental system and the eventual interaction of information they provide might be, the choice of lexical items ultimately reflects a wide range of different, but somehow integrated conditions determining the truth, appropriateness and communicative effect of linguistic utterances. (Bierwisch and Schreuder 1992: 24)

The grammar proper appears to be the simplest of the modules involved (see Section 7.2.2). As far as can be judged, it seems that grammars simply act upon a well-defined semantic input and churn out corresponding surface structures, whereby stepwise, partly parallel processing of successive utterances is taken to be possible (Kempen and Hoenkamp 1987). The inner workings of the phonology module are left undiscussed here.

We also take it that it is the top-down modular complex of grammar and phonology that defines native competence in any specific language or language variety. We take this view because it is clear that any specific language or language variety is defined by competent *production*, not by adequate *interpretation*: adequate interpretation is a necessary but not a sufficient condition for full linguistic competence. As is argued in Section 1.3.3.1 (see also Seuren 2004*b*: 616), to see if a person is a true speaker of English one cannot rely on this person's ability to understand and interpret a, possibly garbled, utterance such as, for example, (7.2a), even though competent speakers of English will have no difficulty understanding it. By contrast, the true test of native command lies in production. A speaker who produces, for example, (7.2b), complete with an authentic pronunciation, is naturally taken to possess a native or near-native command of some known variety of English:

(7.2) a. Me not water in. Me no can swim!

b. Please don't push me into the water. I can't swim!

The fact that a language or language variety is defined by competent production and not by competent interpretation is an important point, which is, on the whole, not appreciated in the community of linguists.

It is assumed that the top-down linguistic modular complex receives its input from a mental compartment that is directly anchored to the pre-existing discourse domain and contains—in a form that has so far remained elusive—an *intent* consisting of a *commitment*, *appeal*, or *rule setting* meant to be socially valid (see Chapter 4), and a *proposition* with regard to which the commitment/appeal/rule-setting becomes valid the moment the intent is realized as actual speech. The combined proposition and commitment/appeal/rule-setting are taken to be cast into the form of a linguistic structure, the *semantic analysis* or SA, containing at least one speech-act operator and at least one L-proposition. The SA underlies the *surface structure* of a sentence, which is to be realized as an actual utterance. SAs are taken to be structures in a *logical language* L_L, which comes close to the language of standard modern predicate calculus. (But note that the similarity to the language of modern predicate calculus is restricted to its *syntax* and excludes its *semantics*. The semantics we have for L_L differs radically from

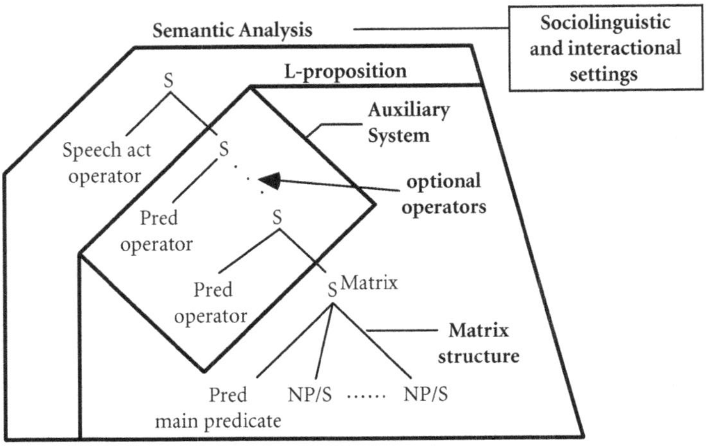

FIGURE 7.7 Internal structure of a VSO Semantic Analysis (SA)

that adhered to in modern logic.) The syntax of L_L assigns to its expressions, the SAs, a general format whereby the predicate comes either first or last (depending on the language type) and the arguments occur in the order SUBJECT–INDIRECT OBJECT–DIRECT OBJECT. For verb-first languages, as well as for languages characterized by surface NP–VP structure, L_L places the verb in initial position. For verb-last languages, L_L places the verb in final position.

Figure 7.7 shows the general format of a right-branching (verb-first or VSO) SA-structure without any topic–comment modulation. It contains a speech-act operator and an L-proposition. The matrix-S (S^{Matrix}) ('nucleus' in Seuren 1969) is so called because it contains the main predicate and the argument terms describing the situation intentionally focused on. One notes that matrix-S argument terms can themselves be S-structures, known as complement-Ss. Their grammatical treatment constitutes the *complementation system* of a language. What is not shown in Figure 7.7—to avoid symbol crowding—is that lexically filled NP-argument terms contain one or more Ss under a *determiner*. Such Ss are realized either as lexical nouns with or without adjectives and/or relative and other clauses or participials. Moreover, operator predicates in the Auxiliary System may themselves contain embedded NP- or S-structures. Embedded Ss may be of any kind, SAs, L-propositions, matrix-Ss, depending on the grammar of the language concerned, but the first S after the Auxiliary System must always be a matrix-S (for details see Seuren 1996).

In most cases, however, the SA will have a discourse-determined topic–comment structure of the form shown in Figure 7.8 (again in VSO format), which is richer than Figure 7.7. The non-topic–comment-driven SA-structure is embedded as S_3, with S_5 as the matrix-S, under the subject-NP of the

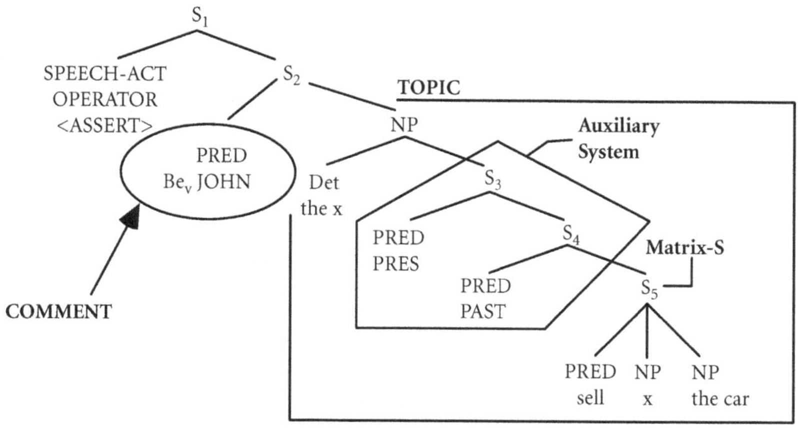

FIGURE 7.8 Example of underlying SA with topic–comment structure

scope-bearing operator Be_v.[6] (The implications of this analysis for the general theory of syntax are left open here.)

Through a relatively simple transformational system, the sentence in question will find a surface expression in (7.3a), (7.3b), or (7.3c):

(7.3) a. It is JOHN who has sold the car.

b. The one who has sold the car is JOHN.

c. JOHN has sold the car.

The grammars of natural languages show a universal tendency, which we call *matrix greed*, for non-matrix elements to be incorporated into the matrix-S by means of a handful of standard transformational rules, the nature of which does not concern us here directly (see Seuren 1974*a*: 183–90; 1996: 317–23). It is clear that Matrix Greed has been maximally operative in (7.3c), where the emphatic accent on JOHN is the only trace of the underlying topic–comment

[6] The operator status of the comment predicate Be_v appears from the fact that it is scope-sensitive, as is shown by the interaction with the negation operator in a language like German:

(i) Nicht HERBERT hat gelacht (sondern sein Sohn).
not Herbert has laughed (but his son)
It wasn't HERBERT who laughed (but his son).

(ii) HERBERT hat nicht gelacht ((*)sondern sein Sohn).
Herbert has not laughed ((*)but his son)
!!It was HERBERT who didn't laugh (*but his son).
√ It wasn't HERBERT who laughed (but his son).

To signal small scope of the negation in (ii), a falling intonation on *gelacht* is required, making the addition of *sondern sein Sohn* ungrammatical. A rising intonation on *gelacht* makes (ii) synonymous with (i), where the negation has large scope. (See also ex. (7.14).)

structure. In (7.3c), the comment-predicate PRED[Be$_V$ JOHN] has been incorporated into the matrix-S by means of the well-known process of LOWERING, which, in this case, leads to the detachment of PRED[Be$_V$ JOHN] from its original position below S_2 and its placement into the position of the corresponding variable x in the matrix-S_5. In the process, the S_2-node as well as $_{Det}$[the x], along with its dominating NP-node, are deleted, as is the predicate Be_v .

Generative syntactic theories tend to ignore any underlying topic–comment structure and generate sentential structures directly as matrix structures with their higher operators for tense, quantification, negation, or what not. In doing so these theories ignore the *lowering* of the comment constituent into the matrix-S and thus fail to account in a principled way for the heavy accent on comment-constituents, which is usually dealt with by positing a phonological rule whereby any constituent can be assigned comment-accent. (The processes leading to cleft and pseudocleft structures like (7.3a, 7.3b) are left in an even vaguer limbo). Yet it was already shown in great detail in Seuren (1974*a*) (see also Seuren 2004*a*: 177–85) that there are all kinds of restrictions on the assignment of comment-accent and that these restrictions are naturally accounted for by assuming that comment-accent is 'deep' in that it is fixed at the level of semantic analysis. Yet, despite the valid criticism that generative (and other) theories of syntax tend to neglect the topic–comment structure of sentences, the best among them still appear to capture most of the essential elements of syntax, precisely because of the phenomenon of matrix greed: matrix structures appear to define most of the syntax of a language.

In contexts in which (7.3c) lacks topic–comment structure, a sentence like (7.3c) has an underlying SA as shown in Figure 7.9. Such SA-structures are justified only in cases where no topic–comment structure is present, as in the opening sentence of a discourse, typically a so-called 'feeding line' or 'feeder' or a 'hot-news' item (see Section 3.2).

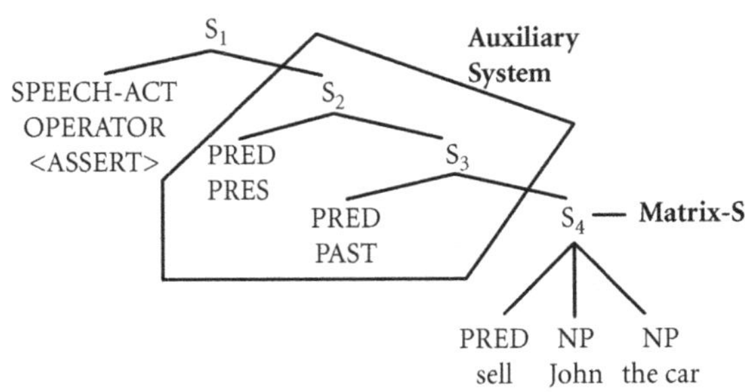

FIGURE 7.9 SA for (7.3c) without topic–comment structure

It is useful, in this context, to consider the wider ecological environment that embeds language, in particular as regards cognition and the world spoken about and spoken in. This is the more useful as relatively little attention is normally paid to such issues in either the philosophical, the logical, or the linguistic literature, including the literature on semantics.

Obviously, when an intent is to be expressed linguistically, it must be expressed in a given language. Therefore, in case a speaker has more than one language or language variety at his or her disposal, the language, or its variety, has to be specified first. This has direct consequences for the sentence to be generated, because different languages (language varieties) impose different requirements as to what must or may be expressed in their well-formed sentences. In all European languages, for example, tense must be expressed for every finite verb form, whereas Chinese, Malay, and many other South-East Asian languages form their sentences without any obligatory indication of tense. The progressive (or *be+V-ing*) form is obligatory in English for the expression of durativity on nondurative verbs but remains unexpressed in many other languages. Evidentiality as an indication of the epistemological source of an assertion, is an obligatory category for assertions in some languages but not in others (see Aikhenvald and Dixon 2003). Such phenomena require a 'scanning' of the cognitive content to be expressed in the sentence under production. It is easy to see that this is one of the major obstacles to machine-translation programs: when a category that remains unexpressed in the source language must be expressed in the target language, the translator must consult his or her comprehension to decide on the proper category in the target language—a process that has so far resisted any formal modelling or machine treatment. We therefore assume, for each specific language or language variety L, the existence of a SEMANTIC CHECKLIST whose parameters must be provided with values before any proposition can be turned into a sentence of L.

Then, as one recalls from Chapter 4, any grammatical sentence *S*, and hence its underlying L-proposition, must stand under a speech-act operator which defines the socially binding aspect, supported by a system of social sanctions, under which the speaker presents the proposition expressed in *S*. Natural language does not allow for the expression of a mere proposition as a sentence. *Me leaving early* may express a proposition, but it is not a well-formed English sentence. Every well-anchored and well-keyed utterance of a sentence, and indeed every conventional sign produced by a signifying subject, whether human or animal, constitutes a socially binding position-taking, creating an interpersonal relation that involves some form of commitment or appeal with respect to the proposition expressed, vis-à-vis one or more social partners who form the *force field* of the speech act's commitment or appeal and who may either exercise or undergo sanctions when the liability conditions are violated. I have called the mental state of deciding to commit oneself in a socially binding way with respect to a proposition the *intent* expressed by the actually ensuing utterance. And I have argued that it is this socially binding position-taking with respect to a given context-bound proposition in a given force field that is the essence of linguistic communication, and not, as is often assumed, the transfer of information. This central aspect is, therefore, fully integrated into the model presented here of language and its position in its ecological environment.

Moreover, as everyone knows but theoretical linguists too often fail to take into account, the utterance-to-be needs a specification for a set of sociolinguistic and interactional parameters. In Seuren (1982, 2001: 88–118) it is argued that there are three main sociolinguistic/interactional parameters determining phonological, lexical, morphological, and syntactic choices, and perhaps even 'deeper' choices to do with the planning of the text—that is, the cutting up of the overall intent into individual sentences and clauses. These are the parameters of geographical area, socio-economic status, and interactional situation. Whether this does indeed exhaust the set of parameters involved in the selection of variable formal expressions may be a matter of dispute. But it is certain that values must be selected for at least these parameters before an utterance is produced. And it is known from a large number of sociolinguistic studies that although such choices are sometimes made consciously or at least recoverably in short-term memory, most of the time they are made without the speaker having any awareness or memory of them.

On these assumptions, we take it that there is a lexical search procedure, whereby the proper lexical item for the comment is selected (the lexical material, as well as the syntactic structure, of the topic are already available). For the opening sentence, or feeder, of a discourse, this means that, first, the main matrix predicate is selected from the mental lexicon and subsequently the items for the argument terms and other elements in the sentence-to-be. The main matrix predicate is listed in the lexicon as requiring a specific syntactic structure with a specific number of argument terms occupying the semantic positions defined for the predicate in question. It is at this early level of lexical selection that the main features of the matrix-S are defined.

A few examples will show that lexical selection of the matrix predicate may make a vast syntactic difference from one language to another. Consider a speaker intending to say that legal punishment for murder is subject to a time limit of twenty years between the deed and the imposition of the punishment.

A German speaker will say something like (7.4a), literally 'murder superannuates after twenty years', but a speaker of English, which has no equivalent of the German verb *verjähren*, will have to come out with something like (7.4b), whose syntactic structure is totally different:

(7.4) a. Mord verjährt nach zwanzig Jahren.
b. There is a statutory limitation of twenty years for murder.

The reason for the syntactic difference is clear. The German speaker's intent directs him or her to the verb *verjähren*, with its specific meaning and argument structure, while the English speaker is directed to the nominal expression *statutory limitation*, which requires a very different syntactic structure.

Or consider the following two translation equivalents:

(7.5) a. You miss him already.
b. Il te manque déjà.

The English subject term corresponds to the indirect object in the French sentence, while the French subject is the direct object in the English sentence. Such differences follow from the lexically defined predicate–argument structure for each predicate. (Little is known about the universal restrictions that no doubt exist regarding the distribution of the semantic or thematic roles over the various argument terms.)

Sometimes, whole sentence forms are lexically selected, in which case one enters upon the territory of idioms. In Dutch, when one accidentally pours a little too much rum into one's guest's tea, one says *Sorry, m'n hand schoot uit* ('my hand slipped'). But in English, the standard expression is *Sorry, I tipped the bottle.* Obviously, cases like this, or like (7.4) or (7.5), form a huge obstacle to machine-translation programs.

The immediate result of assigning the linguistic and sociolinguistic/interactional values, specifying the speech act type, filling in the checklist, putting in any operators, and selecting the appropriate lexical predicates is a structure which we have decided to call the Semantic Analysis (SA) of the sentence (utterance) in the making. Given its sociolinguistic/interactional settings, the SA contains a speech-act operator, symbolizing the type of social binding inherent in the underlying intent, and a type-level L-proposition, symbolizing the underlying token proposition, complete with all the paraphernalia dictated or admitted by the checklist.

Once the SA has constituted itself, it is fed into the grammar module of the language in question, as shown in Figure 7.5. The grammar is an almost fully deterministic module, inaccessible to introspection, conscious control, or interference, which 'prints out' a given SA input as a well-formed surface structure. The grammar is taken to be transformational, in fact, tree-transformational, by definition, since it takes SA tree structures as input and produces surface tree structures as output.[7] The surface structure is subsequently fed into the phonology module, which provides it with a recipe for pronunciation (or writing). When this recipe is implemented, a physical token utterance is produced, which, being physical, is part of the physical world and thus constitutes a legitimate object for extensional reference.[8]

The decision to take a regimented variety of the language of modern predicate calculus (LPC) as the vehicle for the specification of semantic content was taken during the late 1960s, and likewise by McCawley and Seuren at about the same time.[9] These authors were beginning to see not only that LPC is probably the best available instrument for semantic description but also that the rules that map LPC structures onto surface structures in different languages—that is, the grammars of these languages—make for an optimally simple and adequate system of grammatical analysis and description. One may say with full justification that the syntactic descriptions that have been presented in terms of a system transforming regimented LPC structures into

[7] That SAs and their L-propositions must be linguistic tree structures was, although perhaps obvious nowadays, in fact a discovery for linguists. McCawley made this discovery in his (1967); Seuren did so independently (Seuren 1969: 85–8, 219–24). Jerry Sadock has proposed a theory of *Autolexical Syntax* (Sadock 1991), according to which syntax consists in a matching of, on the one hand, autonomous compound linguistic forms (ultimately words) which are defined not only for their phonological form but also for their formal combinatorics and, on the other, the autonomous compound meanings carried by sentences. Such a matching, or mapping, would be nonderivational and nontransformational. This originally Saussurean notion of grammar, which was further elaborated by Louis Hjelmslev during the 1930s under the name of *Glossematics* (Hjelmslev 1953/1943; see also Seuren 1998: 161–7), would make a great deal of sense, were it not for the fact that the meaning representations are themselves syntactic tree structures, which makes any mapping between the two 'planes' transformational by definition, and thus derivational.

[8] There is no linguistic obstacle to the use of metalanguage mixed up with object language (see Section 4.6). Speakers do it all the time, often in order to correct themselves or others, as in:

(i) He was an unhappy, or rather unfulfilled, man.

which is unlikely to mean that the person spoken about was either unhappy or unfulfilled, but rather means that the WORD *unhappy* is less appropriate than the WORD *unfulfilled* to describe the person in question. How grammars deal with such phenomena is a question that has not so far received much attention in the linguistic literature. There is some philosophical literature on the use-mention distinction (token-reflexivity) in language, mainly because of the well-known Liar paradox which threatened to upset the whole of logic and to which Russell presented his solution in the form of a prohibition to mix up object language and metalanguage in the same propositional structure. In more recent times the question has been taken up in the pragmatics literature, where attempts have been made to derive token-reflexivity from the standard 'referential' or 'descriptive' use of language, especially as regards the use of negation (Horn 1985; Carston 1998; for a critique see Seuren 2000). For the question of how natural language manages to avoid any logical paradoxes resulting from the mixing of object language and metalanguage, see Seuren (2001: 119–35).

[9] For more detailed comment on the developments that took place during that period, see the Introduction in Seuren (2001).

surface structures have achieved extraordinary success as regards explanatory adequacy.[10]

The entire modular process, from intent to sound or writing, is taken to consist of serially ordered modules—that is, a modular cascade. Between the modules involved there are interfaces or 'gates', at which the speaker is able to monitor the output of a given component of his or her own symbolization machine, and, if that is felt necessary, to correct its input so that the desired output is produced. Such output self-monitoring is, in principle, within conscious access. It is important to note that self-monitoring as a general psychological process is always output monitoring, not input monitoring. Self-monitoring always retraces the top-down path, which suggests again that the language-modular complex is made primarily for production, not for comprehension.

The analysis presented so far implies the postulation of four serially ordered modules:

(i) for converting intents into SAs via the lexicon;
(ii) for converting SAs into SSs by means of the grammar;
(iii) for converting SSs into recipes for sound or writing via the phonology;
(iv) for converting recipes into sound or writing via the production apparatus.

These modules are automatic and screened off from any possible awareness, conscious access, or interference. In principle, each module simply 'prints out' its output given its input.

The four modules correspond to five interfaces for, possibly conscious, self-monitoring:

(i) the INTENT INTERFACE, where the speaker checks the content of the intent;
(ii) the LEXICAL INTERFACE, where the speaker checks the lexical selections made;
(iii) the SURFACE-STRUCTURE INTERFACE, where the speaker checks the surface structure;
(iv) the INTERNAL PERCEPTUAL INTERFACE, where the speaker checks the instructions for phonetic or graphic production (the 'phonetic plan');
(v) the EXTERNAL OUTPUT INTERFACE, where the speaker checks acoustic or graphic correctness of the actual output.

[10] But no success at all in the community of professional linguists. Owing to an unheard-of defamation campaign, which was led by Noam Chomsky and has had its effect over the past thirty years, it is still so that this theory, to use Oscar Wilde's famous phrase, 'dare not speak its name', that name being Generative Semantics (see Huck and Goldsmith 1995; Seuren 1998: 493–512, 2006, 2007).

Close analysis (see Levelt 1989: 470, 1999; Cutler and Clifton 1999: 152–55) reveals that these five interfaces are probably the only places in the procedure of utterance production where the speaker can exert control or make a choice. This means that the machinery cannot do without feedback controls to earlier input stages, as is emphasized by the authors cited. Obviously, a speaker is, in principle, in control of the semantic content expressed, which gives the *intent interface.* Moreover, given the overall semantic content to be expressed, lexical choice is likewise open to some variation, where choices often have emotional, interpersonal, or social values. Such choices are open to the approval or disapproval of the speaker, which gives the *lexical interface.* Then, a speaker may be assumed to have a, possibly conscious, idea of the combined words and morphemes of the utterance that is about to be produced, which makes us assume the *surface–structure interface.* Furthermore, a speaker has some control over the way an utterance is to be pronounced, just as any other bodily act is preceded by a neuro-mental representation of the intended output. This control is carried out at the *internal perceptual interface.* Lastly, and almost trivially, a speaker monitors the phonetic (graphic) output for acoustic (graphic) quality, which gives the *external output interface.* In the psycholinguistic literature, corrections made after the onset of the physical realization of the element to be corrected are called 'repairs'.

7.2.2 *Evidence and possible counterevidence for grammars as top-down modules*

So far, we have, more or less simply, postulated the machinery described, not without reason perhaps, but still without a full critical examination of the arguments that can be raised for and against such a description. The question is, therefore: does the production machinery of language really constitute a modular complex of the kind described in the preceding section? What is the evidence and how about any possible countervidence? In this context, the actual working of the modules in question (for syntax, see Seuren 1996) is of less interest than their modular character. What has to be shown is that they operate without any external interference from the wider cognitive context. In general, this means that any stylistic, emotive, sociolinguistic, interactional, or semantic shade or nuance of meaning that may or must influence phonetic form, complete with intonation, must have its origin in the underlying intent and must find a place in the input to the grammar. All rules or processes which have a bearing on the condition of proper expression for a given intent—a condition that includes 'grammaticality on a reading' but is not exhausted by it—and which have been proposed in the past as optional insertions in the grammar or the phonology must, therefore, either be shown to be empirically untenable or, at the very least, to be replaceable without empirical loss with a 'deep' treatment at SA-level. In short, all elements that have been regarded in the past as being part of 'surface' semantics must be eliminated in favour of 'deep' semantics. In arguing for this claim, I leave the phonology and the phonetics aside, that being mostly beyond my competence. The main emphasis will be on grammar, in particular syntax.

The question of 'surface' versus 'deep' semantics in relation to grammar has a history in modern linguistic theory. During the early 1970s, Chomsky and his followers abandoned the then current paradigm of Generative Semantics (see note 10), which held that all semantic content is fixed at SA-level and that the grammar of a language is merely an automatic, self-contained, machinery turning SA-structures into surface structures. Although the term was not yet in use, Generative Semantics in fact defended the position that a grammar is a modular piece of top-down cognitive machinery without any influence on the semantic properties of the resulting surface structure—precisely the position we defend here.

The Chomskyans, by contrast, held that surface structure elements may co-determine meaning. Later they went further and claimed that surface structure is the only determinant of meaning. For them, underlying syntactic structure is not identical with the SA or its L-proposition but either partially overlaps with it or, in the later version of the theory, is entirely distinct from it. A position of this nature would be reasonable if the rules and procedures for the assignment of the relevant surface-semantic features were naturally, and with sufficient explanatory power, expressible in terms of surface structure. Over the years, however, it has become clear that this is not so: the distribution of those surface-structural features that signal semantic content of any kind turns out to be naturally and explanatorily definable in terms of underlying SA-structure. Therefore, it is time to reopen the discussion, even if this very substantive issue is no longer considered topical, owing to the fickle sociological fashions that seem to rule the discipline.

When the discussion did become topical, in the period after 1970, it was already clear that matrix structures and sentential complementation features are naturally expressible in terms of SA-structure. Therefore, those who, during the late 1970s, defended full surface semantics were compelled to posit semantic interpretation rules that simply copied the already well-known top-down syntactic rules of sentential complementation in reverse. The same goes for the semantic determination of operator scope. Whereas generative semanticists had to posit rules for the lowering of logical operators into the matrix-S, the Chomskyans, in particular May (1977), proposed the same rules in reverse to distil operator scope by raising the operators in question from surface structures. Yet, despite this obvious redundancy, Chomsky and his followers simply carried on, impervious to rational argument. This being so, we will no longer consider full surface semantics a serious option and limit ourselves to the view that only some semantic features are assigned at surface-structure level—that is, the position defended by Chomsky during the early 1970s.

The discussion with regard to this latter view centred mainly around three issues: (a) grammatical and intonational focusing strategies expressing topic–comment structure; (b) presuppositions; and (c) operator scope. It is shown in a moment that further issues are likewise relevant with regard to a decision on the question at hand, but we will discuss these three issues first. As regards (a) and (b), Chomsky commented:

> To summarize these remarks, we seem to have the following situation. Rules of phonological interpretation assign an intonational contour to surface structures. Certain phrases of the surface structure may be marked, by grammatical processes of a poorly understood sort, as receiving expressive or contrastive stress, and these markings also affect the operation of the rules of phonological interpretation. If no such processes have applied, the rules assign the normal intonation. In any event, phrases that contain the intonation center may be interpreted as focus of utterance, the condition being perhaps somewhat different and more restrictive when the intonation center involves expressive or contrastive stress, as noted. ... The notions 'focus,' 'presupposition,' and 'shared presupposition' ... must be determinable from the semantic interpretation of sentences if we are to be able to explain how discourse is constructed and, in general, how language is used. (Chomsky 1972: 99–100)

Over the years, this position was considered more plausible to the extent that comment or focus accent and presuppositions were considered pragmatic, rather than semantic, phenomena. This, however, is clearly false: comment accent and other focusing phenomena, as well as presuppositions, are semantic phenomena in the full, truth-conditional sense.

As regards focusing strategies, the clinching argument is that there are clear truth-conditional differences between different focus or comment distributions in intensional contexts under emotive intensional verbs. This appears from cases like (7.6a, 7.6b) or (7.7a, 7.7b) (cp. the examples (3.10) and (3.11) in Section 3.2 and (6.12a, 6.12b) in Section 6.2.3.2):

(7.6) a. What worried Sue was that not even JIM laughed at the joke.

b. What worried Sue was that Jim didn't even laugh at the JOKE.

(7.7) a. Sue was angry that JIM had sold the car.

b. Sue was angry that Jim had sold the CAR.

It does not take much reflection to see that (7.6a) may be true while (7.6b) is false and vice versa, and similarly for (7.7a) and (7.7b). This suffices to show that focus or comment accent (and the corresponding particle placement) is not freely substitutable *salva veritate*, a fact which requires a semantic explanation in the full, truth-conditional sense of the word, just as with the lack of free substitution *salva veritate* of co-referring terms—Frege's discovery which gave rise to the entire development of twentieth-century formal semantics.

Presuppositional phenomena are likewise clearly not pragmatic but semantic, in the truth-conditional sense of the word. This appears from the fact that negation is sometimes exclusively presupposition-preserving and sometimes exclusively presupposition-cancelling (see Section 10.4 in Volume II). A couple of examples will suffice to show the point. Consider the following set of sentences ('√' indicates grammatical correctness for the reading intended, '!' semantic incoherence, '*' grammatical ill-formedness on any reading):

(7.8) a. √ Harold not only read the first page. He read the whole book.

b. !*Harold NOT only read the first page. He didn't even look at the book.

c. √ Harold did NOT only read the first page. He didn't even look at the book.

The adverb *only* marks the comment of a sentence and induces the presupposition that the sentence without *only* is true. The sentence with *only* asserts that nothing more than the comment is true. Negation is allowed to occur in construction with *only*, as in (7.8a), and not, as is normally the case, in construction with the finite verb form. Since a negation that does not occur in construction with the finite verb form is necessarily presupposition-preserving and denies only the assertive content of the sentence (see Section 10.4 in Volume II), it follows that a subsequent sentence cannot cancel the presupposition induced by *only* when the negation is not in construction with the finite verb. This is borne out by example (7.8a), where the negation is not in construction with the finite verb but with *only*, so that merely the assertive content of the first sentence can be denied, namely that Harold did not read more than just the first page of the book. The second sentence of (7.8a) confirms that denial, saying that Harold read the whole book and not just the first page. By contrast, the presupposition that Harold did read the first page of the book is cancelled by the second sentence of (7.8b) even though the negation in the first sentence is in construction with *only*, the result being that the two sentences make for an incoherent text, as indicated by the exclamation mark. The first sentence of (7.8b), moreover, is ungrammatical under heavily accented NOT, as is indicated by the asterisk. Example (7.8c) is again fully coherent, which is made possible by the fact that here the (heavily accented) negation is in construction with the finite verb form *did*, so that it is free to cancel presuppositions.

Cases like these show not only that the English negation word *not* can be used in truth-conditionally different senses, but also that attempts at a pragmatic explanation of the origin of presuppositions are doomed to fail, as there is nothing in pragmatics that could account for the fact that negation in construction with the finite verb may cancel presuppositions, whereas negation in other structural positions cannot. Moreover, presuppositions play their part in consistency or coherence relations, which means that they must be incorporated into the logic of language, which rules out a pragmatic status and requires their being recognized as a genuine semantic phenomenon.

The same conclusion follows from cases like the following, involving preposed factive clauses:

(7.9) a. √That Roderick had fallen in love with the stable boy did not surprise his mother. She had seen it coming for some time.

b. !That Roderick had fallen in love with the stable boy did NOT surprise his mother. Nonsense! Roderick didn't even know the boy.

c. √It did NOT surprise Roderick's mother that he had fallen in love with the stable boy. Nonsense! Roderick didn't even know the boy.

The structure of the argument runs parallel to that relating to (7.8a–7.8c), except that instead of *only* under a negation that is or is not in construction with the finite verb, we have to do with preposed *that*-clauses of a factive verb (in all three of (7.9a–7.9c), the negation is safely in construction with the finite verb). The point is that factive presuppositions cannot be cancelled when the factive *that*-clause is preposed, as in (7.9a, 7.9b) (see Seuren 2001: 340). For that reason, (7.9b), where the second sentence implies a denial of the presupposition expressed in the factive *that*-clause, is incoherent, even with heavy accent on the negation. But in (7.9c), the factive *that*-clause is not preposed, so that presupposition-cancelling is possible under heavy-accent negation. Again, it is unclear how pragmatic factors can account for the truth-conditionally different negation in (7.9a) and (7.9c). The grammatical machinery, of course, can: one can formulate structural conditions under which negation is unable to cancel presuppositions.

This having been said, it remains to be shown that focus and presupposition phenomena cannot, with good reason, be taken to be generated at surface-structure level but are best taken to be coded in the SAs that are input to the grammar module.

As regards presuppositions, this is simple. It has become clear, over the years, that presuppositions are, at least for the most part but probably across the board, generated as a result of lexical choice. Since (a) lexical meanings are defined, in principle, by satisfaction conditions, (b) there are two kinds of satisfaction conditions, namely PRECONDITIONS which generate presuppositions, and UPDATE CONDITIONS which generate what is often called 'assertive content',[11] and (c) lexical selection can only take place at a deep semantic level, it follows that most presuppositions are generated at the deep semantic level at which lexical selection takes place. The same goes for presuppositions that are generated as a result of topic–comment structure: the topic is presupposed and the comment is the assertive content. Since, as has been shown, topic–comment structure can only be defined at a deep semantic level, the same applies to the presuppositions it generates.

As regards focusing strategies, it has already been shown that they may make a truth-conditional difference, which brands them as being semantic, and not just pragmatic, phenomena. It remains to be shown that, as semantic phenomena, they cannot reasonably be assumed to be assigned at surface-structure level but must be taken to originate at a deep semantic level. Focusing particles posit no problem, since they are lexical items, and lexical selection is uncontroversially a 'deep' process. But heavy accents and forms of clefting are not so quickly dealt with.

It has been suggested (for example Chomsky 1972: 99–100) that phonological rules may assign heavy accent to any arbitrary element in the sentence. The semantic interpretation of this accent would then not be traceable to the SA underlying the sentence in question. Likewise for cleft and pseudocleft constructions. It has been proposed (for example Akmajian 1970) that clefting processes result from the transformational extraction of any arbitrary nominal or prepositional constituent and are thus not derivable from a semantically given underlying structure.

To show that such proposals are incorrect one needs (a) cases where heavy accent assignment or extraction should be possible, according to the proposals in question, but in fact is not, or (b) cases that are ungrammatical without a given heavy accent or clefting. A clear example of the latter is given in (7.10),

[11] A misnomer, since the term is only applicable when the proposition in question stands under an assertive speech-act operator. The term I normally use is 'update'.

which requires a heavy contrastive accent on the two contrasted elements *the article* and *the book*:

(7.10) Not the article but the book was published last year.

There does not seem to be a viable way of accounting for the syntax of this sentence without assuming an underlying form such as (7.11) expressing its topic–comment structure and read in ordinary English as *What was published last year was not the article but the book*:

(7.11) NOT $_S$[$_{PRED}$[Be$_V$ the x [Article(x)]] $_{NP}$(the x [Published last year(x)])]

—

$_S$[$_{PRED}$[Be$_V$ the x [Book(x)]] $_{NP}$(the x [Published last year(x)])]

Any attempt to account for the syntactic structure of (7.10) either by the imposition of extra structure at surface level or by surface composition rules is bound to fail on grounds of gross ad-hocness. One notes, for example, that, in English (but not in, for example, German or Dutch), the contrastive preposing of the negation word *not* in (7.10) is possible only when the corrected NP is immediately followed by *but* followed by the correcting NP. Failing that, the sentence is ungrammatical:

(7.12) *Not the article was published last year.

An example of the opposite case—heavy accent making a sentence ungrammatical—is (7.13b), as opposed to (7.13a) and (7.13c):

(7.13) a. √John isn't in the least interested.

b. *JOHN isn't in the least interested, PETER is.

c. √JOHN isn't in the least interested, nót PETER.

Here, sentence (7.13a) is grammatical, but it becomes ungrammatical with contrastive accent on *John* while the negation applies to the comment JOHN, as in (7.13b) to be read with rising intonation on *interested*. Sentence (7.13c), with the negation as part of the topic, signalled by falling intonation on *interested*, is all right again. The reason lies in the NEGATIVE POLARITY ITEM (NPI) *in the least*, which belongs to the (strong) class of NPIs requiring a full negation directly over the clause in which it occurs at SA-level. When it is assumed that (7.13b) is derived from a topic–comment structure, that structure must be something like (7.14a), which is ill-formed due to the fact that the negation is too far away from the clause in which *in the least* occurs. By contrast, (7.14b), underlying (7.13c), is all right, because here the negation does not deny the comment but is part of the topic and stands directly over the clause in which *in the least* occurs:

(7.14) a. *NOT [Be$_V$ JOHN (the x [Be in the least interested(x)])]

b. √ Be$_V$ JOHN (the x [NOT [Be in the least interested(x)]])

This suffices to show that an adequate description of presuppositions and focusing strategies requires that they be treated as originating at the SA-level of description, so that they are part of the input to the grammar (sub)module.

The same remains to be shown now for operator-scope assignments. That the scope of logical and semantic operators may make a crucial difference to the meaning of an L-proposition appears from cases such as the following:

(7.15) a. One policeman did not use violence.

b. Not one policeman used violence.

(7.16) a. Most students speak no foreign language.

b. No foreign language is spoken by most students. (ambiguous)

(7.17) a. Every morning Simon reads two poems.

b. Simon reads two poems every morning. (ambiguous)

(7.18) a. Because it was raining, Tom did not leave.

b. Tom did not leave because it was raining. (ambiguous)

Clearly, (7.15a) may be true while (7.15b) is false. The pairs (7.16a, 7.16b), (7.17a, 7.17b) and (7.18a, 7.18b) likewise differ in their truth conditions. The difference hinges, in principle, on the position of the quantifiers or negation in the surface structure. Left-to-right order is significant in that it is always possible to apply the principle that whichever operator comes first in the surface sentence has larger logico-semantic scope. Structural position, however, may override this principle, leading to ambiguity. What appears to be the case is that when a scope-sensitive operator finds itself, in surface structure, in a *peripheral* constituent—that is, a constituent originating from high-up in the SA tree and characterized by being attached in surface structure to the top-S rather than to the VP—scope-sensitivity is diminished. Apparently, peripheral attachment is allowed to override the left-to-right principle, which accounts for the scope ambiguity of sentences (7.16b), (7.17b), and (7.18b).[12]

These facts can be accounted for either by positing a top-down SCOPE ORDERING CONSTRAINT (SOC), which, starting from the lowest, blocks the lowering of a scope-sensitive operator across a scope-sensitive operator

[12] See also the examples (8.12)–(8.15) in Section 8.6.1. The relation between, on the one hand, the left-to-right order of operators in surface structure and their scope relations in logico-semantic structure was the topic of lively discussions during the late 1960s and the 1970s. See, for example, McCawley (1967, 1972), Lakoff (1971), Seuren (1969, 1972*c*, 1984*b*).

lowered earlier (except for operators that land in peripheral position) or by a corresponding bottom-up constraint (May 1977), which, starting from the left, blocks the raising of a scope-sensitive operator across a scope-sensitive operator raised earlier, except for operators to be raised from peripheral position.[13]

The argument is that, although it may look as if the top-down and the bottom-up treatments are equivalent, or notational variants, closer inspection reveals that the bottom-up treatment is, in the end, likely to be untenable. This is so because there are sentences whose grammaticality, or ungrammaticality, crucially hinges on scope configurations in SA-structure and not in surface structure. Therefore, scope configuration cannot be merely a question of semantic interpretation given a surface structure, but must be taken to be part of syntax, as syntax is meant to account for well-formedness phenomena. SA-structure thus not only represents the meaning of sentences in a logically and semantically regular form, it also plays a part in the top-down machinery of syntax. Examples are (Seuren 2004*a*: 188):

(7.19) a. √ Barbara is often not in the least interested.
b. *Barbara is not often in the least interested.
c. √ The teacher did not expect Barbara to be in the least interested.

(7.20) a. √ Either Harry or Fred was late.
b. *Either Harry or Fred were late.
c. √I don't think either Harry or Fred were late.

As was shown in connection with (7.13), the strong NPI *in the least* requires a full negation directly over the clause in which it occurs in SA-structure. This condition is fulfilled in (7.19a), but not in (7.19b), in whose SA the operator *often* stands between *not* and *in the least.* One might say that this can be formulated as a surface structure constraint to the effect that *not* should not be separated from *in the least* by any intervening element representing a scope-sensitive logico-semantic operator, as in (7.19b), but that bid fails in the light of (7.19c), which is fully grammatical even though *not* stands over *expect* and not over *be in the least interested.* The explanation is that *expect* is a negative-raising predicate. Sentence (7.19c) thus has the underlying structure (7.21a), which is turned into (7.21b) by NEGATIVE RAISING.[14]

[13] When May (1977) came into circulation, it gave rise to some hilariousness resulting from the fact that no mention was made of the well-known rules posited by Lakoff, McCawley, and myself for the lowering of operators into the matrix-S, whereas everybody could see that May's rules were exact copies in reverse of the rules posited by these authors.

[14] See, for example, Robin Lakoff (1969), Seuren (1974*b*), Horn (1978) for detailed analyses and arguments.

(7.21) a. Expect (the teacher, [NOT [Be in the least interested (Barbara)]]) ⇒

b. NOT [Expect (the teacher, [Be in the least interested (Barbara)])]

The SA-condition on the position of NOT with respect to 'in the least' is met by (7.21a). The top-down grammar does the rest.

As regards (7.20a–7.20c), the point here is that (7.20a) requires the singular finite verb form *was*: (7.20b) is ungrammatical. Yet (7.20c), with (7.20a) embedded as an object clause under the negative-raising predicate *Think*, is again grammatical. This apparent anomaly is explained if it is assumed (a) that the conjunction AND is a negative-raising predicate just as *Expect* or *Think*, so that the negation standing over the argument clauses of AND is raised over AND, which is converted into OR, so that AND(NOT-S_1, ..., NOT-S_n) is converted into NOT-OR(S_1,..., S_n), meaning 'not a single one of the component L-propositions', and (b) that the plural number assignment in (7.20c) results from *conjunction reduction* originating from the cumulative operator AND. The derivation of (7.20c) will then, in principle be as in (7.22a–7.22c). First, *not* is raised out of both argument clauses of *and* giving *not-either-or*. Then, the negative-raising predicate *think* raises *not* out of its object clause, giving (7.22c). Then, REDUCTION gives (7.22d), with the plural form *were* of the verb *be* owing to the fact that the reduction is originally a reduction under *and*. Final LOWERING of *not* gives (7.20c).

(7.22) a. Think (I, [AND ([NOT[Harry be late]], [NOT[Fred be late]])]) ⇒
NEGATIVE RAISING

b. Think (I, [NOT[EITHER-OR (Harry be late, Fred be late)]]) ⇒
NEGATIVE RAISING

c. NOT[Think (I, EITHER-OR (Harry be late, Fred be late))] ⇒
REDUCTION

d. NOT[Think (I, [EITHER Harry OR Fred be_{PLUR} late])]

Even though this analysis is not without technical complications as regards the transformational procedure,[15] it seems difficult to deny that the plural form *were* in (7.20c) must in any case be due to an underlying AND-conjunction. This implies that any grammatical proposal intended to account for this plural will somehow have to fall back on the semantics of the propositional operator 'and'. This is, in itself, a strong and principled argument in favour of the syntactic theory of Generative Semantics—the more so because it very much

[15] It seems, in particular, that NEGATIVE RAISING applies in a precyclic part of the grammar, where SA-trees are refashioned before being admitted to the transformational cycle. This notion, however, has not, so far, been elaborated with sufficient precision.

looks as if this remarkable fact is unaccountable for in any other theory of grammar.

There is thus plenty of reason to take seriously the view that operator scope as a central element in semantic interpretation cannot be defined on the basis of surface structure, but must be defined at SA-level, which makes it part of the input to the top-down grammar module.

Apart, however, from focusing phenomena, presuppositions, and operator scope, there are other phenomena that may cast a light on the theory that all semantic content must be taken to be represented in SA-structures. One such set of phenomena, showing that conventionalized nonliteral or idiomatic lexical meaning may play a role in syntax, is drawn from standard modern German, which has a lexically and sociolinguistically restricted syntactic regularity whereby prepositions of prepositional objects or adjuncts of place are reduplicated as particles of a compound verb. The syntax of this process is not entirely transparent, but it is clear that we have to do with a single recognizable phenomenon:

(7.23) a. Du solltest **auf** dich **auf**passen.
you should on you on-care
You should be careful (about yourself).

b. Ich kann **in die Garage** nicht **hinein**gehen.
I can into the garage not there-into-go
I can't enter the garage.

c. Ich möchte mich **da** nicht **drauf**setzen.
I would like myself there not there-on-put
I wouldn't like to sit down on it.

d. Er lief **durch** die Strasse **hindurch**.
he walked through the street on-through
He walked all the way down the street.

e. Wir können **mit** Ihnen **mit**kommen.
we can with you with-come
We can come with you.

f. ... weil ich dich denn **aus** der Gruppe **heraus**geschmissen hätte.
... because I you then out of the group out-of-it-thrown had
... because then I would have thrown you out of the group.

g. Jetzt aber her**an an** die Arbeit!
now but there-to to the work
But now get to work!

The point is that, in most cases and in most dialects, this rule or regularity does not apply when the verb is used in a conventionalized idiomatic sense, as appears from the following examples, where the (a)-sentences represent the nonliteral or idiomatic, and the (b)-sentences the literal reading:[16]

(7.24) a. Wie bist du denn auf DIESE Idee gekommen?
how are you then on this idea $\text{come}_{\text{past participle}}$
What has given you THIS idea?

b. Wie bist du denn **auf** das Dach **drauf** gekommen?
how are you then on the roof there-on $\text{come}_{\text{past participle}}$
How did you get up the roof?

(7.25) a. Er darf keinen Blick in das Zimmer werfen.
he may no glance into the room throw
He mustn't look into the room.

b. Er darf keinen Stein **in** das Zimmer **hinein**werfen.
he may no stone into the room into-it-throw
He may not throw a stone into the room.

(7.26) a. ... weil wir großen Wert darauf legten.
... because we great value on it placed
... because we set great store by it.

b. ... weil wir ein großes Buch **da drauf** legten.
... because we a big book there there-on placed
... because we placed a big book on it.

Such examples show that, in some cases at least, the choice between literal and idiomatic or nonliteral meaning is not simply a free decision that can

16 Sometimes, the reduplication is lexically fixed also for nonliteral meaning, as in:

(i) Wie bist du **an** das Geld **heran**gekommen?
how are you at the money at-it $\text{come}_{\text{past participle}}$
How did you get at (put your hands on) the money?

But note that (i) is more literal and less idiomatic than (ii):

(ii) Wie bist du **an** das Geld gekommen?
how are you at the money $\text{come}_{\text{past participle}}$
How did you get (acquire) the money?

Note also that when *auf* DIESE *Idee* in (7.24a) is pronominalized as *da* in (iii), reduplication is again normal, besides the nonreduplicated version given in (iv):

(iii) Wie bist du da drauf gekommen?

(iv) Wie bist du darauf gekommen?
(both: What made you think of that?')

be taken by the listener on pragmatic grounds given any surface structure, but is forced by surface structure. The nonliteral use may have started as a free choice, but then language took over (as it does in so many other areas—such as nominal gender—where lexical or grammatical distinctions are clearly reminiscent of an original cognitively based distinction but where this origin has become subject to linguistic regulation).

This means that, in cases where a nonliteral or idiomatic reading is subject to linguistic regulation, the choice must be encoded in the syntax. And if, as is generally assumed, the syntax determines the conditions for the proper expression of given meanings, the nonliteral or idiomatic character of the lexical meanings concerned must be encoded in the underlying SA-structure and thus be carried along through the top-down transformational process.

An analogous conclusion, this time with regard to THEMATIC FUNCTIONS, is to be drawn in the case of the PREDICATE RAISING (PR) construction in a language like French, best demonstrated with the causative verb *faire* (do, make). When the clause embedded under *faire* is transitive, this PR-construction appears in two forms: one in which the embedded verb is left in the active voice and one in which it is passivized (though without passive morphology). In the former case, the original subject term becomes indirect object (dative), as shown in (7.27a), while in the latter case it appears—if at all—in surface structure as a passive Agent Phrase under the preposition *par* (by), which is the normal preposition for passive Agent Phrases in French. An example of this is shown in (7.27b):

(7.27) a. Luc a fait voir la lettre à/*par sa sœur.
Luc has made see the letter to/*by his sister
Luc has let his sister see the letter.

b. Luc a fait corriger la lettre par/*à sa sœur.
Luc has made correct the letter by/*to his sister
Luc has had the letter corrected by his sister.

As is observed in Seuren (1972*a*, 1996: 192, 198–9), the difference depends on the *thematic function* assigned by the lexical verb of the embedded clause to its subject term: when the lower verb is of the recipient type—that is, it assigns the thematic function *recipient* or *beneficiary* to the subject term, as in (7.27a)—the active option must be taken, with the underlying subject becoming a dative indirect object. But if the lower verb is actor-oriented—that is, it assigns the thematic function *actor* to the subject term, as in (7.27b)—the embedded clause is passivized and the original subject is either left out or expressed under the preposition *par*. If the lower verb is neither of the recipient type nor actor-oriented, the PR-construction cannot be used at all, as shown in (7.28a), which is better expressed as (7.28b):

(7.28) a. *Luc a fait atteindre son but à/par l'article.
Luc has made achieve its purpose to/by the article
Luc has made sure that the article achieved its purpose.

b. Luc a fait de sorte que l'article atteigne son but.
Luc has made so that the article achieves its purpose
Luc has made sure that the article achieves its purpose.

This shows that, just like nonliteral or idiomatic lexical meanings, thematic functions must, at least in some cases, be encoded in the underlying SA-structure and thus be allowed to play a role in the top-down transformational process—even if the relevant machinery has not so far been developed.

A further possible counterargument to the modularity thesis for grammars is provided by the blocking of CONJUNCTION REDUCTION in cases of DOMAIN UNIFICATION versus DOMAIN SPLIT—notions that are discussed more fully in Section 7.2.2.3 in Volume II. Relevant examples are the following:

(7.29) a. Paul may be at home and he may be having his breakfast.

b. Paul may be at home and/or he may be in hospital.

In the normal interpretation, (7.29a) conveys the information that it is possible that Paul is at home, having his breakfast. That is, (7.29a) is interpreted in such a way that only one possibility domain is involved. By contrast, (7.29b) involves two possibility domains, one in which Paul is (possibly) at home and a different domain in which he is (possibly) in hospital. The reason for the domain split in (7.29b) is, of course, the fact that one cannot, at the same time, be at home and in hospital.

The curious fact is that (7.29a) allows for the syntactic process of *conjunction reduction*, as in (7.30a), while (7.29b) does not, as shown in (7.30b) (which, incidentally, becomes fully acceptable with *or* instead of *and*):

(7.30) a. Paul may be at home and be having his breakfast.

b. !! Paul may be at home and in hospital.

This means that conjunction reduction is subject to a semantic condition—the condition of domain unification versus domain split—which is not formally encoded in terms of any of the existing theories of grammar.

It seems that there are two possible answers to this difficulty. One may say, as many do, that conjunction reduction is not a rule of syntax because all conjunctive phrases are generated directly as such. If this view can be upheld, the problem disappears because then both (7.30a) and (7.30b) allow for only one interpretation, namely the one in which only one possibility domain is involved, which makes (7.30b) anomalous. The second possible answer would be to say that, in relevant cases, input SA-structures must be encoded for domain split versus domain unification. It is not possible, at this point, to decide which of the two answers, if either, is more likely to be the correct one. For the time being, we must allow the question to rest.

As with all inductive theories, it is impossible to PROVE the semantic invariance of the top-down transformational system of grammar and hence the modularity thesis. All we can do is rebut possible counterexamples. We have done so for focusing phenomena, presuppositions, operator scope, nonliteral lexical meanings, thematic functions, and domain split, and we are ready to face possible further challenges.

7.2.3 *Is comprehension also modular?*

The reverse, bottom-up, process of comprehension, from sound or writing to the underlying intent, which involves the grammatical *parsing process* as a central element, appears to be less determined by a purely mechanical grammar module than the production process (see also Seuren 2004*a*: 83–5). The reasons for thinking that, other than the top-down input, which largely determines the final output, the bottom-up input does so to a far lesser extent are mainly the following.

Top-down we have a one-to-almost-one mapping, once the parameter settings for sociolinguistic, interactional, and stylistic variation have been set. Bottom-up, however, we have to do with a one-to-many mapping. Even within syntax, the result of parsing methods is far from uniquely determined by the phonetic or graphic input.

It is true that some elements in the parsing process are fully or largely deterministic in that they deliver a unique or almost unique output. This holds in particular for complex word forms created by systems of flectional morphology, such as the Turkish form *yaz-dır-a-ma-malı-y-dı* ('it should not have been possible to get someone to write') quoted in Section 7.1.3. Less dramatic examples are provided by verbal tenses, which are, on the whole, both expressed morphologically and uniquely reconstructible, so that large chunks of the auxiliary system are in place immediately and simply and independently of the generative processes defined in the grammar. In general, assuming that the phonological discrimination system allows a competent listener to specify all the different morphemes involved, a simple algorithmic routine leads straight up to the corresponding chunk of SA structure, with hardly any need for feedback control. As has been said, the straightforwardness of parsing procedures for possibly complex forms created by the flectional morphology of a language may well be the functional *raison d'être* for the systems of flectional morphology found in the languages of the world.[17]

Case morphology, by and large, allows for the unique reconstruction of argument functions. And if that fails, the number of alternatives tends to be extremely limited. For example, the German sentence (7.31), with the accusative NP *den Jungen* (the boy), allows for both the readings (7.32a) and (7.32b), both of which include the lexical ambiguity of the German verb *lassen* meaning either 'allow' or 'cause':

(7.31) Er ließ den Jungen töten.

(7.32) a. He allowed the boy to kill. / He made the boy kill.
b. He allowed the boy to be killed. / He had the boy killed.

Here the accusative case is not a unique determinant of the semantic argument function of the NP in question. But the number of possible readings for the accusative *den Jungen* is minimal: just two. This is still within the range of what may be assumed to be manageable within a practical mechanical application. To understand what is at issue, one should know that the German verb *lassen* induces PREDICATE RAISING for its object clause, which is either active, or passive but without passive morphology, just like the French *laisser* (allow) and *faire* (make), or Dutch *laten*, the equivalent of German *lassen*. All the parser has to do, in such cases, is try out the production system for both the active and the passive version and check if the result matches the utterance in question. Available discourse or long-term memory will, in most cases, determine, or preempt, the choice to be made.

However, other types of construction, especially those that go under the name 'syntactic', produce such numbers of possible analyses as can no longer be accommodated within a plausible realist theory of linguistic processing (or within a practicable computer program). Such constructions are frequent enough to justify the general statement that the bottom-up parsing process is

[17] One might wonder why, if this is true, one does not witness a universal trend in the languages of the world to develop complex flectional morphologies. The answer would appear to be that complex morphological systems, while delivering a substantial profit in terms of quick and certain parsing algorithms, require a heavy investment in terms of linguistic neural machinery, risking atrophy unless activated early. L2 learners of such languages know from experience how difficult it is to gain an adequate and fluent command of them at a later age. The maintenance of such machinery in a language community requires a stable social structure with adequate opportunities for young children to acquire the language of their environment in a natural, organic manner and without too many upheavals. Typically, the morphological system is the first to go and the syntax needs to be reinforced accordingly, under conditions of enduring socio-political and cultural instability, as is vividly shown by the transition from Latin to the Romance languages.

heavily underdetermined by the products of the top-down machinery. It is for that reason that Figure 7.6 has the look of a maze of feedback processes.

Scope differences are a typical example of the kind of phenomena that burden the parsing process beyond tolerable limits. Consider the sentence:

(7.33) I had not written to the man I wanted to succeed in 1985 because of his blue eyes.

Here the number of possible readings makes a mechanical parser operating on just the sentence theoretically futile and practically unmanageable. The sentence may mean any of the following, and more:

(7.34) a. 'in 1985 the situation was such that I had not written to the man I wanted to be the successor, and this was because of his blue eyes'

b. 'my writing to the man I wanted to be the successor had not been in 1985, and this was because of his blue eyes'

c. 'it was not because of his blue eyes that in 1985 I had not written to the man I wanted to be the successor'

d. 'it was not because of his blue eyes that my writing to the man I wanted to be the successor had been in 1985'

e. 'it was not because of his blue eyes that I had written to the man I wanted to be the successor in 1985'

f. 'because of his blue eyes, in 1985 the situation was such that I had not written to the man I wanted to be the successor'

g. 'because of his blue eyes, my writing to the man I wanted to be the successor had not been in 1985'

h. 'I had not written to the man who, in 1985 and because of his blue eyes, I wanted to be the successor' and so on

The number of possible readings is doubled when the verb *succeed* is read as 'be successful'. The number is tripled when the readings are added where the phrase *the man I wanted to succeed* is read as 'the man such that I wanted to succeed him', rather than as 'the man such that I wanted him to succeed (be the successor)'.

The conclusion must be that the parsing process gone through by a listener is not a straightforward inversion of the procedure followed during production. But if it is not that, what is it? We know that those elements that lead to an explosion of the parsing process must find the proper interpretation with the help of nonlinguistic factors, in particular world knowledge, situational knowledge, and information stored in the discourse domain at hand, possibly reflected in topic–comment structure.

Suppose the discourse has been about a man whose successor I want to be. This settles the threefold ambiguity of *the man I wanted to succeed.* Suppose further that preceding discourse contains the information that the succession I desired was due in 1985. That places *in 1985* in construction with the relative clause *I wanted to succeed.* Assume, furthermore, that the discourse contains the information, introduced by previous text, that prior to 1985 I had written to the man I wanted to succeed in 1985, and that, at the point in the discourse where (7.33) is uttered, the question has come up why I had written to the man I wanted to succeed in 1985. Now sentence (7.33) has a topic–comment structure reflecting the fact that it serves as an answer to that question—a negative answer: it was not because of his blue eyes! In the context described, the sentence has now suddenly lost its multiple ambiguity and is effortlessly and immediately interpreted with just one reading. This shows that the programming of the parsing process cannot be successfully carried out without a formal theory of discourse incrementation and of the anchoring of new utterances in given discourse, possibly reflected in topic–comment structure and the appropriate intonation pattern. No such theory, however, has so far been developed to a sufficient degree of explicitness and completeness. But starts have been made (Seuren 1972*b*, 1980, 1985; Kamp 1981; Kamp and Reyle 1993).

Then, there is evidence from psycholinguistics (Ritchie 1983; Konieczny *et al.* 1997; Cutler and Clifton 1999: 124, 142) that comprehension involves an early identification of key lexical items, complete with the corresponding concept—evidence that is confirmed by studies in second language acquisition (and by ordinary experience). Ritchie writes:

> Psycholinguistic evidence seems to suggest that people absorb much (or all) of the semantic content of a sentence on a word-by-word basis, but it is very hard to design a clean, systematic, theoretically defensible way of programming such a mechanism. (Ritchie 1983: 217)

Early lexical identification suggests that listeners jump to a reconstruction of the most likely relevant scope and predicate–argument relations in the underlying intent structure. Any such reconstruction is constrained, as we have seen, by the information contained in the given discourse structure, supplemented by encyclopedic world knowledge and also by whatever is readily reconstructible on grammatical grounds, including syntactic, morphological, and intonational clues. The early availability of the key lexical items in an utterance entering the comprehension system, together with a plausible scenario giving rise to an initial semantic structure in which they have found a place, automatically sets into motion the grammar module, which cannot help delivering a mental print-out of a corresponding surface structure.

This again suggests that the comprehension process is essentially a process of reconstruction-by-hypothesis, or analysis-by-synthesis, involving tacit appeals to general knowledge, combined, of course, with strict constraints imposed by the sound or writing input and by the grammar, which now functions mainly as a feedback or control mechanism. It would also put the parsing process beyond present limits of algorithmic (modular) modelling, which would explain the limited success of the massive efforts made in computational linguistics to produce mechanical parsers.

There is, furthermore, the fact that adequate real-life comprehension of uttered sentences often takes place despite defective input or despite insufficient knowledge of the language in question on the part of the listener—a fact that has invariably plagued all formal parsing theories on the market. Subjects, especially second language learners, often understand the sentences of a language L perfectly well, picking up all sorts of semantic, stylistic, and sociolinguistic subtleties, but are not able to produce correct sentences or to distinguish correctly between well-formed and non-well-formed sentences of L. Such persons can hardly be said to be fully competent in L. To be considered fully competent in a language, one must have full *active* competence, in that one is able to *produce* flawless sentences that express precisely what is intended, and to avoid flawed sentences. When that is achieved, the speaker will automatically also have full *passive* competence, that is, competence with regard to comprehension. It thus seems that competence is to be measured by the capacity to *produce* syntactically correct sentences for given meanings, not by the capacity to *comprehend* sentences, whether or not they are syntactically correct.

It is fairly obvious that the phenomenon of correct comprehension despite defective input should be seen in the light of the principle of OPTIMIZATION OF SENSE, TRUTH, and ACTUALITY, or the OSTA Principle, introduced in Section 3.4.2. This principle, however, though explanatory in its own right, does not lend itself to formalization, given available techniques, and is thus well beyond the modelling powers of present-day language technology. This being so, it should come as no surprise that available formal models for sentence parsing fail to achieve acceptable levels of success.

In addition, one notes that an analysis of the points of conscious access and control (self-monitoring) during the processes of production and comprehension indicates a top-down, rather than a bottom-up flow. This is implicit in the fact that self-monitoring, as a known psycholinguistic process, is OUTPUT monitoring, not INPUT monitoring. As has been said, the monitoring always consists in checking a top-down output result, given a particular input, which may be an intent, as in the case of production, or, as in the case of comprehension, sound or writing. If the input consists in sound or writing and self-monitoring is top-down, it follows that the parsing process involves a top-down route, which is only possible if it is at least partly defined as a process of reconstruction-by-hypothesis.

Finally, recent neurophysiological research has revealed (e.g. Rizzolatti *et al.* 2001; Arbib 2002) that the pre-motor area F5 in the monkey brain as well as the corresponding Broca area in the human brain contain a 'mirror system' activated for both the execution and the observation of manual actions. Highly localized probings ('single unit recordings') have shown that the F5 area in the monkey brain, which is used for the pre-motor planning of actions, is also activated, as if by a 'mirror' effect, when actions executed by other monkeys or by humans (such as the grasping for an object) are perceived and interpreted. The authors in question assume that the F5 area, like the corresponding Broca area in humans, contains so-called mirror neurons. Similar findings, using transcranial magnetic stimulation (TMS), have been reported for the perception and interpretation by humans of human speech sounds (Fadiga *et al.* 2002; Watkins *et al.* 2003). This appears to indicate that generally, in higher vertebrates, perception and interpretation of actions, including speech acts, are processes that are somehow dependent on, or secondary to, the planning and execution of corresponding actions, which are mirrored in the interpretation process. Not until the perceiver has gone through the mental process of preparing for the action perceived can he or she reliably identify and interpret the action.

There are, however, a couple of important caveats. If the 'mirror' requirement is taken in too absolute a fashion, it becomes impossible to explain why interpretation always has an edge on active competence. People watching a football match would, on the whole, never be able to emulate the players in the field but may be very good at judging what they do. And competent speakers of a language can always understand more than they can say. Moreover, the 'mirror' requirement lands one in a predicament regarding the acquisition of manual and linguistic skills. For it would follow that if the perceiver is lacking in competence regarding the execution of an action, whether manual or linguistic, (s)he would be unable to interpret it. Since it is clear that the 'mirror' requirement does not *fully* determine interpretation, it remains to be seen *to what extent* it does.

There is the further interesting fact that the 'mirror' effect seems closely related to the principle argued for in Section 2.2.2, with regard to the perception and interpretation of sense data in general, that there can be no interpretation without mental representation and hence prior virtual reality. That is, interpretation in general requires the creation of a virtual object, state of affairs, or event before any actual object, state of affairs, or event can be seen as instantiating it. Yet the 'mirror' effect phenomenon goes further in that it also activates the very neurological structures that are required for the actual production of the action at hand. This further step appears to be limited to the observation and interpretation of *intentional acts* performed by human congeners or by members of a closely enough related nonhuman species.

If this is correct, we have an extremely interesting neurological confirmation of the hypothesis that higher vertebrates (and perhaps other species as well) have a special sensitivity with regard to any sense data that may play a role in processes of personal interaction with members of the same or a closely related species. Such a finding would be of great importance to any theory of personal interaction, as it would show a neurologically innate proclivity to empathize with, or project oneself onto, other beings presumed to be similar in certain cognitive and emotional respects. De Waal (2006) defends the thesis that there can be no social strategy and no morality without empathy. What transpires from studies on language comprehension would suggest that there can be no comprehension of linguistic utterances without empathy. Likewise, it would seem that the qualia problem, discussed in Section 8.2.2, cannot be solved without the assumption of an innate capacity of empathy. These, however, are aspects that must be allowed to rest in the present context.

Townsend and Bever (2001: 160–72) present the view that utterance comprehension is, in principle, a process of reconstruction-by-hypothesis or analysis-by-synthesis, harking back to theories of motor-driven phoneme perception developed in the Haskins Laboratory during the 1960s, and extending these to early recognition of lexical items and hence to modular grammatical generation. In their view, the reconstruction or synthesis is driven *lexically* by the lexical items found but *configurationally* by knowledge of plausible scenarios.

These authors present their theory of utterance comprehension as 'entirely consistent with the minimalist syntactic model' of Chomsky (1995), which they call 'a rather compelling model' (Townsend and Bever 2001: 178–9). The rationale underlying this view is that the arbitrary set of lexical items, which is the starting point of a minimalist syntactic derivation, is, in these authors' eyes, naturally provided by the comprehension system once the lexical items have been identified. This may be so, but that does not mean that the 'configurational' part of the process of reconstruction-by-hypothesis is driven only by plausible scenario knowledge. If it were, it would be impossible to identify sentences that do not fit into a given 'scenario': for such an identification to be possible the configurational part of the comprehension process must receive the final approval of the grammar, regardless of contextual factors. The grammar, therefore, cannot be done without. But Chomsky's initial set of lexical items does not contain any grammar. The set of items {*hit*, *John*, *Bill*} gives both *John hit Bill* and *Bill hit John*, both of which may be equally plausible in a given 'scenario'. Worse, if it is assumed that *do*-support is added during the derivation, the set of items {*not*, *many*, *all*, *players*, *referees*, *hate*} may give (7.35a–7.35h), all of which may be equally natural in a given context:

(7.35) a. Not many players hate all referees.
b. Not all players hate many referees.
c. Not many referees hate all players.
d. Not all referees hate many players.
e. Many players do not hate all referees.
f. All players do not hate many referees.
g. Many referees do not hate all players.
h. All referees do not hate many players.

The minimalist model thus fails as a production model for basic reasons. It may provide the lexical items for a derivation, but it fails to provide the 'configurations'. By contrast, Generative Semantics (or Semantic Syntax) will foot the bill, as it provides for both a lexical and a configurational input to the grammar while also leaving room for contextual and other cognitive input supporting successful comprehension (Seuren 2004*a*: 85; 2004*b*: 598).

In the light of these considerations, in particular the fact that parsing is co-determined by general and situational (including discourse-stored) knowledge, it is at least doubtful and probably just false that the parsing process is adequately describable in terms of full modularity. On the other hand, the production process is successfully described as a self-contained production algorithm once the sociolinguistic and interactional parameters have been set, lexical selections have been made and the checklist has been gone through, and hence as modular in the sense indicated. For this reason, we prefer to see the grammar module as designed for production and thus naturally modelled as a production algorithm. The process of comprehension is taken to be only partly algorithmic, a large part being of a higher order, involving access to a knowledge base and to given discourse structure. One knows, of course, that the mechanisms underlying these latter aspects of comprehension have so far escaped any formal treatment that could lay claim to a realist account of the mental processes involved. Even so, we feel that the perspective of analysis-by-synthesis, or reconstruction-by-hypothesis, provides a fruitful basis for further research in this area. These matters will not, however, be pursued any further here, since they are marginal to the main topic of this book.

7.3 The machinery of truth-value assignment

Figure 7.10 shows the overall structure of the truth-value assignment machinery applied to a token occurrence of the tenseless L-proposition Graze (the horse), corresponding to the sentence *The horse is grazing.* The analysis is limited to the matrix-S, which represents the grammatical 'main frame' of the sentence. Full L-propositions contain, in addition to the matrix-S, a great deal of higher auxiliary structure, expressed in the form of higher operators (abstract predicates), such as tense, modality, negation, conjunction, or disjunction operators, quantifiers, adverbial modifiers, and the like, which define modes of anchoring and keying as well as other delimitations of the application range of the proposition concerned. In the present context, however, we neglect the auxiliary structure of the L-propositions concerned and limit the discussion to the matrix-S containing the main lexical predicate of the sentence in the making.

First we consider the machinery for truth-value assignment to the mere proposition as a mental occurrence, without any corresponding L-proposition. The proposition assigns the property of grazing to a particular horse intentionally singled out for reference, say the horse Sirius. The assignment takes the form of adding the satisfaction conditions defining the concept 'graze' to a given object representation, say discourse address **d–8** (though it may be an object representation unconnected with any on going linguistic discourse, as in the case of the sun, the moon, the world, and so on), which mentally represents (is intentionally keyed to) the horse Sirius (step 0 in Figure 7.10). In order not to complicate matters, let us take it that Sirius is an actually existing horse, and not a virtual horse, and that **d–8** is keyed to Sirius (step 3 in Figure 7.10). Now Sirius is tested for satisfaction of the conditions defining the concept 'graze' (step 4). When he passes the test, the proposition is true, but when he fails it, the proposition is false (step 5a). So far, the whole procedure takes place in the mind of the thinking individual: there is no speaker and no listener.

But the proposition can be fed into the modular complex for language and thus be linguistically expressed as an L-proposition and eventually as an utterance. The matrix-S of this L-proposition consists of the predicate *Graze* and the definite nominal subject term the x (horse(x)), corresponding to the surface form *the horse.*

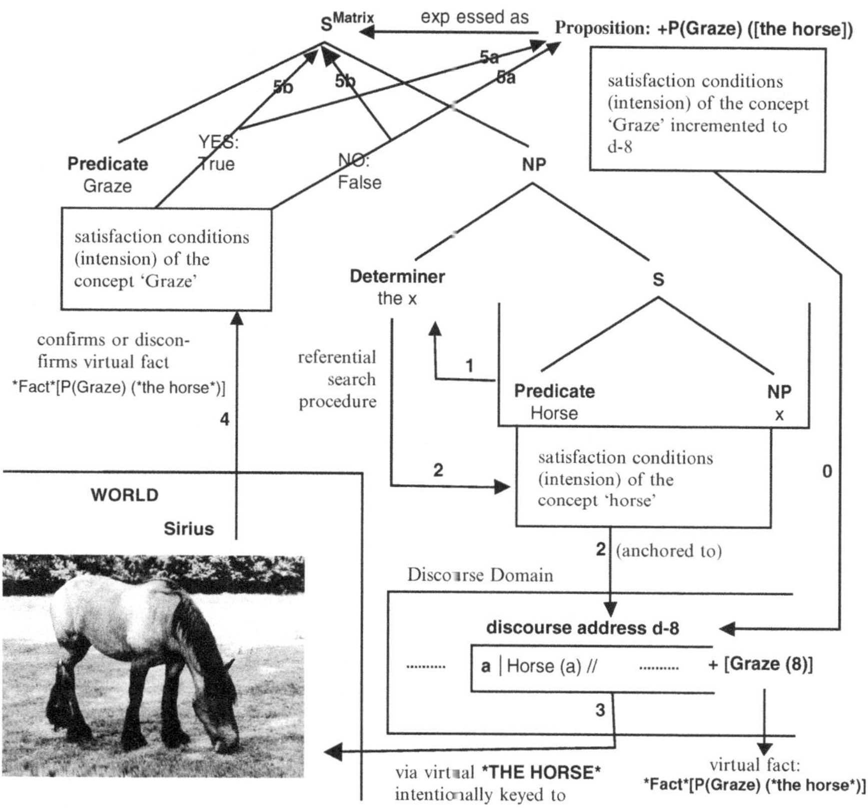

FIGURE 7.10 Truth-value assignment for the L-proposition Graze (the horse)

Assuming that the listener has grasped the structure of the matrix-S, the propositional function Horse(x) is fed into the Determiner function (step 1 in Figure 7.10), which triggers the *referential search procedure.* The first step in this procedure (step 2) is to locate the corresponding DA, say **d–8**, which is required for further processing.[18] Often the corresponding DA is lacking. The

[18] Some highly relevant research has been done recently, with the help of neuroimaging techniques, into the brain processes involved in DA-assignment procedures (see in particular Nieuwland *et al.* 2007 and Van Berkum *et al.* 2007 and the references cited there). Although these articles speak about 'reference', it should be borne in mind that they deal, in fact, with the cerebral procedures involved in the identification (or setting up) of DAs, not with the processes involved in the intentional keying of definite noun phrases and their corresponding DAs to actually existing objects in those cases where such reference objects are available. What these findings show is that certain brain areas (in particular the medial frontal and parietal cortices) are involved in *problem solving* when a pronoun fails to land immediately at any of the DAs given. This does not necessarily say something about the brain areas involved in *nonproblematic* cases of DA assignment, since it may well be the case that *general* problem-solving procedures are invoked only when the *specific* assignment procedure runs aground. When speaking about DA assignment, Nieuwland *et al.* state correctly:

normal procedure, in such cases, is to slip in the missing address post hoc in virtue of a process known as *accommodation* or *post hoc insertion*. However, any form of post hoc insertion must be screened by cognition. For example, when there is a DA for a human face, a DA for a nose is naturally supplied post hoc, but not for two noses, or for a lamp post, or whatever does not normally belong to a face.

Once the DA has been located, the object to which it is intentionally related (keyed), the actual horse Sirius, is fixed upon (step 3). The reference object, Sirius, is then tested for the satisfaction conditions associated with the main predicate of the matrix-S, the predicate *Graze* (step 4). When Sirius passes the test, both the (token utterance of) the matrix-S (step 5b) and the underlying proposition (step 5a) are true, otherwise they are false.[19]

In either case, the matrix-S is added ('incremented') to the DA involved, say **d–8** still in the form of (7.36a). After incrementation, the address **d–8** has the form (7.36b), to be read as 'the horse grazes':[20]

(7.36) a. **d–8** [a | Horse(a)]

b. **d–8** [a | Horse(a) // Graze(**8**)]

Now suppose someone has two DAs, say **d–8** and **d–9**, both intentionally related (keyed) to the same horse Sirius. A good reason for there being two different addresses may be that the person in question believes there to be two distinct horses, the horse Sirius and the horse Bronco, not knowing that they

However, with the exception of a few studies focusing on cohesion/coherence processing in discourse comprehension..., there has been no functional magnetic resonance imaging (fMRI) work on the neural substrate for referential [=DA-assigning; PAMS] processing. As a result, we know virtually nothing about how the brain subserves a core aspect of language comprehension, how people work out who's who and what's what in text or conversation. (Nieuwland *et al.* 2007: 994)

[19] The process is more complicated in cases like:

(i) The first Americans landed on the moon in 1969.

(ii) Most people get chicken pox before the age of ten.

Both sentences are ambiguous, but their most natural or prominent readings are structurally peculiar. Sentence (i) preferably does not say that the President of the United States and his or her spouse landed on the moon in 1969, but rather that the first Americans who landed on the moon did so in 1969. Similarly for sentence (ii), which preferably does not say that of all the people in the world most get chicken pox before the age of ten, but rather that most people who get chicken pox do so before they are ten. The analysis of this kind of sentence is left open here.

[20] The canonical way of introducing addresses is by means of existential quantification. The existentially quantified address for the case at hand is represented as (7.36a), to be read as 'there is/was a horse' with Horse(a) standing for the propositional function Be a horse and '**a**' for the (Russellian) existential quantifier. For reasons discussed in Section 8.1 in Volume II, existential addresses are 'closed' (symbolised as '//') as soon as they are used as a reference anchor. The result of address closure is that a changes from being an existential quantifier to being a definite reference functor. The sentence *The horse grazes* then extends (7.36a) to (7.36b), to be read as 'the-horse-introduced-by-means-of-**d–8**-before-closure grazes'.

are the same animal, which is officially registered as 'Sirius' but usually called 'Bronco' on the farm. Let the two addresses, after incrementation, be rendered as (7.37a) and (7.37b), respectively, to be read as 'the object *a* such that *a* is a horse and *a* is called ["Sirius"/ "Bronco"] grazes':

(7.37) a. **d–8** [**a** | Horse(a), Called 'Sirius'(a) // Graze(**8**)]

b. **d–9** [**a** | Horse(a), Called 'Bronco'(a) // Graze(**9**)]

Whether the property of grazing is added to **d–8** or to **d–9** makes no difference for the truth-value of the proposition or of its linguistic (L-propositional) expression Graze(Sirius) or Graze(Bronco), because the truth testing is carried out on the horse, not on the DA involved or on the term used. Yet in either case there is a different underlying proposition. And since it is this underlying proposition that is the ρ-value for object terms under predicates like *believe*, and since believing is defined not as being about truth *per se* but about *being willing to accept* a proposition as being true, it follows that the embedded L-proposition Graze(Sirius) cannot be freely substituted *salva veritate* for Graze(Bronco), or vice versa, in sentences like:

(7.38) a. Henry believes that Sirius grazes.

b. Henry believes that Bronco grazes.

In conclusion, a word must be said about the spirit in which the analysis spelled out above should be taken. It is most certainly not meant to be the final word about the matters discussed. On the contrary, it is almost certainly deficient in a number of ways. But it is intended as a stimulus for the reader to engage in a conceptual analysis of the problems involved. Such an analysis, if properly carried out, should be a useful tool for psycholinguistic experimenters and brain-imagists who are looking for the psychological and physiological substrate of linguistic functions. In this respect one should not be overoptimistic and expect the conceptual analysis to be confirmed immediately by psycholinguistic experiments and neuroimaging techniques. On the contrary, whatever little is available in the way of experimental results appears to suggest at best some degree of incompatibility and at worst a straightforward clash with most or all theoretical models proposed so far. It seems to me that the proper response in such situations is to be patient and reflect and experiment further on the issues involved. It is not only from the clash of opinions but also from the clash of results that light tends to spring.

8

Sentence meaning and lexical meaning

8.1 Signs, sentence meaning, and token interpretation

It is high time, meanwhile, for us to make explicit what we mean by 'meaning'. In answering this question we shall look at the notion of meaning in a wider context than just language. But not in too wide a context. For example, the verb *mean* is frequently used, in common speech, to denote the all-important general phenomenon in human and animal life of having a degree of importance, as in *Only then did she realize what her family had meant to her.* Obviously, the phenomenon itself of something or somebody having importance for someone antedates human language by a long shot, but the use of the English word *mean* to refer to it is relatively recent.[1] This use of *mean* is the result of normal processes of extension of lexical meaning, in this case of the word *mean*, but we do not consider this use in our analysis of linguistic meaning. Instead, we concentrate on meaningful forms or percepts as consequences of nonperceptible causes, which, in the case of linguistic meaning, are the nonperceptible intents causing utterances to come about according to some specific algorithmic system—that is, the lexicon and the grammar of the language in question.

We take it that MEANING IS A PROPERTY OF SIGNS. This may seem a platitude, but the importance assigned to the notion of sign in earlier ages may well make us think otherwise. Yet, for no apparent reason, this notion has, strangely, vanished from the theoretical literature on language and meaning. This disappearance can perhaps be understood better when one takes into account the history of the notion of sign, which not only is long and chequered but also bears the burden of intellectual and academic shadiness. Whereas the ancient Stoic philosophers still treated the notion of sign in a very serious manner which is still relevant today, the highly influential St Augustine and his theological followers in the Middle Ages and later introduced so much

[1] The multivolume *Oxford English Dictionary* (Supplement) places its first occurrence in 1888: '... all that the squire had meant to him during the past year'.

fluidity that it became unfit for a professional, disciplined theory of language. Nowadays the notion is current in various kinds of holistic or otherwise nonanalytical and sometimes highly subjectivistic philosophies, and notably also among followers of the wayward but original late-nineteenth-century American philosopher Charles Sanders Peirce.

It is probably because of this background that more analytically oriented philosophers and semanticists look upon 'the sign' with more than a little suspicion. We do not share that suspicion, because it is not inherent in the notion of sign to be the object of mystifications. On the contrary, we agree with the ancient Stoics, who, for good reasons, found the notion entirely serious. Any theory of knowledge and meaning, whether in philosophy or in psychology, needs a theory of signs as an important basic element. It is a mere accident of history that the notion of sign has come to find itself under a cloud of disrespect in modern times (see Seuren 1994*a* for further comment).

Our definition of sign runs as follows:

> SIGN:
> At token level, a sign is a perceptible form F that the perceiver knows can have come about only in virtue of a necessary cause C, which, at the occurrence of F, is itself not perceptible. Signs are mentally classified as type-level units. At type level, a sign S is a mental procedural unit allowing a knowing perceiver to infer C from any token occurrence of F.[2]

Accordingly, meaning in a general sense can now be defined as follows:

> MEANING:
> The meaning of a sign S is the systematic property of S of being able to bring about knowledge of the present or past occurrence of its necessary cause C, despite the imperceptibility of C.

This applies not only to so-called NATURAL SIGNS, such as smoke being a sign of fire, or a footprint in the sand being a sign of some animate being having trodden on it, but also to so-called CONVENTIONAL SIGNS. Conventional signs differ from natural signs in that the cause C of a natural sign is a natural cause, whereas the cause C of a conventional sign is the decision taken by a thinking individual to express, and by expressing it to enact, through a socially shared system of signs, an intent, consisting of a socially binding operator and a proposition. Such signalling of intents usually takes place in terms of a symbolic system, which calls for some degree of conventionalization. That is why the tradition has always spoken of conventional signs in those cases. And

[2] When it is reliably known that C is not a necessary but merely a sufficient condition for S to occur, or that S and C regularly co-occur, we tend to call S a *symptom*, not a sign.

since linguistic utterances are a subcategory of conventional signs, a linguistic utterance is a sign of the speaker's decision to express, and by expressing it to enact, an underlying intent, and thus of the presence of that intent.

In giving these definitions of 'sign' and 'meaning' we place ourselves in the CAUSAL TRADITION of defining signs, thereby putting aside the better known ASSOCIATIVE TRADITION, which sees signs as 'standing for' or 'representing' either an object or class of objects or, more plausibly, a corresponding concept or a mental representation. The reason why this latter tradition is better known is probably the fact that it is more directly useful for the description of lexical meaning, and lexical meaning has always appealed more to the popular mind than sentence meaning.

While the usefulness of this second notion of sign, in so far as it applies to lexical meaning, is not in doubt, it would seem that the associative notion fails to provide a satisfactory account covering the overall notion of sign. It is often said, in the associative tradition, that, for example, smoke 'stands for' fire, but it is unclear in what sense this can be taken to be the case. Surely, to be able to interpret smoke as a sign of fire it is not necessary to replace a perception of smoke by a perception of fire. Firemen would be in serious trouble if they confused fighting smoke with fighting fire. Other interpretations of 'standing for' meet with similar problems. It seems that the only reasonable answer is to adopt a causal theory of signs, as defined above. It is in that vein that we now proceed.

For a natural or conventional sign S to function as a sign in any given token situation, it must have been learned: the interpreter must not only interpret the token occurrence of S in terms of a type-level category of occurrences, but must also have knowledge of causal relations.[3] Since knowledge is organized in types, a token occurrence of a (natural or conventional) sign S can only function as a sign because it belongs to a *sign type.* With natural signs the type–token distinction is usually taken for granted: it is equally natural to say, at type level, that the meaning of smoke is that there is fire as it is to say, at token level, that the meaning of that smoke over there is that there is fire now in the spot the smoke is coming from.

The type–token distinction applies with equal force to linguistic signs, but it has become customary in linguistic semantics to use different terms for the meaning of sentence types and of utterances tokens. Since the middle of the twentieth century it has been normal practice, in circles of nonformal, structuralist, semanticists, to speak of *meaning* as a type-level property of

[3] Sometimes what passes for causal knowledge is, in fact, belief in magic. In such cases we, who know better, only speak of signs with the necessary qualification.

sentences, and of *interpretation* as a token-level process involving utterances. This terminological distinction seems useful and we adopt it.[4]

Linguistic utterances can only be signs (expressions) of underlying token intents in virtue of being tokens of a given sentence type, analysed as representing an intent schema or intent type, which consists of (a) a schematic proposition p in which the parameters that are to be filled in by token values have been left unspecified, and (b) a superordinate type of social binding with regard to p. We thus define the notion of sentence meaning as follows:

> SENTENCE MEANING:
> The meaning of a sentence S as a potential perceptible utterance token is the system-driven, type-level property of S to reveal the speaker's decision to make public an underlying intent (consisting of a socially binding operator and a proposition) and thus to enter a commitment, or put out an appeal, or set a rule of behaviour, with respect to the proposition expressed at any token utterance occurrence of S.

Thus the meaning of the type-level sentence *The policeman caught the thief* is its property of being usable for the expression of an assertion of a proposition conforming to the schema in which the property of catching is assigned to whoever is referred to by the noun phrase *the policeman* as the catcher and to whoever is referred to by the noun phrase *the thief* as the catchee. The application of the linguistic meaning of a sentence type to a token state of affairs is called the *interpretation* of the sentence uttered in the discourse/situation at hand. Any actual token utterance of the sentence *The policeman caught the thief* is interpreted as revealing the speaker's decision to utter the intent in which the speaker vouches for the truth of the proposition in which the property of catching is assigned to the actual person referred to by the phrase *the policeman* as the catcher and to the person referred to by the phrase *the thief* as the catchee.

The more mathematically minded might prefer a formulation in terms of functions, such as, for example:

> SENTENCE MEANING:
> The meaning of a sentence S is a function from the token utterances of S to the intent expressed in S.

Yet although such a formulation is probably ultimately correct, one should realize that the interpretation of a given sentence meaning is not a fully

[4] The distinction between type-level meaning and token interpretation has been known at least since the days of European structuralist semantics (e.g. Antal 1964). It was elaborated in Sperber and Wilson (1986), although these authors make exaggerated and merely promissory claims about the formal treatability of the factors at play in interpretation over and above the recognition of meaning (see Seuren 1987).

compositional process. Interpretation depends not only on the linguistic material in any given utterance but also, and in ways that are essential to the interpretation process, on situational, contextual, and other cognitive factors requiring processing in nonlinguistic parts of cognition, which are not, as yet, open to formal treatment. That there is likely to be a function of some sort, probably a very complex one, appears from the fact that humans mostly understand utterances in their language unambiguously: interpretations are, on the whole, uniquely determined. But the function itself, if any, is, as yet, beyond definition.

The interpretation is normally much richer than what one would expect a token instantiation of the sentence meaning to be, owing to the fact that utterances are keyed to a given state of affairs and cognitively anchored in a given context, an integral part of which consists of general and episodic knowledge. Moreover, utterances contain lexical items whose meanings are also strongly codetermined by world knowledge and other cognitive factors. These factors tend to contribute to the interpretation of an utterance in ways that go beyond a simple filling in of ρ-values for the argument terms and values for the temporal parameters that specify time. No theory has been made available so far that provides a systematic account of exactly what is supplied in utterance interpretation over and above the recognition of sentence meaning and how this happens, although attempts have been made in pragmatics, in particular in Sperber and Wilson (1986) and other publications by the same authors (see note 4).

Besides full sentence meaning, there is also the meaning of the propositional part P of a sentence *S*, the propositional meaning of *S*, defined as follows:

> PROPOSITIONAL MEANING:
> The meaning of the propositional part P of a sentence *S* is the system-driven, type-level property of P to reveal the underlying mental proposition *p* at any token occurrence of P.

Or, if one prefers:

> PROPOSITIONAL MEANING:
> The meaning of the propositional part P of a sentence *S* is a function from any token utterances of P to the proposition expressed in *S*.

But the reservation expressed above with regard to a definition of the concept of sentence meaning in terms of mathematical functions is as valid here as it is there.

8.2 Lexical meaning

8.2.1 *What is lexical meaning?*

Whereas we speak of linguistic meaning as a systematic property of sentence types, interpreted in actual token contexts, this is not what is commonly found in the professional literature, where one speaks of meaning primarily as LEXICAL MEANING or WORD MEANING, which is also the normal use of the term in popular, nonprofessional contexts. Correspondingly, in whatever little there is in the way of literature on signs, one commonly finds that signs are equated with words, rather than with uttered sentences. But this is not the perspective developed here. In our causal analysis of signs, linguistic meaning is primarily sentence meaning, that being the only viable explication of the notion of meaning in a general sense, which includes both natural and conventional signs, and, within the latter category, both linguistic and nonlinguistic signs.[5]

This was also the view held by the Oxford philosopher John Austin and expressed in his 1940 paper 'The meaning of a word', posthumously published in Austin (1970: 55–75), where we can read:

> A preliminary remark. It may justly be urged that, properly speaking, what alone has meaning is a *sentence*. Of course, we can speak quite properly of, for example, 'looking up the meaning of a word' in a dictionary. Nevertheless, it appears that the sense in which a word or phrase 'has a meaning' is derivative from the sense in which a sentence 'has a meaning': to say a word or phrase 'has a meaning' is to say that there are sentences in which it occurs which 'have meanings'; and to know the meaning which the word or phrase has, is to know the meanings of sentences in which it occurs. All the dictionary can do when we 'look up the meaning of a word' is to suggest aids to the understanding of sentences in which it occurs. Hence it appears correct to say that what 'has meaning' in the primary sense is the sentence. (Austin 1970: 56)

Austin then proceeds to pose the question of what it is that makes words help build the meanings and interpretations of sentences as efficiently as they do. In his own words:

> Leaving historical disputes aside, it is a matter of urgency that a doctrine should be developed about the various kinds of good reasons for which we 'call different things by the same name'. This is an absorbing question, but habitually neglected, so far as I know, by philologists as well as by philosophers. Lying in the no man's land between them, it falls between two schools: to develop such a doctrine fully would be very

[5] Interestingly, in English the preposition *of* is used for signs as defined here, but *for* for lexical predicates. Thus the word (predicate) *shoe* is a sign *for* a particular class of objects made for covering and protecting feet, but smoke is a sign *of* (there being) fire, and an utterance is a sign *of* (a decision to express) an underlying propositional intent.

complicated and perhaps tedious, but also very useful in many ways. It demands the study of *actual* languages, *not* ideal ones. That the Polish semanticists have discussed such questions I neither know nor believe. Aristotle did to a quite considerable extent, but scrappily and inexactly. (Austin 1970: 70)

He thus not only stresses the need for an adequate theory ('doctrine'), but also emphasizes that all those current formal and less formal theories that are based on the assumption of one or more fixed meanings for any given word must be taken to fail to fit the facts: 'We are using a working-model which fails to fit the facts that we really wish to talk about' (Austin 1970: 63).

The question of what 'working-model' might perhaps fit the facts more adequately he, somewhat dodgingly, answers as follows, suggesting that his concern for an adequate theory might not be as pressing as he just said:

To summarize the contentions of this paper then. Firstly, the phrase 'the meaning of a word' is a spurious phrase. Secondly and consequently, a re-examination is needed of phrases like . . . 'being a part of the meaning of' and 'having the same meaning'. On these matters, dogmatists require prodding; although history indeed suggests that it may sometimes be better to let sleeping dogmatists lie. (Austin 1970: 75)

This amounts to saying 'We don't know'—an answer to which I have no hesitation to subscribe. Yet it also suggests that 'dogmatists' (Austin's term for what I prefer to call 'theoreticians' or 'theorists', even though these terms do not have *dog-* as their first syllable) were asleep around 1940 and that it might be better to leave them in that state—perhaps because he felt that the chances of doing better than Aristotle, whose efforts in this direction he calls 'scrappy' and 'inexact', are slim indeed.

Unfortunately, it is this a-theoretical, or perhaps even anti-theoretical, slant in Austin's, otherwise sharp, thinking that has caught on among the pragmatically inclined linguists and philosophers who deal with these questions today. A tendency has developed lately to rake up the old view, current during the 1950s and represented by such diverse people as the philosopher Ludwig Wittgenstein and the linguist John R. Firth, that the concept of word meaning should be abandoned altogether and replaced with a notion of the 'uses' to which a word is put, where these 'uses' are taken to be determined, or at least co-determined, by linguistic and extralinguistic context.

Regrettably, it still appears to be necessary to point out that this leads nowhere. At best, such views merely state the problem (and not in a very useful way at that). At worst, they lead straight into inconsistency, since if linguistic and extralinguistic context fully determine the 'use' of a given word, then this word cannot, in turn, contribute to the meaning or interpretation of the sentence in which it occurs, and thus change the linguistic and extralinguistic context. The central question is and remains:

How do words contribute to sentence meanings and utterance interpretations and, in particular, what are the limits of possible 'uses' or, rather, meaning applications, of words in a language?

This is the question we are facing and to which we must admit having no satisfactory answer as yet. The inevitable result of our ignorance in this respect is that talk about lexical meaning quickly lands us with the bric-à-brac of a lexical curiosity shop, as is amply evident from, for example, Firth's writings on word meaning, but also from Chapter 9 in the present book, whose purpose it is to give the reader an idea of the phenomena concerned and thus of the magnitude of the question at hand. (Little wonder that Aristotle's treatment of the phenomena concerned may well strike one as 'scrappy' and 'inexact'!)

Although I agree with Austin that the variability and adaptability phenomena of lexical meanings pose a huge problem that is still largely unresolved, my inclination is not to shy away from that problem, but rather to be patient and see what can be done in the circumstances. If the circumstances are such, as I believe they are, that the failure to come up with an adequate solution is due precisely to the fact that we lack sufficient knowledge of the cognitive structures and processes underlying concept formation and concept application, all we can do is try to prepare the ground as best we can—and not be too impatient.

Despite the uncertainties regarding lexical meaning, it still seems best to follow tradition in that a distinction is drawn between, on the one hand, the lexically fixed, type-level meaning (or, more strictly, meaning function) of any given lexical item (predicate) and, on the other, its token-level possible applications in any given linguistic and extralinguistic context.

This position is risky for at least two reasons. First, because it has so far proved impossible to develop a systematic and adequate method for the analysis and description of type-level lexical meanings, and second because there isn't even the beginning of a theory of their possible token-level applications. Yet the distinction at hand seems both theoretically inevitable and intuitively right. As regards type-level lexical meanings fixed for each word (predicate) in the lexicon of a language, we pin our hopes on cognitive research into concept formation. As regards token-level applications, our hopes are placed in a theory of the composition of concepts into larger wholes, in particular propositional structures. Until such theories have been properly developed, the reader must bear with us, as we must all bear with the situation as it is.

8.2.2 *The lexicon and cognition: the qualia problem*

One thing, however, is clear. The lexicon is inextricably and essentially linked up with general cognition through unique private experiences that owe their identity to their specific phenomenal or experiential quality, not effable through analytic descriptions or paraphrases. In philosophy, one often finds the term QUALIA in connection with this category of phenomena, sometimes to refer to qualia sensations, but sometimes also to refer to whatever physical properties are taken to underlie these sensations. I adopt this term, but only to refer to the sensations, which provide the concept, not to the physical properties, which are too diverse to be captured in the terms of the phenomenal qualia categories, as is stated by Musacchio:

> The evidence from neuroscience indicates that the relationship between experiences and their realizers is that of a contingent identity. (Musacchio 2005: 422)

Qualia are the basic irreducible elements of experience.[6] They define what it is for an itch to be an itch, or for a sour taste to be sour. In some cases, qualia can be communicated by describing a typical situation in which they occur. For example, Dutch has a word *tintelen*, whose closest equivalent in English is *tingle*. It denotes the typical feeling one has in one's hands, feet, or cheeks when one enters a warm room from the freezing cold. In general, the communicative transfer of qualia meanings requires recognition by fellow humans of the experience in question. Description in terms of their physical correlates is useless, since, in most cases, what may cause a quale does not form a coherent physical category, even though qualia predicates *denote* the physical correlates, not the experiences or sensations. The satisfaction condition of a qualia predicate such as *red* thus refers to the ontological property 'causing the experience of red in human beings', no matter how this experience is caused: the experience itself is to be retrieved from phenomenal recognition.

Some meanings, by contrast, are composite in that they allow for a precise analysis of their satisfaction conditions in terms of other lexical items. The German noun *Ermittlung*, for example, is adequately rendered as 'judicial investigation', as it denotes not just investigations but, more precisely, investigations carried out or ordered by the authorities for legal reasons. The word *Ermittlung*, therefore, does not express a single quale, even though judicial investigations may well carry special connotations and associations in the experience of some speakers.

[6] I remain neutral as to the issue of what it is to have an experience and what experiences consist of in neurophysiological terms, although I am tempted to take sides with those, such as Jackson (1982) or Block (2002), who claim that there is an 'explanatory gap' in the sense that what specialists know about neurophysiological processes is (still) insufficient to explain the sensation of private experiences (and hence every person's certainty regarding his or her own existence—Descartes' cogito-argument).

Apart from the qualia, cognition makes its presence felt in the lexicon through the device of OPEN PARAMETERS (often called 'free variables'; see Sections 1.3.2.2 and 7.4 in Volume II) as elements in the specification of the satisfaction conditions of predicates. The values of the open parameters are to be retrieved from shared situational, contextual, and world knowledge. Moreover, cases of polysemy and of ambiguity, in whatever sense and on whatever level, are resolved with the help of available nonlinguistic knowledge. Combined with the fact that available nonlinguistic knowledge is likewise required for adequate utterance incrementation in discourse (see Chapter 8 in Volume II), this forces the conclusion, much as one would wish to avoid it, that bottom-up grammar (parsing) cannot be considered modular but needs the help of shared situational, contextual, and world knowledge—precisely the conclusion reached in Section 7.2.3.

8.2.3 *All lexical words are predicates*

One should realize that at L-propositional level all lexical words are predicates of some sort. It is shown in Section 2.3 in Volume II that the logical operators are profitably treated as predicates. This illustrates the general principle valid for the lexicon as a whole that all lexical elements are labelled PRED(ICATE) at L-propositional-level (Seuren 1996: 28–9). This principle is endorsed in Bierwisch and Schreuder (1992: 29): '... lexical items are *n*-place predicates assigning properties or relations to *n*-tuples of entities (or sets of entities).'

At surface level these PRED-nodes are relabelled as Noun, Verb, Adjective, Preposition, and so on, or appear as particles or morphological elements (as often with tenses and sometimes with modalities)—depending on the grammar and lexicon of the language in question. The strategy to treat all lexical elements as predicates (McCawley 1973) has many advantages, paramount among which is the fact that it drastically simplifies and unifies L-propositional structure, making it possible to keep the transformational system of the grammar simple and compact. Another advantage is that it contributes to the building of a common semantic basis for translation equivalents in different languages even though they differ radically in surface structure. Consider, for example, the English sentence *He has just left*, which corresponds to French *Il vient de partir*, or *He is reportedly ill*, which can be rendered in French as *On dit qu'il est malade*. Or *He likes to sing*, which translates into German as *Er singt gerne*, with *singt* (sings) as the finite verb form, and the adverb *gerne* (with pleasure) as an adverbial modifier. Such cases are not exceptional. The languages of the world show vast structural differences for the expression of identical or near-identical meanings. To give just one more example, the Turkish sentence

(8.1) translates literally as 'the man in-s the house'. Here, the locative inflection suffix (postposition if one likes) *de* (in), combined with its object *ev* (house), acts as the main surface predicate taking the verbal tense suffix *dir*:

(8.1) Adam ev- de- dir
man house- in-3SG- PRES
The man is in the house.

8.2.4 *Word meanings as satisfaction conditions*

Given that all lexical items are treated as predicates at L-propositional level, word meaning or lexical meaning, can be defined as follows:

> LEXICAL (WORD) MEANING:
> The meaning of a lexical item (predicate) F is the type-level property of F that defines, for any token occurrence of F, the satisfaction conditions of the property expressed by F in the proposition at hand.

And again we might wish to use the following, more mathematical, formulation:

> LEXICAL (WORD) MEANING:
> The meaning of a lexical item (predicate) F is a function from *n*-tuples of ρ-values (term extensions) to propositions.

Again, however, this should not be taken to suggest that the functions that constitute lexical meanings are open to actual formalization. As is stressed in the remainder of the present and the whole of the following chapter, word meanings are notoriously slippery and are arguably the least amenable to formal treatment of all phenomena occurring in natural language.

Talk of functions tends to make one forget the ecological vagaries of lexical meanings and to suggest a degree of formal precision which, in fact, is not there. One might think that function talk helps to see the overall architecture of language and its uses. Yet an architectural sketch in terms of a system of interrelated functions is also to some extent arbitrary. We have just defined a predicate, or predicate meaning, as a *function from* ρ*-values to propositions.* This is not incorrect but it is a little arbitrary, because a proposition has, by definition, a unique truth-value (in whatever system of truth-values), which means that a predicate is also, in a sense, a *function from* ρ*-values to truth-values*—the standard, formal semantics way of defining predicates. Moreover, since a proposition necessarily, by definition, defines a unique virtual fact, a predicate is also a *function from* ρ*-values to virtual facts.* And since virtual facts uniquely define valuation spaces, we may add that a predicate is also a *function from* ρ*-values to valuation spaces.* And this is only with regard to the semantic aspect. But then there is also the syntactic aspect. Syntactically, a predicate (defined as a morphophonological unit) can be viewed as a *function from terms to L-propositions.* Therefore, definitions in terms of a functional architecture may be useful for specific purposes, they are not a panacea to make up for lack of insight. What is needed is an adequate analysis of the overall structure of the language machinery and its ecological embedding in relation to the world and the cognizing mind. The rest follows from that.

Yet even the less mathematical-looking definition, which does not speak of a function but only of satisfaction conditions to be satisfied by the term referents of any given predicate, glosses over the problems actually encountered when one analyses and describes lexical meanings. When we speak of the meaning of a predicate F as the *satisfaction conditions* of F, we may be taken to promise more than we can give. Strictly speaking, it is, of course, correct to say that lexical meanings are defined by satisfaction conditions, but these conditions vary so much from case of use to case of use, and in such apparently whimsical ways, that one must give up the simple thought that lexical satisfaction conditions only need to be formulated and listed for an adequate lexical description to come about. Much of this and the following chapter is devoted to an illustration of the difficulties actually encountered when one tries to specify lexical meanings—difficulties that tend to drive lexicologists and lexicographers to despair.

Even though it is probably correct to say that lexical meanings are defined by their satisfaction conditions, it is almost certainly wrong to say that these can be given in the familiar, logically manageable, format of necessary and sufficient conditions. Jackendoff (2002: 351) criticizes the view, adopted by the entire Frege–Tarski tradition, that lexical meanings are definable by means of a system of satisfaction conditions that are 'individually necessary and collectively sufficient to guarantee truth of the sentence'. This criticism is entirely to the point. Jackendoff mentions two kinds of phenomena, the vagueness of many lexical predicates (e.g. colour terms), and what he calls 'cluster' concepts (2002: 352–6), covering, in principle, the phenomena usually caught under the term polysemy (see Section 9.6), as well as those of prototypicality (see Section 8.8). Pustejovsky (1995) voices a similar opinion, mentioning polysemy and related phenomena of dynamic meaning application as his grounds, but leaving prototypicality and vagueness out of account. Both authors are right in principle. If we continue to speak of satisfaction conditions as defining lexical meanings, it should be understood that these conditions, in all probability, cannot be captured in the format familiar from logic and set theory, of 'individually necessary and collectively sufficient' conditions. What other way of specifying the satisfaction conditions of lexical predicates will do the job is simply not known.

In this and the following chapter we deal with these complications in a cursory manner, just stating the problems as clearly as is possible given the constraints of the present text, but without trying to be complete or to force solutions. Since we take logic to be a form of lexical semantics (see Chapter 1 in Volume II), we must probe into the question of the semantic specification of the logical predicates (constants) as well as of the logical properties of those predicates that are not traditionally considered 'logical'. In this perspective, it is necessary to look at the general question of lexical semantic specification from a purely linguistic point of view and without any distortion possibly imposed by previous experience with logical matters. For this purpose an overall theory of lexical meaning specification is not an absolute requirement, although it would clearly be of great advantage if we had one. But one must at least have an idea of the range and nature of the problems that arise when one tries to specify lexical meanings. And also of the specific properties of those predicates that form a logical system of some sort and of the possible format and metalanguage that can be used to specify their meanings. For this reason we present a cursory and inevitably incomplete discussion of the problems encountered, without aiming at an overall theory of the analysis and description of lexical meanings in natural languages.

Yet even without an overall lexical theory, it will become clear that the theoretical and philosophical consequences of our cursory discussion are potentially momentous. Our discussion shows that any form of model-theoretic semantics, including the varieties of 'update' semantics that are currently being developed in various centres, must be deemed basically inadequate for natural language—even though the insights gained may turn out to be useful for a proper understanding of what language is and how it works. Then, even more seriously, since truth and falsity result from the application of the satisfaction conditions of predicates to their ρ-values, it will be seen that the Aristotelian definition of truth as correspondence will have to be subjected to a thorough overhaul, again a task we only broach in the present book, leaving fuller discussion to future philosophical, linguistic, and cognitive research.

8.3 The metalanguage problem: the circle of synonymy

In this section we will take a more detailed look at the question of how to specify lexical meanings—that is, the satisfaction conditions of predicates. The first problem that presents itself is the problem of the DESCRIPTIVE METALANGUAGE. Commercial dictionaries use ordinary language, occasionally supplemented with the technical jargon of some discipline or area of public life. But although this is largely satisfactory for practical purposes, it leads directly to a theoretical quandary that blocks an adequate understanding of what actually goes on in language and language use. The quandary consists in what we call the CIRCLE OF SYNONYMY: any language used for the description of meanings will again contain expressions that have a meaning, which will then again have to be described in terms of some language containing expressions that have a meaning, and so on *ad infinitum*. When ordinary language is used for the specification of lexical meanings, one will be forever caught up in a circle of synonymy.

Various attempts have been made to escape from this circle. On the whole, their aim is to achieve the escape by treating lexical satisfaction conditions as cognitive criteria to be satisfied by the ρ-values of the predicates in question. That being so, it is felt that the specification of lexical meanings should be given in terms of a theory that analyses and describes the criteria at work in cognition when phenomena are identified and distinguished. The problem is, however, that empirical access to these aspects of human cognition is largely limited to what can be culled from the lexical study of natural language predicates. Independent empirical evidence is extremely hard to come by. One keeps being caught in circles.[7]

The attempts of this nature are divided over two main schools, the REDUCTIONIST and the GLOBAL schools.[8] The difference between the two schools amounts, in essence, to the following. According to the reductionists, the bundles of satisfaction conditions of each separate predicate in whatever language are the result of a language-internal process of composition of cognitive primitives, each specific composition serving the needs in any given community for the labelling of certain phenomena, given the community's system of values and beliefs, its social and political organization,

[7] There have been attempts, still rife in anthropological linguistics, to show that cognitive criteria are, if not determined, at least influenced by the satisfaction conditions of the predicates of the language of a given community—a point of view dubiously ascribed to Humboldt and Sapir and properly ascribed to Benjamin Lee Whorf (1897–1941), a chemical engineer and fire insurance inspector with an amateur interest in linguistics. Since no cogent empirical evidence has been presented for this view, I shall disregard it in the discussion that follows.

[8] I disregard here the largely German movement of 'lexical field' or 'Wortfeld' semantics, initiated by Trier (1931, 1934, 1973) and taken over by a small school of followers. The main idea was to split up the lexicon of a language into a 'mosaic' of 'fields', within which lexical meanings are distinguished by oppositions in a more or less Saussurean sense. For adequate critique see Lehrer (1974: 15–19) and Geeraerts (1994*a*). The movement suffered from a serious lack of notional clarity and theoretical foundation and did not lead to any tangible results.

Nor will I pay further attention to the idea, developed by Anna Wierzbicka in a large number of publications since (1972), that word meanings can be explicated in terms of a 'metalanguage' consisting of sixty-odd so-called 'conceptual primitives' all corresponding to the presumably universal words (1998: 114–15 and elsewhere): *I, you, someone(person), something(thing), people, body, this, the same, other, one, two, some, many/much, all, good, bad, big, small, think, know, want, feel, see, hear, say, word, true, do, happen, move, there is, have, live, die, not, maybe, can, because, if, when(time), now, after, before, a long time, a short time, for some time, moment, where(place), here, above, below, far, near, side, inside, touching, very, more, kind of, part of, like*. According to this author (1998: 115), 'evidence suggests that the 60 or so words listed as conceptual primitives do match in meaning across languages'. The author

its technical development, and so on. The globalists, by contrast, reject any idea of language-internal composition. They hold that each lexical meaning corresponds directly and in total to a pre-existing set of cognitive satisfaction criteria, whose possible internal analysis is entirely a matter of cognitive research. Language has nothing to do with the internal analysis of lexical meanings; it merely provides a phonological reflection of the cognitive units.

In the reductionist school the conviction prevails that lexical meanings are the product of a language-driven system allowing for the combination of a finite number of (hopefully universal) lexico-cognitive primitives into a possibly infinite variety of larger structures. According to some practitioners, the lexico-cognitive primitives are theoretical constructs with technical terms as their labels. According to others, they have to be terms occurring as such in natural language(s), although it is not made clear why this should be so.[9]

An argument for reductionism may be derived from a well-known difference between verbs of locomotion in, for example, French and English. The difference consists in the fact that one can say (8.2a) in English, but not (8.2b) in French, although (8.2b) is the French word-by-word equivalent of the English (8.2a). Sentence (8.2c), with the French verb *aller* (go), is the, or a, correct French translation of the English (8.2a):

claims (1997: 130) to be able, by using just the words mentioned, to explicate the subtle semantics of a word like *freedom*, rendered as:

freedom
(a) someone (X) can think something like this:
(b) if I want to do something I can do it
(c) no one else can say to me: 'you can't do it because I don't want this'
(d) if I don't want to do something I don't have to do it
(e) no one else can say to me: 'you have to do it because I want this'
(f) this is good for X
(g) it is bad if someone cannot think this

Apart from the deplorable lack of precision and sophistication of this 'theory', the first thing one notices is this author's blindness to the qualia problem (see Section 8.2.1). The satisfaction conditions of qualia predicates can only be defined in a yet to be developed technical language enabling one to identify and analyse the elements of cognition (Musacchio 2005).

[9] Wierzbicka (1998: 116) argues as follows:

Because this metalanguage is carved out of natural language (any natural language), the semantic explications constructed in it are intuitively meaningful and have psychological reality. Consequently, unlike semantic formulae based on various artificial formalisms, NSM [=Natural Semantic Metalanguage] formulae are open to verification (they can be tested against native speakers' intuitions).

This author represents extreme ecologism, in the sense discussed in Section 1.3.3. She seems to hold that formalization in linguistics leads to unverifiability (unfalsifiability?) because native speakers' intuitions have nothing to say about formal theories or formal terminology. In the physical sciences this would amount to saying that formalization leads to unverifiability because native observations and perceptions reveal nothing about formal physical theories and their terminology. This view illustrates the deep-seated aversion to technicalities and formalization all too often encountered in the humanities, linguistics included.

(8.2) a. John walked from the house to the shop.
b. *Jean marchait de la maison à la boutique.
c. Jean allait de la maison à la boutique en marchant / à pied.
lit.: Jean went from the house to the shop while walking / on foot.

A reductionist may argue that the explanation lies in the fact that the suitability of local adjuncts of the form 'from A to B' is inherent to the concept of going, but not to the concept of using. He will thus analyse English *walk* with the help of the general (abstract) predicate GO and French *marcher* with the help of the (abstract) predicate USE, as shown in (8.3a, 8.b), where the format used is the standard format for the specification of the extensions of predicates by comprehension (see Section 3.3.2):

(8.3) a. ⟦Walk⟧ = {x | x GOES by the use of x's legs according to pattern W}
(the extension of the predicate *walk* is the set of all objects *x* such that x goes by the use of x's legs according to the prototypical 'walking' pattern)
b. ⟦Marcher⟧ = {x | x USES x's legs for locomotion according to pattern W}
(the extension of the predicate *marcher* is the set of all objects x such that x uses x's legs for locomotion according to the prototypical 'walking' pattern)

Obviously, a fully developed argument for analyses of this kind requires a great deal of lexicological, grammatical, and crosslinguistic research—a task I am happy to leave to others.

An early initiative (perhaps the first) to set up a reductionist lexicon in this sense is Leibniz's well-known *Characteristica Universalis*, a fledgling attempt at universal lexical analysis, which never got beyond the programmatic stage (Seuren 1998: 368–9). The twentieth century saw many more such attempts, which, however, for reasons of feasibility were mostly restricted to certain limited areas of related word meanings. Predicates of perception, for example, have provided a fertile ground for such analyses.[10] And so have lexical studies in kinship relations, especially in anthropological linguistics. Other areas have been studied according to this method as well.

It is a curious fact that this approach has been relatively successful when applied to restricted areas of meaning or to partial semantic descriptions, but clearly fails when applied to the lexicon of any natural language as a whole or to individual meanings in their totality. And the failure is always due

[10] See, for example, Grigorjewa (1973), Rogers (1973) for perception verbs in general, Gruber (1967), Bülow (1972), Schepping (1982) for visual perception verbs, Vliegen (1988) for auditory perception verbs.

to the fact that, besides and beyond the 'systematic' meaning elements that look as if they can be captured by the theory at hand, there is almost always an unalienable remainder of extrasystematic or perhaps even unsystematic, unique, irreducible criteria to be satisfied, the sort of criteria that are satisfied by phenomenal or experiential sensations not open to further philosophical, cognitive, or linguistic analysis, such as the criteria that make one recognize colours or distinguish between apples and pears, between swans and geese, or between the smell of roses and that of hyacinths—in short, the criteria defining qualia (see Section 8.2.2).

8.4 Katz and Fodor (1963)

Let us look at a few notable attempts. In Katz and Fodor (1963) a system of' 'semantic markers', 'distinguishers', and 'selection restrictions' was introduced for the specification of lexical satisfaction conditions. Unfortunately, the range of examples is restricted to just four words, the nouns *bachelor* and *ball*, the adjective *colourful* and the transitive verb *hit*. The noun *bachelor* is specified semantically as follows:

(8.4) bachelor → (Human) → (Male) → [who has never married]
or: → (Human) → (Male) → [young knight serving under the standard of another knight]
or: → (Human) → [who has the first or lowest academic degree]
or: → (Animal) → (Male) → [young fur seal when without a mate during the breeding time]

The elements between round brackets are the semantic markers; those enclosed by square brackets are the distinguishers. The difference between the two is that the semantic markers are part of a descriptive system, whereas the distinguishers are ad hoc and extrasystematic or unsystematic—that is, the qualia just mentioned. Each semantic marker or distinguisher is to be read as stating a condition to be fulfilled by whatever object is assigned the predicate in question. The arrows represent passages from the more general to the more specific.

This particular proposal regarding the specification of lexical meanings was subjected to a thorough critique by Weinreich (1972: 16–43), to which the reader is referred. Pustejovsky (1995: 33–54) rejects the method followed by Katz and Fodor on the general grounds that it is based on the notion of sense enumeration and thus fails to do justice to the creative use of words in sentences and texts. In addition to this fully justified critique, we may observe that the Katz and Fodor system begins to crumble when nonnominal predicates are tried out. Consider the specifications given by these authors for the adjective *colourful* and the transitive verb *hit* ('∨' stands for 'or'):

(8.5) a. colourful → (Colour) → [Abounding in contrast or variety of bright colours] <(Physical Object) ∨ (Social Activity)>

or: → (Evaluative) → [Having distinctive character, vividness, or picturesqueness] < (Aesthetic Object) ∨ (Social Activity)>

b. hit → (Action) → (Instancy) → (Intensity) → [Collides with an impact] <SUBJECT: (Higher Animal) ∨ (Improper Part)[11] ∨ (Physical Object), OBJECT: (Physical Object)>

or: → (Action) → (Instancy) → (Intensity) → [Strikes with a blow or missile] <SUBJECT: (Human) ∨ (Higher Animal), OBJECT: (Physical Object), INSTRUMENTAL: (Physical Object)>

Here we have, besides semantic markers and distinguishers, also SELECTION RESTRICTIONS, given at the end between angled brackets. The selection restrictions state conditions to be satisfied by the reference values of the argument terms of the predicates in question. But isn't that exactly what the semantic markers do for nouns, that is, nominal predicates? A nominal predicate like *bachelor* has one single argument term, which must refer to a (male) human or animal for truth to come about. This being so, one wonders why the semantic markers in (8.4) do not figure as selection restrictions, since they express conditions to be satisfied by the one argument term of the predicate *bachelor*.[12] A further difficulty is that it remains unclear whether the selection restrictions are to be specified again in terms of systematic markers, or in the ad hoc terminology of the distinguishers. Moreover, there is no exact interpretation of markers like 'Colour', 'Evaluative', 'Instancy', 'Intensity', whose meaning is a matter of guessing rather than of precise specification.

8.5 Pustejovsky (1995)

The system proposed in Pustejovsky (1995) may be seen as a continuation of the line initiated by Katz and Fodor (1963). Beyond a specification of argument

[11] In a footnote, Katz and Fodor (1963) add the not very enlightening explanation: 'the marker *Improper Part* is assigned to lexical items that represent wholes which the language contrasts with their parts'.

[12] Pustejovsky's rendering (1995: 35 and elsewhere) of this marker-type mode of lexical description gives rise to the same question. In his rendering of this mode of description verbs impose selection restrictions on their argument terms, but nouns have what he calls 'genus' conditions. Later, when presenting his own analysis, the two kinds of conditions are correctly conflated into the one category of selection restrictions on argument terms.

structure with concomitant selection restrictions, Pustejovsky adds specifications for event structure and for what he calls 'qualia structure', covering constitutive, formal, functional and causal properties of whatever is denoted by the predicate in question. The 'qualia structure' may be seen as Pustejovsky's attempt at capturing in more systematic terms, in particular in terms of Aristotelian ontology, the largely ad hoc combinations of properties and conditions that Katz and Fodor put up in their 'distinguishers'. This theory, which looks more formal than it is in reality, is called 'generative' because it aims to establish rules for the dynamic interaction of these levels of semantic representation in order to account for the infinitely varied play of applications of lexical meanings in larger structures. Much as this is to be applauded, however, no really formal generative system seems to be getting off the ground, which means that this particular aim of the enterprise must be taken not to have been achieved and to have remained largely promissory. A further problem is that, as in the case of Katz and Fodor (1963), the 'markerese' used for the specification of the argument, event and 'qualia structures' is, on the whole, so opaque that hardly any clear and testable predictions in terms of semantic judgements by native speakers appear to follow. If this work is of value, it is because of the many examples given and discussed, and because of its wide coverage.

8.6 Lexical decomposition

The question of whether lexical meanings should be seen as undividable wholes (the 'holistic' view) or as consisting of one or more primitive elements (the qualia) put together into hierarchical structures (the decompositional view) has dominated the psycholinguistic literature of the past few decades. And it is far from having been resolved. The main problem is that, on the one hand, there are strong indications that meanings have an internal structure and should therefore be seen as decomposable units, while, on the other hand, experimental results (see Bierwisch and Schreuder 1992: 46–7 and the references cited there) have shown that the processing of more specific predicates, and thus predicates with more internal structure, is as fast as, and sometimes even faster than, that of more general predicates which must be taken to have a simpler internal structure. For example, the more specific verb *rise* was processed faster than the more general verb *move* in an experiment in which the subjects had to verify descriptions containing these verbs against pictures (Schreuder 1978). Even so, Bierwisch and Schreuder support the decompositional view:

We will endorse the decompositional over the holistic view for three reasons:

(a) There is growing evidence, ranging from linguistic analyses to speech errors resulting in lexical mis-selection, and to language acquisition, that strongly supports decomposition. . . .
(b) There are clear cases of systematic lexical relations that are captured in a straightforward way by decomposition, but cannot plausibly be expressed with meaning postulates. . . .
(c) The problem of lexical access cannot reasonably be formulated under the view of meaning holism, because that would require a pre-established congruity between concepts and the semantic form of lexical items. Under this (clearly counterfactual) assumption, the conceptualizer would have to operate essentially in terms of lexical items, so that the real problem of lexical access is trivialized. (Bierwisch and Schreuder 1992: 28)

Whether these arguments are sufficiently convincing to make one opt for the decompositional view is open to doubt. What seems to me to be a potentially fruitful stance is to see if the two views cannot be reconciled with each other, where the notion of prototypicality may well be of considerable help. In Section 8.8 it is argued that a basic distinction should be made between prototypes and concepts (Fodor 1998: 94). Prototypes are to be taken as direct and undivided psychological units—qualia in their own right—even though they are subject to truth-conditional conceptual analysis in terms of a conceptual system still to be worked out by cognitive psychologists. If it is true, as I am inclined to argue it is, that prototypes figure as cognitive criteria in the satisfaction conditions of predicates, one may think of a theory that unites the direct-prototype view—that is, the holistic view—with the decompositional view which likes to see lexical meanings as consisting of satisfaction conditions. It is in this light that the following subsections should be read.

8.6.1 *Prelexical syntax*

During the mid-1960s, in the context of what was then called GENERATIVE SEMANTICS, Jeffrey Gruber[13] and Jim McCawley[14] developed, each in their own way, a specific, and so far unique, variety of the decompositional view, called PRELEXICAL SYNTAX. Prelexical syntax is based on the thought that at least some lexical predicates are the result of a process whereby chunks of syntactic

[13] See Gruber (1976), which contains his unpublished MIT doctoral dissertation *Studies in Lexical Relations* of 1965 and his unpublished technical memorandum *Functions of the Lexicon in Formal Descriptive Grammars*, written in 1967 for the System Development Corporation, Santa Monica, California.

[14] Jim McCawley wrote a number of papers on prelexical syntax between 1968 and 1972, all of which are bundled in his book *Grammar and Meaning* of 1973.

structure, at some suitable level of syntactic representation, are replaced by a single surface lexical item (lexeme). The parts of syntax preceding the insertion of a surface lexical item were called 'prelexical syntax'. At the time, the notion of prelexical syntax provoked wide interest in the linguistic community, no doubt partly due to the intellectually exciting character of the proposals that were put forward. Unfortunately, however, their empirical basis was weak, which made this approach even more vulnerable than the other ideas that arose in the context of the Generative Semantics movement. This intrinsic vulnerability was soon ruthlessly exploited by Chomsky and his followers, with the result that for a few decades prelexical syntax fell into disrepute, a state of affairs that is now slowly being redressed (see, for example, Harris 1993; Huck and Goldsmith 1995; Seuren 1998: 493–527). Given the genuine intellectual value of the approach, it seems appropriate to devote some detailed attention to the flurry of activity around prelexical syntax during the late 1960s and early 1970s and to present some arguments that support at least a restricted validity of this theory.

It was argued, in particular by McCawley (1968), that the assumption of prelexical syntax meant that one can no longer posit a 'deep structure' level of syntactic representation at which all lexemes are in place, as the insertion of surface lexical items replacing chunks of syntactic structure was supposed to take place gradually throughout the syntactic derivation. However, since McCawley (other than Gruber) insisted at the same time that the structures to be replaced by a single lexeme should form one categorial constituent created by the application of syntactic rules, it would seem to follow that all prelexical syntax may at least be taken to be already defined within the lexicon. If that is so, one may still assume, for each sentence, a level of deep structure with all surface lexical items filled in at the proper places, each item representing a bit of lexically defined syntactic structure. The argument against lexically defined deep structure thus comes to nothing.

The crucial point is not the question of whether or not a deep-structure level of representation should be postulated for sentences, but rather the thesis that at least some lexical predicates are to be analysed as chunks of structure put together according to rules that are independently motivated in the syntactic descriptions of natural languages. McCawley's celebrated example is the analysis of the English verb *kill* as CAUSE-BECOME-NOT-ALIVE (McCawley 1973: 157–8 and elsewhere). The sentence schema *x kill y* is analysed by McCawley in the following way. The deepest underlying structure is (8.6a), where 'S' stands for 'sentential structure', 'Pred' for 'predicate', 'NP' for 'noun phrase', and where the order of constituents is Verb–Subject–Object, or VSO (see McCawley 1970).

(8.6)

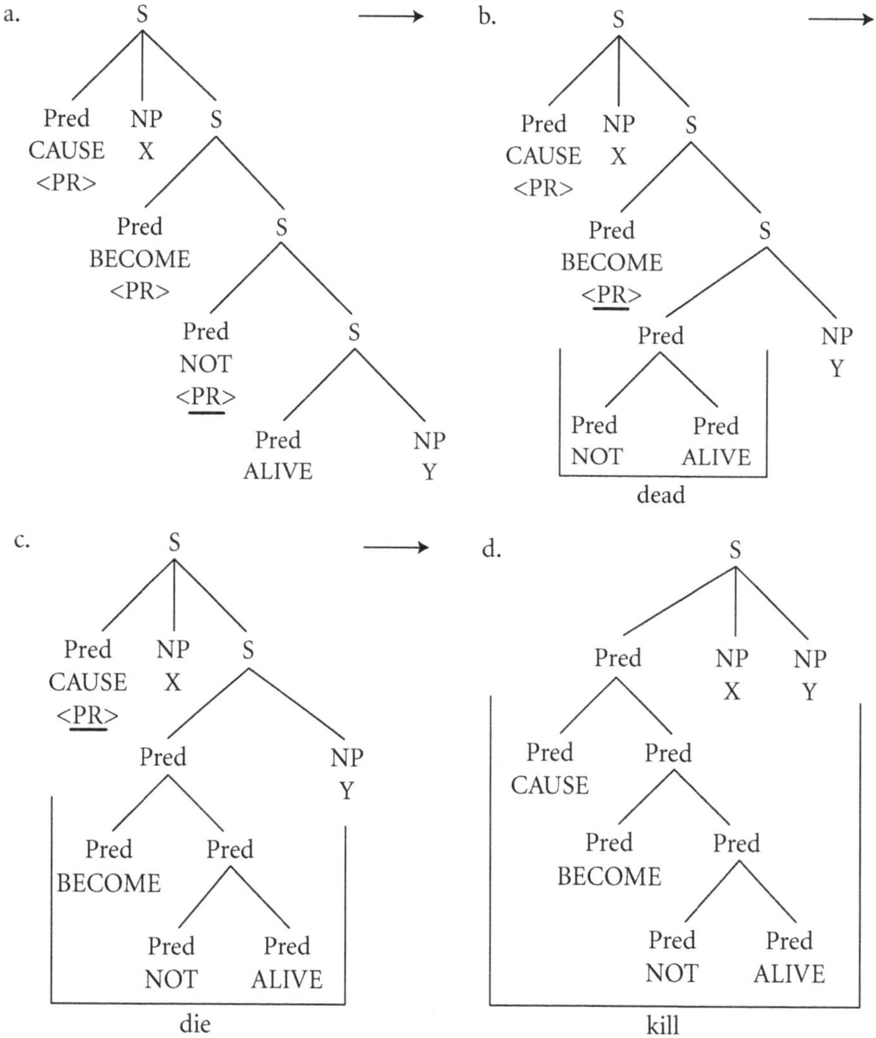

The predicates CAUSE, BECOME, NOT, and ALIVE are written in capital letters to distinguish them from the items occurring in surface structures. They may be taken to represent semantic primitives corresponding directly to cognitive satisfaction conditions, although that aspect was largely left undiscussed at the time. It is also problematic, in that CAUSE may still be analysed as either axiomatic physical causation or as 'do something which causes', and BECOME may be analysed further as BEGIN-BE, with BEGIN and BE as primitives. But be that as it may, CAUSE, BECOME, and NOT are taken to be specified in the lexicon as inducing the cyclic syntactic rule of PREDICATE RAISING (PR), indicated in the tree structure as the rule feature <PR> associated with the predicates in question. PR consists in the unification of the lower predicate with the inducing predicate as illustrated in (8.6b–8.6d). It applies cyclically—that is, at each S-level starting from the bottom. The first application of PR results in (8.6b), the second in (8.6c), and the last in (8.6d) (each instance of PR up for application is underlined). The predicate cluster [CAUSE[BECOME[NOT[ALIVE]]]] in (8.6d) is then replaced with the single verb *kill*, although earlier replacements may also take place, such as *dead* for [NOT [ALIVE]], or *die* for [BECOME[NOT[ALIVE]]]. This analysis has become known as the analysis of *kill* as 'cause to die'.

8.6.2 *Criticisms of prelexical syntax*

McCawley's inspiring and highly original idea of prelexical syntax immediately came under unusually heavy fire, which led to an unjustified wholesale rejection of this approach. Yet, although this rejection was far too drastic, it must be admitted that there are serious problems, most of which still await a solution.

The first and most obvious problem with prelexical syntax is its empirical motivation. Chomsky, who strongly opposed prelexical syntax, wrote:

> This is the approach taken by McCawley in the case of words such as *kill* = 'cause to die'. In the proposed underlying structure, *John caused Bill to die* (or *John caused Bill to become not alive*), the unit that is replaced by *kill* is not a constituent, but it becomes one by the otherwise quite unnecessary rule of predicate raising. Such a device will always be available, so that the hypothesis that Q [i.e. the prelexical tree taken to be replaced by the surface lexical item; PAMS] is a constituent has little empirical content. (Chomsky 1972: 142)

However, Seuren (1972*a*) and Evers (1975) found that the rule of Predicate Raising, far from being 'quite unnecessary', is in fact one of the best motivated and most widespread rules in natural language syntax, although it does not occur in English—other than in rare transparent lexemes like *hearsay* (from the V-island $_V[_V[\text{hear}]\ _V[\text{say}]]$), or in equally rare semilexicalized predicates like *let go* (from the V-island $_V[_V[\text{let}]\ _V[\text{go}]]$) or *make do* (from the V-island $_V[_V[\text{make}]\ _V[\text{do}]]$). This led Seuren to conclude that if McCawley's theory of prelexical syntax is considered tenable, linguistic theory must stipulate that, for each language L, a distinction must be made between rules allowed to occur in the 'open' syntax of L and those allowed to occur in its prelexical, lexicon-internal, syntax. Unfortunately, McCawley himself failed to capitalize on the findings of Seuren and Evers and more or less gave up on the whole Generative Semantics enterprise (see Seuren 2006).

Other, more serious, criticisms came from Fodor (1970), where three reasons are given why McCawley's analysis of *kill* as 'cause to die' should be rejected.[15] The first reason, to do with *do so* anaphora, is not very stringent. Fodor observes that (8.7a) and (8.7c) are good English but (8.7b) is not:

(8.7) a. Floyd melted the glass though it surprised me that it would do so.
b. *John killed Mary and it surprised me that she did so.
c. John caused Mary to die and it surprised me that she did so.

Now, he says, the verb *melt* in (8.7a) is causative and should, therefore be read as 'cause to melt', incorporating the intransitive *melt*, which is then the antecedent of the *do so* anaphor. Analogously, he says, if (8.7b) were to be read as (8.7c), there would be no reason why (8.7b) should be ungrammatical. The answer to this is simple. Even if (8.7a) is considered acceptable (many native speakers dislike it), the argument hinges on the conditions of *do so* anaphora, which at the very least requires a preceding phonologically and semantically identifiable element as antecedent. (For further illustration, see (8.26a, 8.26b) and note 18.)

So let us pass on to the second and third arguments, which are more substantial. The second argument is based on the observation that (8.8a) is fully acceptable but (8.8b) is semantically or pragmatically unacceptable, and the third argument on the observation that (8.9a) is ambiguous in a way that (8.9b) is not:

(8.8) a. John caused Bill to die on Sunday by stabbing him on Saturday.
b. !John killed Bill on Sunday by stabbing him on Saturday.

(8.9) a. John caused Bill to die by igniting the charge.
b. John killed Bill by igniting the charge.

As regards (8.8a), one observes that *on Sunday* in (8.8a) can only be read as modifying *to die*: Bill's death took place on Sunday as a result of John's stabbing him on Saturday. This reading, however, is impossible for (8.8b), a fact which, in Fodor's view, shows that *kill* cannot be derived from CAUSE-DIE. But that conclusion is premature, since there appears to be a scale of restrictions, about which more in a moment, on the occurrence of operators within PR-islands in open syntax as well as within fully lexicalized PR-islands. These restrictions do not apply in SUBJECT-RAISING constructions, manifest in (8.8a) and (8.9a),

[15] For an extensive discussion of Fodor's critique, see Seuren (1985: 204–9).

but they do apply to verbal PR-islands, which are the syntactic constructions assumed to underlie lexicalizations like *kill.* In (8.8b), *on Sunday* would have a position internal to the semantic analysis of *kill*, as shown in (8.10a), and this is what Fodor takes to be excluded for fully lexicalized PR-islands.

(8.10)

a.

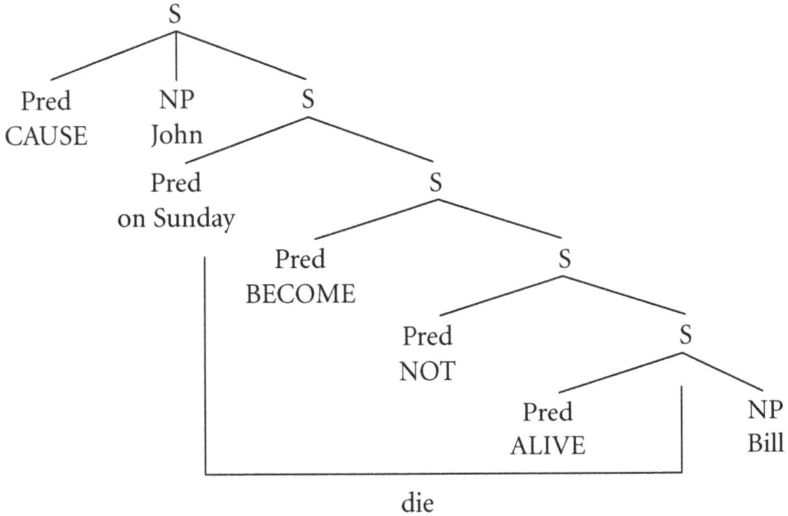

b.

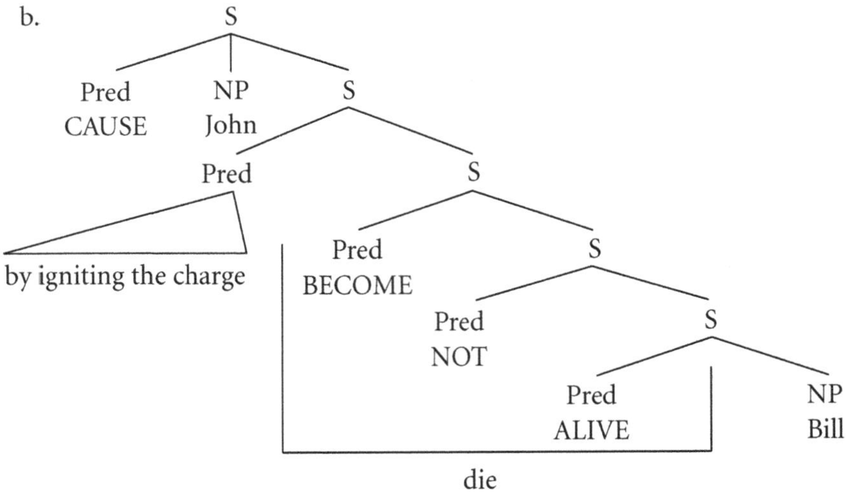

As regards the ambiguous (8.9a), in one of its two readings the operator *by igniting the charge* modifies the whole structure '$_{S}$[$_{V}$[CAUSE]$_{NP}$[John] $_{S}$[$_{V}$[BECOME-NOT-ALIVE] $_{NP}$[Bill]]]'. One notes that *by igniting the charge* is itself the result of syntactic processing involving the deletion of the subject term of *ignite* (Seuren 1996: 70–2), which makes John the one who ignited the charge. In the other reading, however, shown in (8.10b) and excluded for (8.9b), *by igniting the charge* modifies only the sub-island $_S$[$_V$[BECOME-NOT-ALIVE] $_{NP}$[Bill]], which makes Bill the one who ignited the charge. This would suggest that 'internal' operators such as *on Sunday* or *by igniting the charge* block unification into one lexeme. This suggestion is supported by the fact that such restrictions clearly do exist with regard to PR-islands in open syntax, as is shown in examples (8.16a–8.16d) below.

But this cannot be a general principle, because there are many cases where internal scope-bearing operators do not prevent lexical unification. Morgan observed (1969: 62–3) that the sentence:

(8.11) John almost killed Bill.

is open to at least three readings, one in which John (was so angry that he) almost killed Bill, one in which John did something, perhaps inadvertently, which almost killed Bill, and one in which John reduced Bill to a state of near-death: he caused Bill to be almost dead. These readings are differentiated by the different scopes of *almost*, at least one of which is internal to *kill*. Moreover, as was shown in Section 5.5, sentences with predicates like *look for*, *look like*, or *need* whose object terms are existentially quantified show systematic ambiguities that are naturally accounted for by assuming an internal prelexical structure allowing for the insertion of the existential quantifier in two different positions corresponding to the scope differences concerned.

Likewise, as observed by Morgan (1969: 61), sentence (8.12a), with the causative verb *jail* analysable as CAUSE-BE-IN-JAIL, is ambiguous between a large scope reading in which it was in Reading that the Sheriff of Nottingham decided to jail Robin Hood, no matter where the jail was, and a (more natural) reading with internal scope, in which the Sheriff decided to put Robin Hood in the Reading jail, no matter where the decision was taken. Interestingly, (8.12b) strongly favours the former, less natural, reading. Likewise for (8.13a, 8.13b): (8.13a) and (8.13b) share the less natural reading in which the Sheriff kept jailing Robin Hood repeatedly for four years, whereas (8.13a) also has the more natural reading in which Robin Hood was forced to stay in prison for four years—a reading that can only be forced on (8.13b) with artificiality. Likewise again for (8.14a, 8.14b), both of which have the reading in which for two days the speaker kept lending his bicycle to Harry, whereas (8.14a) also has the more natural reading in which Harry could have the bicycle for two days—a highly artificial reading for (8.14b). Similar observations apply to (8.15a, 8.15b): in the natural reading of (8.15a) *for the rest of his life* operates on the internal BE-IN-RUINS of the prelexical analysis of *ruin*. The conclusion must be that one cannot simply impose a general rule prohibiting lexicalization when there is an internal operator. Yet the conditions under which lexicalizations are allowed to unite material across an internal operator are unclear.

(8.12) a. The Sheriff of Nottingham jailed Robin Hood in Reading.
b. In Reading the Sheriff of Nottingham jailed Robin Hood.

(8.13) a. The Sheriff of Nottingham jailed Robin Hood for four years.
b. For four years the Sheriff of Nottingham jailed Robin Hood.

(8.14) a. I lent Harry my bicycle for two days.
b. For two days I lent Harry my bicycle.

(8.15) a. The scandal ruined John's reputation for the rest of his life.
b. For the rest of his life the scandal ruined John's reputation.

A tentative answer may be that internal operators are allowed in the prelexical analysis of lexical items (predicates) only if these operators do not involve any form of subject deletion in the operator in question. This accounts for (8.9a, 8.9b), where the subject of *igniting* has been deleted in both readings. The deletion operation would thus block lexical unification into *kill*. But it fails to account for (8.8b), which does not involve any subject deletion in the higher operator. Perhaps the event character of the lower sub-island is at play here, in the sense that operators over events—such as DIE—are not allowed to cut into a PR-island. Note that in the examples (8.12)–(8.15) the sub-island under CAUSE is not an event but a state (BE-IN-JAIL, HAVE-MY-BICYCLE, BE-IN-RUINS). Further research is obviously called for.

In none of the other cases of internal operators—that is, the examples (8.11)–(8.15)—is there any form of subject deletion in the operator in question. It might thus be tentatively concluded that if the prelexical analysis of a lexical predicate contains an internal operator not involving subject deletion, then lexical unification is allowed for adverbs (*almost*) and also for preposition phrases but only over stative, non-event sub-islands. In the latter case, the normal position of the prepositon phrase in surface structure will be at the end of the sentence, not at the beginning. It is obvious, however, that this whole question requires a great deal of further research. That no such research is taking place is entirely due to the sociological cloud that has been hanging over the notion of prelexical syntax ever since Generative Semantics was brought into disrepute by the defamation campaign emanating from the MIT linguistics department.

Even so, we will indulge in some preliminary investigation. A closer look at some French sentences strongly suggests that the blocking power of subject deletion in internal operators applies not only to the lexical unification of prelexical structures but has a wider validity in that it generally blocks application of the syntactic rule PR in open syntax. By contrast, internal operators not involving subject deletion appear not to block PR in open syntax. French, which has open-syntax PR, provides the data: (8.16a), (8.16b), and (8.16d) only allow for the reading in which Alain did not say a word and exclude the reading in which Paul didn't say a word, but (8.16c), like its English version, also allows for the reading in which Paul died without a word:

(8.16) a. Alain a fait mourir Paul sans dire un mot.
Alain has made die Paul without say a word
Alain caused Paul to die without (Alain) saying a word.

b. Alain a tué Paul sans dire un mot.
Alain has killed Paul without say a word
Alain killed Paul without saying a word.

c. Alain a fait mourir Paul sans un mot.
Alain has made die Paul without a word
Alain caused Paul to die without a word.

d. Alain a tué Paul sans un mot.
Alain has killed Paul without a word
Alain killed Paul without a word.

The French construction *a fait mourir* in (8.16a) and (8.16c) is the result of PR in open syntax, unlike the English construction *caused Paul to die*, which is the result of subject RAISING (Seuren 1972*a*, 1996: 191–203). Apparently, in cases that involve SUBJECT DELETION of the type in question, the OPEN SYNTAX unification of FAIRE (cause) and MOURIR (die) into one open-syntax categorial island $_V$[$_V$[FAIRE] $_V$[MOURIR]] is subject to the same restriction as the corresponding LEXICAL unification (see Seuren 1985: 206–7 for more details). Yet it cannot be denied that (8.16a) is built up from the elements FAIRE and MOURIR. This means that if Fodor's third objection, expressed in (8.9a, 8.9b), were valid, then French *faire mourir* in (8.16a) would not be derivable from FAIRE-MOURIR, whereas in (8.16c) it would be!

The question remains why *sans un mot* in (8.16d) cannot be read as an internal operator, so that Paul died without a word, whereas *for the rest of his life* in (8.15a) can be, so that John's reputation is ruined for the rest of his life. Again, the event character of the PR-sub-island would seem to make the difference. In any case, the data provided should be sufficient to show that there are principles at work here which have not, so far, been brought to light with sufficient clarity.

Similar problems, equally hard to solve, occur with predicates that imply a change of state, such as *have forgotten*, *grow*, or *age*, when they take a subject or object term with variable reference (De Rijk 1974; Morgan 1969: 57–9). Despite the fact that *have forgotten* is naturally analysed as NO LONGER REMEMBER, *grow* as BECOME TALLER, and *age* as GET OLDER, the (a)-sentences below allow for both a single reference and a variable reference reading, whereas the (b)-sentences allow for a single reference reading only:

(8.17) a. I no longer remember the names of my students.
b. I have forgotten the names of my students.

(8.18) a. Students have become taller over the past twenty years.
b. Students have grown over the past twenty years.

(8.19) a. Cabinet ministers have been getting older over the past fifty years.
b. Cabinet ministers have been ageing over the past fifty years.

It must be admitted that no clear and convincing solution for such cases has so far been provided, no doubt partly owing to the fact that (a) the analysis of comparative constructions in open syntax is still a matter of controversy, and (b) the phenomenon of variable reference has so far remained not only unsolved but also practically undiscussed in the semantics literature.

Further criticisms, voiced by Chomsky (1972), Ruwet (1972), and others, are equally serious. Thus, Chomsky observes:

> If we insist that there be a 'natural' or 'motivated' rule to make Q [i.e. the prelexical tree taken to be replaced by the surface lexical item] a constituent, then the claim that Q is a constituent in a lexical insertion operator seems untenable. Consider, for example, the word *assassinate*, which, in terms of this approach, should replace everything but X in 'kill X by unlawful means and with malice aforethought, where X is human, reasonably important, [...]'. There is no way, without extreme artificiality, to make all of this, except for X, a constituent. Yet it is just this that would be required by the assumption that a lexical transformation inserts an item with only phonological and syntactic features for a subphrase-marker Q that is a constituent. (Chomsky 1972: 143)

Here Chomsky is clearly right. His objection touches on the qualia problem, which Katz and Fodor (1963) tried to solve by means of their semantic distinguishers and which is handled by Pustejovsky (1995) by means of his 'qualia structure': lexical meanings often contain a set of unique conditions that can sometimes be expressed by linguistic means but are sometimes embodied in purely cognitive categories which escape the boundaries of language, as with the predicates *apple*, or *pain*, or *red*.

Similar criticisms were voiced by Ruwet (1972: 139–62). Ruwet shows, with a wealth of examples, that there are many cases in French (some but not all of which translate into English) where a replacement of the construction *faire*+*X*, with an intransitive verb *X*, by a single homophonous causative verb *X* often leads to a change of meaning, so much so that it may lead to absurdity. One remembers that the French *faire*-construction is a clear case of PR in open syntax.

Consider, for example, (8.20)–(8.23). Although both (8.20a) and (8.20b) are acceptable in French, the analogous sentences (8.21a, 8.21b) behave differently. For reasons (Ruwet 1972: 162) still unknown today, the 'synthetic' (8.21b) and (8.22b) are pragmatically unacceptable. The fact that the analytic (8.23a) is also unacceptable (unless read as 'Fritz had the supplies brought up from the cellar by someone'—a reading not at issue here), as opposed to the synthetic (8.23b), can probably be reduced to the fact that (8.23a) requires an intransitive embedded clause *les provisions montent de la cave* ('the supplies rise from the cellar'), which is unacceptable because intransitive *monter* with an adjunct of local origin, like its English counterpart *rise*, appears to require either an animate subject term (*He rose from the grave*) or one that is driven by natural causes (*Smoke rose from the ashes*) (Ruwet 1972: 141).

(8.20) a. Delphine a fait entrer la voiture dans le garage.
Delphine has made enter the car into the garage
Delphine has put the car into the garage.

b. Delphine a entré la voiture dans le garage.
Delphine has entered the car into the garage
Delphine has put the car into the garage.

(8.21) a. Delphine a fait entrer les invités au salon.
Delphine has made enter the guests into the drawing room
Delphine has ushered the guests into the drawing room.

b. !Delphine a entré les invités au salon.
Delphine has put the guests into the drawing room.

(8.22) a. Les pluies ont fait monter le niveau de la rivière.
the rains have made rise the level of the river
The rains have made the level of the river rise.

b. !Les pluies ont monté le niveau de la rivière.
The rains have raised the level of the river.

(8.23) (a) !Fritz a fait monter les provisions de la cave.
!Fritz has made the supplies rise from the cellar.

(b) Fritz a monté les provisions de la cave.
Fritz has brought up the supplies from the cellar.

These criticisms, however, no matter how valid and how useful, are insufficient to force the conclusion that the entire notion of prelexical syntax should be rejected. But they do force the conclusion that the original Generative Semantics notion of lexical insertion during the transformational process, whereby a surface item is put in the place of a single-node subtree without any change in meaning, must be abandoned. This means that if prelexical syntax is viable, it must be seen as an internal lexical affair and not as part of the syntax of the language in question. It also means that the old notion of deep structure representation can be reinstated.[16]

It would seem that prelexical syntax can be saved if one sees the lexemes in question as partially defined by prelexical syntactic structures forming one categorial constituent, while the remainder of the semantic description is cast in terms of item-specific conditions, which include unique qualia or prototype conditions and conditions of a presuppositional nature. This would put prelexical syntax on a par with compounding as a morphological process (see Section 9.2), since the constituent elements of a compound are insufficient to determine the meaning of the compound as a whole. The prelexical analysis of an item thus makes the item semantically transparent, yet it fails to specify the full meaning, just as compounds are semantically transparent but underspecified.

One can almost observe prelexical syntax at work in the case of semantically transparent semilexicalized formations like Dutch *(iemand) leren kennen* or its German equivalent *(jemanden) kennen lernen*. Both expressions mean literally 'learn to know (someone)' but that is not their idiomatic meaning, which can be rendered as 'make (someone's) acquaintance' or 'meet (someone)', as in (8.24a, 8.24b). Yet the literal, compositional, meaning is normally present

[16] A striking example of prelexical syntax is the French verb *rendre* followed by an adjectival complement as in (i). Sentence (ii), which should be the well-formed result of PREDICATE RAISING meaning what (i) means, is, in fact, ungrammatical:

(i) Cela rend Marie heureuse.
(That makes Mary happy.)

(ii) *Cela fait être Marie heureuse.
(That makes be Mary happy.)

Since *rendre* behaves in all respects like FAIRE+ÊTRE, it is very tempting to see *rendre* as a surface manifestation of FAIRE+ÊTRE when followed by an adjective. I am inclined to give in to that temptation.

when verbs other than Dutch/German *kennen* ('know') are used, as in (8.25a, 8.25b):

(8.24) a. Zij heeft Stavros op een vacantiereis leren kennen.
she has Stavros on a holiday trip learn know

b. Sie hat Stavros auf einer Ferienreise kennen gelernt.
she has Stavros on a holiday trip know learned
both: She met Stavros on a holiday trip.

(8.25) a. Zij heeft Stavros op een vacantiereis leren waarderen.
she has Stavros on a holiday trip learn appreciate

b. Sie hat Stavros auf einer Ferienreise schätzen gelernt.
she has Stavros on a holiday trip appreciate learned
both: She learned to appreciate Stavros on a holiday trip.

Here the specific right-branching combination [LEREN [KENNEN]] in Dutch and its left-branching equivalent [[KENNEN] LERNEN] in German, both resulting from PR in open syntax (see Seuren 2003), take on a specific meaning element over and above, or perhaps replacing, the 'skeletal' meaning compositionally expressed by the constituent predicates. The same can be done for cases like English *assassinate*, to be analysed as containing a 'skeletal' internal syntactic structure CAUSE-DIE, enriched with further, more idiosyncratic and partially presuppositional, restrictions.[17]

Once one develops an eye for such cases, they are easily spotted in all kinds of languages. A striking case is the Malay formation *beri-tahu* (inform), which is transparently derived by (right-branching) PR from *beri* (give) and *tahu* (know): 'give to know', but which has acquired the specialized meaning of 'inform'. Note that Malay also has PR in open syntax. Similar cases are found in Turkish, which likewise has PR in open syntax. There one finds, for example, *bil-dir-mek* (literally CAUSE-KNOW, but in fact 'tell', 'inform') formed by (left-branching) PR induced by the infix *-dir-* (cause) and applied to the verb *bil* (know), followed by the infinitival suffix *-mek*. The irregular phonology in forms like *gös-ter-mek* (MAKE-SEE hence 'show'), from the stem *gör-* (see), or *ge-tir-mek* (MAKE-COME hence 'bring'), from the stem *gel-* (come) shows that lexicalization has actually taken place.

[17] Charles Collier from Florida University, who is both a legal expert and a philosopher of language, pointed out to me that in English common law it used to be so that the criminal infliction of injuries to a victim was not 'criminal homicide' unless the victim died within one year, death occurring after one year being presumed to have stemmed from natural causes. It would, of course, be nonsensical to stipulate that such legal details, important though they may be, should be taken to be part of the prelexical analysis of the concept 'homicide', transformed by motivated syntactic rules into one lexical island.

The same is found in French, in a number of semilexicalized compositions with the verb *laisser* ('let, allow'), also borrowed into English, such as the nouns *laisser-aller* ('carelessness'; lit. 'let go'), *laisser-faire* ('non-interference'; lit. 'let do'), *laisser-passer* ('permission to pass'; lit. 'let pass'). French, as we have seen, has PR in open syntax. Yet even English, which has no PR in open syntax, has the semilexicalized *let go* in the sense of 'release' and the noun *hearsay* in its specialized sense.

German is particularly rich in such semilexicalizations. An example is the compound noun *Mekkareise* (literally: 'trip to Mecca'), which, as a compound, has acquired the specialized meaning of 'trip to Mecca for religious purposes, hadj'. The satisfaction conditions of *Mekkareise* can thus be specified partly as the syntactic NP *Reise nach Mekka* ('trip to Mecca'), with the added condition that it must be a trip of the kind prescribed in Islam. It is thus possible to say of a man that he is on a trip to Mecca without it being true that he is making a *Mekkareise*.

Also, one notes that pronominal anaphora to the antecedent *Mekka* is still possible when *Mekka* has been incorporated into the compound *Mekkareise*, as long as the antecedent *Mekka* remains phonologically and semantically recognizable, as in (8.26a). But as soon as the phonological form of the intended antecedent has disappeared, taking away its semantic identifiability, as in (8.26b), pronominal anaphora is no longer possible.[18] This is a further illustration of the point made above in connection with (8.7a–8.c) above:

(8.26) a. Er hat schon zwei mal eine Mekkareise gemacht, aber er weiß noch immer nicht, wo *es* liegt.
He has already been to Mecca twice, but he still doesn't know where *it* is.

b. !Er hat schon zwei mal einen Hadsch gemacht, aber er weiß noch immer nicht, wo *es* liegt.
!He has already done a hadj twice, but he still doesn't know where *it* is.

Such cases illustrate the fact that lexemes typically arise as a result of a social need: when a particular concept has acquired a prototypical status in a given

[18] As regards semantic identifiability, note that both pronominal and epithetic anaphora in (i) is clearly ruled out despite the fact that the word *Jesuit* incorporates the name *Jesus* and does so in a way that maintains phonological identifiability. The reason is that the semantic value of the name *Jesus* is no longer transparent in the noun *Jesuit*. Compared with (i), (ii) is a great deal better, mainly, it seems, because the antecedent *Marx* is not only phonologically but also semantically clearly identifiable:

(i) !Dennis is a Jesuit, but he doesn't believe in him/the man.

(ii) Dennis is a Marxist, but he doesn't know anything about him/the man.

community, it is convenient to be able to express it by means of a single lexeme, complete with argument structure and grammatical and phonological features and ready for processing by the grammar in any sentence in which it can do service. Transparent constructs, such as *laisser-passer* or *Mekkareise*, are useful in such cases as pointers or mnemonics, provided the speakers of the language in question know the specific meaning that has been assigned to the transparent construct in question.

8.6.3 *Case shift and other arguments for prelexical syntax*

Transparent constructs have the advantage of illustrating the distinction between a prelexical syntactic structure and a nonsyntactic part in the satisfaction conditions of a predicate. As such they constitute an argument in favour of a moderate form of prelexical syntax for the semantic description of predicates. But are there any further positive indications that prelexical syntax makes sense? The answer is that there are, but only, as far as we know, for a limited range of lexemes, not for the lexicon as a whole. Yet the indications that we have are intriguing enough to justify a closer look.

For one thing, the examples (8.12)–(8.16) have already shown that scope-bearing adverbial adjuncts (operators) can sometimes modify an element in the internal structure assumed for causative verbs. Although we do not know the precise conditions under which this is allowed, the very fact that it is possible supports prelexical syntax.

A further argument, which we call the CASE-SHIFT ARGUMENT, derives from French and other languages, such as Turkish or Dutch, that have PR in open syntax and assign dative case to the underlying subject of an embedded transitive clause (German does have open PR but assigns accusative case to such argument terms). The point is that in languages that assign dative case, such as French, Turkish, or Dutch, an underlying intransitive subject becomes accusative in the resulting surface structure, whereas an underlying transitive subject becomes dative, provided it denotes an animate being that is the recipient or beneficiary of the process or state expressed by the lower verb, as with the verbs *know*, *see*, *hear*, *eat*, and so on (see Section 7.2.2). The accusative is then reserved for the underlying object term. When, however, the underlying lower transitive subject term denotes an active (animate or inanimate) force, as with causative verbs, the embedded clause is passivized (but without passive morphology), and the transitive subject appears in surface structure as an agent phrase with the appropriate preposition, in French *par* (by).

Consider the following French examples (with their assumed semantic structure):

(8.27) a. Il fera rire les enfants.
he will make laugh the $children_{ACCUS}$
He will make the children laugh.
S[V[FAIRE] NP[IL] S[V[RIRE] NP[LES ENFANTS]]]

b. Il fera voir la lettre à Luc.
he will make see the $letter_{ACCUS}$ to Luc_{DATIVE}
He will make/let Luc see the letter.
S[V[FAIRE] NP[IL] S[V[VOIR] NP[LUC] NP[LA LETTRE]]]

c. Il fera fermer la porte par Luc.
he will make $close_{PASSIVE}$ the $door_{ACCUS}$ by Luc
He will make/have Luc close the door.
S[V[FAIRE] NP[IL] S[V[FERMER] NP[LUC] NP[LA PORTE]]]

In some cases, a verb can be used both transitively and intransitively. The English verbs *write* or *eat*, for example, may occur with and without a direct object, and the same goes for their French equivalents. Typically, therefore the intransitive subject of such verbs will appear, under *faire*, in the accusative (with or without passivization) but the otherwise identical transitive subject in the dative:

(8.28) a. Il fera manger du pain à Luc.
he will make eat $bread_{ACCUS}$ to Luc_{DATIVE}
He will make Luc eat bread.
S[V[FAIRE] NP[IL] S[V[MANGER] NP[LUC] NP[DU PAIN]]]

b. Il fera manger du pain.
he will make $eat_{PASSIVE}$ $bread_{ACCUS}$
He will make/let bread be eaten.
S[V[FAIRE] NP[IL] S[V[$MANGER_{PASSIVE}$] NP[DU PAIN]]]

c. Il fera manger Luc.
he will make $eat_{(PASSIVE)}$ Luc_{ACCUS}
He will make/let Luc eat / be eaten.
S[V[FAIRE] NP[IL] S[V[$MANGER_{PASSIVE}$] NP[Luc]]]

Sentence (8.28c) is now seen to be ambiguous between a reading in which Luc eats and a reading where Luc is eaten (we reckon (8.28b) not to be ambiguous that way owing to selection restrictions: bread does not eat but is eaten).

The argument for prelexical syntax consists in the fact that analogous case-shift phenomena are observed with single-morpheme causative predicates in a large variety of languages. Take English *feed* and assume that its semantic specification contains a prelexical syntactic structure 'cause to eat' united into one single V-constituent in virtue of prelexical PR:

(8.29) a. Anne will feed bread to Luc.
b. Anne will feed bread.
c. Anne will feed Luc.

Just as in (8.28), the subject term *Luc* of the predicate *eat* taken to be hidden in *feed* occurs in the dative in (8.29a) but in the accusative in (8.29c), which is ambiguous depending on whether hidden *eat* is taken to be active or passive (Luc may be fed to the lions). The object term *bread* takes accusative case throughout: in (8.29a) because it is the underlying object, in (8.29b) because it has become the intransitive subject of the passivized embedded clause, as in one of the two readings of (8.29c). This argument thus runs parallel to what was observed above in connection with the examples (8.16a–8.16d), where it was shown that the ban on internal operators involving subject deletion applies equally to prelexical and open-syntax PR constructions.

The case-shift phenomenon occurs regularly, in English and many other languages, as illustrated by (8.30)–(8.33), although many more examples could be provided in all kinds of languages (see Seuren 1985: 178–80).

(8.30) a. teach French to the student
b. teach French
c. teach the student

(8.31) a. serve soup to the client
b. serve soup
c. serve the client[19]

(8.32) a. pay fifty pounds to the doctor
b. pay fifty pounds
c. pay the doctor

(8.33) a. delegate a task to the secretary
b. delegate a task
c. delegate the secretary

One should be warned, however, that it is hardly ever possible to PREDICT that the case-shift phenomenon will actually occur in a particular case, since lexicons mainly provide data for RETRODICTIVE explanations of the kind current in the historical sciences, not for the stronger kind of explanation that allows

[19] Hence the anecdote of the two ladies in a restaurant who are told by the waiter 'Sorry, ladies, we only serve men here', whereupon one of the ladies replies 'Oh good, we'll have one each'.

for prediction. Thus French has the verb *persuader* (persuade, convince), analysable as 'cause X to believe Y', as in *persuader son innocence au juge*, ('persuade the judge of one's innocence'; literally 'persuade one's innocence to the judge'). But although one finds *persuader le juge* ('persuade the judge'), one does not find **persuader son innocence* in the expected meaning of 'protest one's innocence'.

It must be observed that prelexical syntax is the only theory to offer a principled analysis of this striking case-shift regularity: no theory of syntax or of the lexicon has been able, so far, to capture this generalization. This in itself should be a reason for extreme caution when it is felt that the theory of prelexical syntax should be rejected.

A further indication comes from PRELEXICAL PASSIVES. We already encountered 'hidden' passives in (8.27c), (8.28b, 8.28c), (8.29b, 8.29c), (8.30b), (8.31b, 8.31c), (8.32b) and (8.33b). A further striking example is English *refuse*, which may be thought to contain the prelexical structure 'decide not to give', as in (8.34a). The ambiguity of (8.34b) is naturally reduced to the difference between the active and the passive form of internal GIVE:

(8.34) a. He refused the child a cigarette (decide not to give)
b. He refused a cigarette (decide not to give/decide not to be given)

Or consider French *défendre* ('forbid' but without a dative either 'forbid' or 'defend', depending on the interpretation of the internal predicate corresponding to 'take'):

(8.35) a. Il défend le vin aux enfants. (not allow X to take Y)
he forbids the wine to the children
He forbids the children to have wine.

b. Il défend le vin. (not allow Y to be taken)
he forbids/defends the wine
He forbids wine/defends the wine.

c. Il défend le pays. (not allow Y to be taken)
he defends the country
He defends the country.

Here both (8.35b) and (8.35c) involve an internal passive.

8.6.4 *Lexically incorporated negation*

More confirmation for prelexical syntax comes from phenomena that suggest a lexically incorporated negation. This is a particularly intriguing topic. Negativity, in one form or another, is clearly present within the semantic structure of many lexemes, as appears, for example, from the fact that negativity is often manifest as a recognizable morpheme. Sometimes, however, this negativity is not morphologically transparent yet appears from the licensing of a negative polarity item (NPI) like *any*, *at all*, and so on. Then again, there are predicates (items) that clearly require some sort of negativity somewhere in their definition yet do not license NPIs (the 'fourth class' in Janet Fodor *et al.* 1975: 520–1, but see also (8.37d) below, where the quantifier *each* does not allow for *any* but other universal quantifiers like *all* or *every* do). Moreover, the negativity in question sometimes consists in a propositional (logical) negation, but sometimes also in less clearly defined forms of negativity. In the context of prelexical syntax, we are primarily interested in the question of how to determine whether it makes empirical sense to assume an internal propositional negation, that is, the logical predicate NOT, as a primitive in the semantic structure of a lexeme, but that is only a limited aspect of the question of internal negativity as a whole.

Intensive attempts have been undertaken over the past few decades, in the wake of Ladusaw (1979), to reduce the more general notion of negativity to the logical property of DOWNWARD ENTAILMENT (↓⊨):

> DOWNWARD ENTAILMENT(↓⊨):
> A sentence *A* has the property of DOWNWARD ENTAILMENT with respect to the expression *e* in *A* when it is always possible to form a sentence *B* which is entailed by *A* and is identical with *A* but for the substitution of an expression *f* for *e*, where *f* denotes a class that is contained in the denotation of *e*.

Upward entailment (↑⊨) is the converse:

> UPWARD ENTAILMENT (↑⊨):
> A sentence *A* has the property of UPWARD ENTAILMENT with respect to the expression *e* in *A* when it is always possible to form a sentence *B* which is entailed by *A* and is identical with *A* but for the substitution of an expression *f* for *e*, where *f* denotes a class that contains the denotation of *e*.

Upward entailment is illustrated in (8.36a, 8.36b), downward entailment in (8.36c, 8.36d) (the expressions *e* and *f* have been italicized in each case):

(8.36) a. She bought some *red tulips* ↑⊨*tulips* ↑⊨*flowers*
b. Jim *has been murdered* ↑⊨*has been killed* ↑⊨*is dead*
c. She didn't buy any *flowers* ↓⊨*tulips* ↓⊨*red tulips*
d. Jim *isn't dead* ↓⊨*hasn't been killed* ↓⊨*hasn't been murdered*

Yet these attempts at reducing negativity to downward entailment, which is then taken to be the licensing factor for the occurrence of an NPI such as *any*, have also met with serious obstacles. It is true that the propositional negation *not* generally causes downward entailment. In other forms of negativity, however, such as the negativity expressed by *hardly*, there appears to be some variation. In (8.37a) there is downward entailment for *change*, but in (8.37b) there is no downward entailment for *anything*. Despite the correct *any* in *She hardly bought any flowers*, there is no downward entailment, since it may well be true that she hardly bought anything while she still bought a big bunch of flowers. Similarly for the extremely troublesome word *only*, which does license the NPI *any*, but lacks downward entailment, as shown in (8.37c).[20] Sentence (8.37d), finally, shows that the quantifier *each* does not allow for *any* in the expression denoting the class quantified over, unlike *all* and *every* which do allow for *any* in that position. At the same time it looks as if there is also no downward entailment, again unlike *all* and *every*, for which there is. *Every boy (all boys) who had eaten some fish was (were) checked* has the downward entailment *Every boy (all boys) who had eaten some cod was (were) checked.* But (8.37d) does not have downward entailment for *fish*, since (8.37d) may be true while *Each boy who had eaten some cod was checked* is false. Intuitively, the reason seems to be that there may have been one or more boys who did eat fish but never ate any cod, but how this is different for the quantifiers *all* and *every* is simply not clear:

(8.37) a. She hardly noticed any *change* $\downarrow\models$*small change / big change*

b. She hardly bought *anything* $\downarrow\not\models$*any flowers*

c. Only John employs any *immigrants* $\downarrow\not\models$*illegal immigrants*

d. Each boy who had eaten some (?any) *fish* $\downarrow\not\models$*cod* was checked

Moreover, (8.38a), with the nonnegative predicate *like*, is downward entailing when it takes a generic object term for the expression *e*, just like its negative counterpart (8.38b). Likewise for predicates like *love* or *hate*. Also, (8.39a), with the negative predicate *regret*, is neither upward nor downward entailing. Whether its negation (8.39b) is downward entailing depends on whether or not one is prepared to take presuppositional phenomena into account. If one does, there is no downward entailment:

(8.38) a. She likes *pets* $\downarrow\models$*cats* $\downarrow\models$*siamese cats* ($\uparrow\not\models$*animate beings*)

b. She doesn't like *pets* $\downarrow\models$*cats* $\downarrow\models$*siamese cats* ($\uparrow\not\models$*animate beings*)

[20] Interestingly, translation equivalents of *only* in some other languages, such as *móno* in Modern Greek, do not license NPIs (Giannakidou 2004).

(8.39) a. She regrets feeding any *pet* ↓⊭*cat* ↓⊭*siamese cat*
(↑⊭*animate being*)

b. She doesn't regret feeding any *pet* ? ↓⊨? *cat*
(↑⊭*animate being*)

On pain of becoming monotonous we must say again that the question has so far remained unsolved: it is simply not known how precisely direction of entailment is related to semantic properties of predicates or sentences, although we have some idea of the factors that are probably involved one way or another.

Given that downward entailment is insufficient as a diagnostic, the question arises of how negativity in a wider sense can be diagnosed. Often the presence of a negative morpheme, mostly a prefix, such as English *un-*, *in-*, or *dis-*, is an indication, but this criterion is not infallible. For one thing, there often is some form of semantic specialization, as with *uneasy*, which means neither 'not easy' nor 'difficult', and which does not show clear signs of negativity. Sometimes a negative morpheme combines with an element that has no independent status, as in *dissent*, *discomfiture*, *difficult*, *impervious*, *uncouth*, *discrimination*, *disperse*, and so on. Sometimes also an original negative element has lost all semantic negativity, as in *atom* (etymologically 'unsplittable'), or *anecdote* (etymologically 'not to be made public').

In the absence of open morphological criteria, negativity appears from the marked character of the negative element in a positive–negative pair. Thus we say naturally *They came out, one after the other*, and not *!They came out, one before the other*, although the latter is truth-conditionally equivalent to the former. Likewise we say more naturally *How far is it from A to B?* than its presuppositionally marked but otherwise equivalent *How close is A to B?* On this reckoning, *before* and *close* should count as negative.

Then, as has been said, lexically internal negativity often shows up by the appearance of a negative polarity item such as *any* in the following sentences:

(8.40) a. She left *before* (*after) anything untoward happened.

b. It is *hard* (*easy) for her to admit that she did anything wrong.

c. She is *far from* (*close to) admitting that she did anything wrong.[21]

d. This is *different from* (*the same as) anything I've seen so far.

[21] In this sentence *far from* functions as the negative counterpart of the positive *close to*, whereas a few lines earlier it is taken to be the positive element of the pair. The reason is that metaphorical use may invert polarity, as with the nonnegative *distance*, which becomes negative in the metaphorical *take distance from (any allegation)*.

e. She *denied* (*asserted) that she had done anything wrong.

f. She *doubts* (*believes) that she has done anything wrong.

g. She *ignored* (*acknowledged) any hecklers.[22]

h. She came *without* (*with) any children.

i. She *lacks* (*has) any charm.

j. *She *missed* any train.

Yet although this may be a sufficient criterion it does not seem to be a necessary one. For example, (8.40i), with the negative verb *lack*, analysable as 'not have', is a good sentence, whereas (8.40j), with the verb *miss*, is ungrammatical and clearly cannot be used to express the meaning 'she did not catch any train', even though *miss* is equally naturally analysed as 'not hit' or 'not catch':

We hit here on a problem that is similar to the problem signalled above in connection with (8.8)–(8.15). There the problem was whether and how an open-syntax adverbial adjunct, representing a scope-bearing operator at the semantic level, can modify an element in the internal semantic structure of an item. We saw that this is sometimes but not always possible and that the conditions are not entirely clear. Here the problem is whether and how an internal scope-bearing operator, in this case the negation element, can affect elements outside the lexeme concerned. The examples (8.40a–8.40j) show that this is sometimes but not always possible, and we will have to conclude, again, that the conditions are not clear.

Consider the cases (8.41) and(8.42). *Have lost* in (8.41a) is naturally analysed as 'no longer have'. Yet this lexically incorporated negation cannot take scope over *all*, since if it could, (8.41a) would have the reading of (8.41b). And (8.42b) shows that the negative existential quantifier *no-one* can bind a variable pronoun in open syntax, but not from a lexically internal position, as in (8.42a):

(8.41) a. She *has lost* all the jewels she had before her marriage.

b. ≠She *no longer has* all the jewels she had before her marriage.

(8.42) a. !This is a no-go area if *they* want to survive.

b. This an area that *no-one* should enter if *they* want to survive.

In other cases, however, it looks as if it IS possible for a lexically incorporated negation to take scope over the universal quantifier. If, as is reasonable, *deny* is analysed as 'assert that not', then (8.43a) shows that this *not* can take scope over *all*, just as it does in (8.43b), which is synonymous with (8.43a) but for the

22 *Ignore* may be analysed as 'pretend not to notice'. French *ignorer* has the different meaning of 'not know.'

presupposition in (8.43a) that the proposition underlying *All parts were faulty* has been expressed before. Likewise in (8.44a, 8.44b) and (8.45a, 8.45b), where (8.45b) may be rephrased under NEGATIVE RAISING, leading to (8.45c):

(8.43) a. He *denied* that all parts were faulty.
b. He *asserted* that not all parts were faulty.

(8.44) a. He *dissuaded* her *from* eating all of it.
b. He *persuaded* her *not to* eat all of it.

(8.45) a. He *disbelieved* that all parts were faulty.
b. He *believed that not* all parts were faulty.
c. He *didn't believe* that all parts were faulty.

It would seem that the difference is caused by the fact that the negation taken to be incorporated into predicates like *deny*, *dissuade*, or *disbelieve* follows the main verb and thus has scope directly over the subordinate complement clause:

deny ≅ ASSERT THAT NOT ...
dissuade ≅ PERSUADE NOT TO ...
disbelieve ≅ BELIEVE THAT NOT ...

By contrast, in predicates like *have lost* or *have forgotten* the presumed internal negation takes scope over the main verb:

have lost ≅ NO LONGER HAVE
have forgotten ≅ NO LONGER KNOW

If this proves to be correct, it is again a serious argument in favour of the principle of internal lexical analysis. Be all this as it may, further inquiry is clearly needed. Unfortunately, however, such research, no matter how much sense it would make, has been stifled over the past thirty or so years by sociological pressure emanating from Chomskyan linguistics, the only motivation being that it was initiated by the school of GENERATIVE SEMANTICS, whose name dare not be spoken.

A further sign of negativity is provided by the copying of a lexically incorporated negation in the remainder of the sentence under certain structurally defined conditions.[23] This occurs frequently and is a sure sign of internal

[23] Copying of overt negation is rampant in many English dialects, especially substandard ones, as appears from sentences like (i) or (ii):

(i) It ain't no cat can't get in no coop. (Labov 1972: 130)
(ii) 'E's an odd fella. 'E ain't never been no good to no woman, not never. (British TV-play)

negativity. In London Cockney, for example, where negation copying is rampant, one can hear sentences like the following ((8.46a) is taken from a song that was popular in Britain during the depression years):

(8.46) a. There was *hardly* no money nor hardly no hope.
b. She did a better job *than* I never thought she would do.

We take *hardly* to contain some form of negativity (as appears also from the fact that it induces auxiliary inversion when fronted: *Hardly had she left, when the row started*—another sign of internal negativity). The multiple iteration of the negation further down the sentence can only be based on the internal negativity of *hardly*. Likewise with *than* of the comparative in English. Joly (1967) argues convincingly that *than* derives historically from *thon-ne* ('by/to which not'), so that *John is taller than Bill* is, roughly speaking, derivable from 'John is tall to a degree to which Bill is not tall' (for an extensive analysis see Seuren 1973, 1984*a*). Again it is the lexically internal negation or negativity that is taken to trigger the copying.[24]

It should be mentioned, at this point, that on none of the above criteria can the predicate *dead* (a member of the 'fourth class' in Janet Fodor *et al.* 1975: 520–1) be regarded as in any sense containing a negation as part of its lexically internal syntactic structure (if any), as was proposed by McCawley, who analysed *dead* as 'not-alive'. In hindsight it seems more sensible to describe *dead* without any prelexical syntax and place the condition 'not alive' in the metalinguistic satisfaction conditions. The same holds for a host of other items, including the much discussed *bachelor*, usually analysed as 'human male of marriable age but not married'. Such definitions are not wrong or

[24] Interestingly, semantically functional, non-copied, double negation blocks NPIs and licenses positive polarity items (PPIs), such as the PPI *rather* in (i) (Baker 1970: 179):

(i) There isn't anyone in this camp who wouldn't rather be in Montpelier.

Likewise, English *few* is somehow internally negative, as appears from the fact that it licenses negative polarity items(NPIs), as in (ii). Yet, when placed under negation, as in (iii) (Sanford *et al.* 2007: 24), *few* loses its power of licensing NPIs:

(ii) Fewer than ten people believe Jones anymore.
(iii) *No fewer than 10 people believe Jones anymore.

Similar phenomena are observed in the use of lexically incorporated double negation, as in Latin *nonnulli* (literally 'not none', hence 'several'), *nonnumquam* ('not never', hence 'sometimes', 'occasionally') and *nonnusquam* ('not nowhere', hence 'in some places', 'here and there'). In the same way, the Latin PPI *potius* ('rather'), which, being a PPI, does not occur directly under negation, is seen to occur freely directly under the double-negative Latin quantifiers mentioned above. (*Potius* does have an idiomatic occurrence directly under negation in the expression *non potius A quam B* (Kühner and Stegmann 1955, II: 483), which, however, has lost its literal meaning 'not rather A than B' and has come to mean 'A as much as B'.)

illegitimate, but, pending further research, they should not be considered to represent chunks of prelexical, or internal lexical, syntactic structure.

It thus appears that, on the one hand, the notion of prelexical syntax looks tantalizing, while, on the other, there are severe problems which are awaiting a solution. It would be wrong to reject the entire notion on the grounds that there still are so many problems to solve. But it would be equally wrong to ignore the problems and declare prelexical syntax a valid theory. Further research is obviously needed.

8.7 Fodor's global-cognitivist approach to lexical meaning

So much for lexical decomposition. A totally different approach from prelexical syntax has been advocated by Jerry Fodor in Janet Fodor *et al.* (1975), Jerry Fodor (1975, 1998, 1999), Jerry Fodor *et al.* (1980) and other publications. In Fodor's view (1975: 150–1), the distinction that must be made anyway between linguistic and other knowledge is best drawn in such a way that lexical meanings are treated as atomic, undivided wholes, stored in cognition, the 'concepts' of traditional psychology and philosophy, including the philosophy of language. They are external to the language machinery (Fodor's 'sentence understander'), which just carries the lexical meanings (concepts) to and from the surface so that utterances can be cognitively processed by the listener in the form they are produced by the speaker. The phonological representations of lexemes, on the contrary, are part of the language machinery and are therefore profitably studied by linguists. For Fodor, the meaning of a lexical predicate consists in its property of being linked up with a particular cognitive concept, which may exist without any linguistic counterpart, as is the case in animals, which have no language but do have forms of thinking. In Fodor's view, a concept that serves as a lexical meaning is connected with the whole of cognition through a set of meaning postulates as proposed in Carnap (1956) and discussed (and rejected) in Section 2.3.1 in Volume II.

One of Fodor's main points is that the necessary and sufficient conditions taken to define meanings cannot be expressed in natural language terms but must be formulated in a 'language of thought', which is categorially different from any natural language and whose terms and combinatorial properties will have to be established as a result of psychological theorizing. Concepts are taken to be compositional in that they allow for composition into more specific concepts and, accordingly, for decomposition into less specific concepts. This presupposes a basis of 'primitive' concepts, which is taken to be innate:

[T]he theory of *concept learning* consists of a set of inductive procedures...which determines the availability of a complex concept relative to (a) the availability of appropriate basic concepts (in particular, the ones from which the complex concept is constructed) and (b) the experience of the organism.... [T]heories of concept learning *presuppose* the availability of the primitive conceptual basis; they don't explain it. If, however, the primitive basis is presupposed in concept learning, then it cannot itself be learned. If it is not learned, then, presumably, it is innate. The claim, then, is that all standard theories of concept learning require the innateness of the primitive basis and explain *at most* the acquisition of complex concepts relative to the availability of that basis.[25] (Fodor *et al.* 1999: 503–4)

What is 'innate' in language, therefore, is not 'concepts like "airplane" ready formed', but 'the elements into which such concepts decompose, together with the appropriate combinatorial operations defined over the elements' (Fodor 1975: 152). Moreover, children are born with the innate capacity to link chunks of thought structure with chunks of sound representations, so that each word becomes 'an abbreviation for a complex expression of the internal language...the abbreviatory convention [being] stored as a principle of the logic' (ibid.). Fodor's main argument for this position is that utterance comprehension is too fast to allow for the amount of cognitive processing that will be needed if the internal analysis of words (predicates) is considered part of the on-line comprehension process (Fodor 1975: 150–1).

This view makes a great deal of sense. It obviates the qualia problem in lexical meanings, mentioned in Section 8.2.2, since it is impossible to define, for example, the specific property that makes an apple an apple in terms of any form of natural language. The conclusion that it takes nonlinguistic, purely cognitive criteria to capture that sort of property seems inescapable, even though that conclusion is merely programmatic in that it shifts the burden

[25] Jackendoff seems to do Fodor an injustice when he claims (Jackendoff 2002: 334):

Jerry Fodor...takes the position (e.g. 1975; 1998) that all lexical concepts (or at least all morphologically simple lexical concepts) are innate. This includes, for instance, the meanings of *telephone*, *carburettor*, *merlot*, *soffit*, and *yarmulka*: even if some of these words are absent from someone's vocabulary, their meanings are present in his or her f[unctional]-mind, just waiting to be triggered. Fodor arrives at this position because he maintains that lexical concepts are monadic: they *have* no parts, hence no parts that could be previously known. Hence he is forced to say that all lexical concepts are innate.

Fodor does not speak of 'morphologically simple lexical concepts', whatever these may be, but of concepts in cognition, which fall outside the realm of morphology. The misunderstanding continues in Jackendoff:

So here is the answer to Fodor's argument against definitions: although he has shown that lexical meanings cannot be decomposed definitionally, he neglects the possibility that there are nondefinitional forms of decomposition of the sort found in phonology and physics. Since his alternative to nondefinitional decomposition is genetic transmission of the whole meanings of *quark* and *fax*, our choice ought to be clear. (Jackendoff 2002: 336)

of specifying such criteria to future psychological and physiological research (Musacchio 2005).

Critics, however, may well object that Fodor's account still fails to answer the question of what comprehension of an utterance amounts to. Fodor distinguishes between two components, a purely linguistic component that enables a listener to understand or interpret the utterance received, the 'sentence understander', and a 'logical' component which enables a listener to draw inferences (through meaning postulates) and integrate the information 'gleaned' from the utterance into the totality of available knowledge. The first component is taken to use up relatively little time and working space, but one has to pay dearly for the speed and the economy in terms of modular machinery and further infrastructure. The second component may take much longer and is free to use the entire working space of long-term memory, which is a whole lot cheaper in terms of infrastructural investment. In Fodor's words (1975: 150): 'Computing memory is expensive, but long-term memory is cheap'. The relevant passage is worth quoting in full:

> A theory of the hearer can reasonably be expected to contain two distinguishable components. The first of these is concerned with explaining sentence comprehension proper; i.e., with characterizing the computations which effect the correspondence between wave forms and messages; i.e., with specifying those mental operations which eventuate in a display of the information that utterances of sentences convey; i.e., with showing how hearers reconstruct the communicative intentions of speakers. Call this component a 'sentence understander.' The second component is concerned with representing the data processes (including the drawing of inferences) which are *defined over* the information that utterances of sentences convey; i.e. those data processes which mediate the hearer's use of the information he gleans from the utterances he hears. Call this component a *logic*. Then, roughly (abstracting from feedback and the like) the output of the sentence understander is the input to the logic. Equivalently, the (or a) function of the sentence understander is to represent utterances in the normal form for which operations in the logic are defined. (Fodor 1975: 150–1)

There are several problems with this account. One, it seems, is that the notion of 'comprehension-as-such', the intended result of the 'sentence understander', remains cloudy. The 'computations that effect the correspondence between wave forms and messages' appear to be merely the computations of the phonology and the grammar, which do not amount to actual semantic interpretation. Specifying 'those mental operations which eventuate in a display of the information that utterances of sentences convey' is precisely the task at hand, but Fodor fails to tell us how this feat is accomplished, and the same goes for his phrase 'showing how hearers reconstruct the communicative intentions of speakers'.

What seems to be lacking is an account of the on-line process of comprehending the bare proposition. One can rake one's mind as much as one likes, there does not seem to be any alternative to saying that the on-line understanding of an utterance means at least the activation of the satisfaction conditions that define the (possibly complex) concepts involved and their application to the term referents at hand. This not only makes the Carnapian meaning postulates redundant, it also, one fears, takes away the force of Fodor's argument that the near-immediacy of utterance comprehension rests on the fact that on-line processing is fast but requires an expensive infrastructure, while long-term memory is slow but cheap in terms of investment, since the application of satisfaction conditions to term referents surely takes some fast and expensive on-line processing, but it also needs to activate—cheaply but slowly, as Fodor has it—chunks of long-term memory.

Another, directly related, problem with Fodor's account is to do with his strict separation of a purely linguistic level of representation and the use made of that information in further cognitive processes. The assumption of such a strict separation has, until recently, not been controversial in semantics and philosophy. Most linguistic, semantic, and/or pragmatic theories are cast in terms of a two-step interpretation procedure consisting of a linguistic analysis followed by further cognitive processing of whatever kind. However, results obtained by means of recently developed on-line brain imaging techniques indicate with increasing force that such a separation is unwarranted (Hagoort *et al.* 2003, 2004). It appears that the activation of a lexical concept SIMULTANEOUSLY activates just about everything stocked in long-term and short-term memory that is associated with the concept in question. And since, during interpretation, lexical concepts are activated well before the total process of interpretation is completed, the entire cognitive machinery is directly open to access in the interpretation process. In fact, the results strongly support the view that any element in the wider cognitive machinery may be instrumental in achieving the very interpretation process that led to its activation—precisely the point of view defended in Section 7.2.3. With all due caution, one may conclude, therefore, that, in interpretation, there is no strict temporal separation between the activation of linguistic and nonlinguistic elements in the cognitive mind, although there may still be a strict separation between the various machineries that are activated.

A further point that can be made against Fodor's account is the fact, established above in connection with Sections 8.6.3 and 8.6.4, involving case shift, passives, and negativity, that there are serious arguments, of a well-established kind, in favour of the assumption of prelexical syntax for at least part of the meaning description of at least some lexical predicates. It therefore appears that, if Fodor's account deserves support, it does so only for certain aspects of lexical meaning descriptions.

8.8 Prototypicality

Many authors, including Fodor and Jackendoff, recognize the importance of prototypicality for the specification of lexical meanings. Fodor maintains that prototypes and concepts are not identical, although they may co-exist as correlates of lexical predicates:

> Concepts can't be prototypes, *pace* all the evidence that everybody who has a concept is highly likely to have its prototype as well. (Fodor 1998: 93)

Fodor is no doubt right in maintaining that concepts are not the same sort of thing as prototypes, but his account of prototypicality is deficient and fails to integrate the two notions. The matter is important enough to deserve some attention.

Fodor's argument that concepts and prototypes are different things rests on the premiss that complex concepts are built up compositionally out of basic concepts whereas prototypes are not. In his familiar idiom he writes:

> In a nutshell, the trouble with prototypes is this. Concepts are productive and systematic. Since compositionality is what explains systematicity and productivity, it must be that concepts are compositional. But it's as certain as anything ever gets in cognitive science that prototypes don't compose. So it's as certain as anything ever gets in cognitive science that concepts can't *be* prototypes and that the glue that holds concepts together can't be statistical. (Fodor 1998: 94)

Even if we accept this argument (although it may be a trifle too cocksure), Fodor does not tell us what prototypes are and how they come about. Yet this question is crucial, since prototypicality clearly is a powerful determinant of lexical meaning in a large number of cases.

There is one answer that will not do, despite its great popularity. According to many, presumably including Fodor, prototypicality is based on the statistics of experience. Take a class C of objects satisfying a set of conditions that define the concept *C*. Now suppose there is a subclass C_S of objects that share one or more features not shared by other objects of C. Suppose further that the objects of C_S are encountered with significantly greater frequency than those of the complement of C_S. Then, according to this view, the subclass C_S forms a PROTOTYPE for the class C.

This account has something going for it, yet it does not tell the whole story, as the following example shows. Let C be the class of animals and C_S the subclass of flies. One may take it that most people encounter flies in their lives with significantly greater frequency than, say, dogs. Yet it is hardly likely that anyone asked to mention an animal (that sort of question being taken by psychologists to bring out prototypicality) will come up with the answer 'a fly', whereas the answer 'a dog' is clearly within the range of expectancy. Why should that be so? Presumably because there is an autonomous cognitive parameter of *salience*. In the formation of prototypes, salience counts for more than frequency, although frequency undoubtedly helps and may even be a necessary condition. It seems safe to assume that the encounters at issue must, first of all, be SALIENT encounters, a condition usually not met by encounters with individual flies. (A cloud of flies will be salient enough, but it crucially lacks frequency.)

Moreover, the features shared by the subclass C_S must be salient features. Take the class of objects that we call by the collective name *hair*. Hair has the feature of growing on a surface, yet only one kind of surface makes for a salient feature, the top of a human head, called *pate* when it lacks hair. The reason for this salience is probably the fact that human heads are of prime importance in human interaction, but be that as it may, it is this salience that makes the hair that grows on human heads prototypical, in contrast to, for example, the hair that grows on chests, or on the bodies of some animals.[26] Or consider the class of prime numbers, which are defined by the property of not being divisible other than by themselves and the number 1, and which are all odd except for the even number 2. Now the cardinality of a group is salient only in the lower numbers: according to psychologists, the cardinality of a group stands out saliently up to 5, perhaps 6 or even 7, but not beyond that. This is probably why subjects, when asked to mention a prime number, tend to be quick to give the number 3, or 5, but not, for example, 41 (or 2, which is untypical on account of its being the only even prime number). Or when asked to make a drawing of a tower, most subjects will not draw something resembling the famous painting of the Tower of Babel by Pieter Bruegel the Elder, since that tower is saliently squat, is surrounded by a spiralling walkway and contains enough living space for a whole town—features that are distinctly untypical of run-of-the-mill towers.

What makes for salience is hard to say. The honest answer is that we do not know. What we do know is that, up to a point, salience can be measured

[26] It is no doubt this prototypical distinction that lies at the bottom of lexical distinctions such as between English *hair* and *fur*, or between French *cheveu* (Italian *capello*), which is 'hair growing on a human head' (but not beard hair) and French *poil* (Italian *pelo*), which covers other varieties of hair. Correspondingly, French *chauve* (Italian *calvo*) means 'deprived of *cheveux* (*capelli*)' and not 'deprived of *poils* (*peli*)'.

in terms of brain activity or psychological function. Thus, for individual occurrences to be salient they must raise the attention, which is a measurable parameter, and they will normally have a high rate of memory retention, likewise a measurable aspect. For features of objects to be salient they must satisfy conditions of prominence, and also be measurable. But this amounts to a recording of symptoms, not to a specification of what it is that makes for salience, and why. To answer those questions one needs an adequate theory of cognitive structures and processes, together with an appropriate theoretical language, as in the case of qualia.

If, on the other hand, the objectively measurable aspect is left out of account, all we have is an explanation of the obscure in terms of the equally or more obscure. Then one can say, for example, that salience is closely related to 'significance' or 'meaningfulness' in a nonlinguistic sense, which does not bring us any closer to an answer. So let us follow accepted practice and accept the obscurity along with the reality of the phenomena concerned. This way we have at least the advantage of a cognitive parameter that plays an explanatory role in a larger theoretical set-up—as far as it goes. And it goes as far as the study of human (or animal) experience: in the physical sciences, salience is not a valid concept, though statistical preponderance is.

The criteria for prototypes seem to centre around a certain level of specificity. Prototypicality typically makes its appearance at that level of specificity at which humans classify objects when no special conditions prevail. It is a well-known fact that unconditioned naming tends to float at some level of specificity halfway between the very general and the very specific. For example, when placed before a picture of an oak and asked to tell what they see, subjects will normally say 'a tree' and not, for example, 'an oak' or 'a Lebanon oak' (if that is what is represented), such answers being too specific. Nor will they say 'a vegetable object', that being too general. Or when someone asks me how John got here, I will normally say 'by car', if that is the case and I will say 'in his Rolls' only when that detail is especially relevant. Otherwise such a reply is overinformative. An answer along the lines of 'by some means of transportation' is, although true, definitely underinformative. The question of what determines the level of specificity that characterizes unmarked naming is as obvious as it is hard to answer. In all likelihood, the answer is to be found in the complex mental structure or structures that may be taken to represent and specify relevance values in social interaction, but too little is known about those aspects of human mental life to draw any firm conclusions regarding the question at hand. Yet it does seem to be the case that criteria for prototypicality are typically found at that level.

Conditions of prototypicality play a part in the satisfaction conditions of at least some predicates. We say this on a variety of grounds. First, prototypes have the property of attracting *spurious objects.* When many or most objects that are called a name satisfy a set of prototypical conditions, then objects that satisfy all of those but fail to satisfy one or two of the nonprototypical conditions, are likely to be slipped into the extension of the concept or predicate meaning at hand. This, it seems, is what makes people treat whales and dolphins as kinds of fish, although that is not their biological class. Professional taxonomists will not be so indulgent, as salience and prototypicality are alien to their business.

Then, although, as Fodor puts it, 'prototypes don't compose', they do shift, along with experience, whether individual or socially shared. And, as we know, shifting is a fixture in lexical semantics, leading to specialization and metonymy, to polysemy phenomena, and even to the kind of transfer we find in conventionalized metaphors (see Section 9.5). Such phenomena cannot be understood without an appeal to prototypicality, one way or another.

Finally, as has been said, it seems a useful approach, given the strong cognitive reality of prototypes, to assume that conditions of prototypicality play a part in the satisfaction conditions of at least some predicates. The interesting thing is that, as long as they stay put, the part they play is not truth-conditionally defined. But when they begin to make themselves felt by exerting an influence on the lexical meanings they are part of, in the form of specialization, metonymy, polysemy, or conventionalized metaphor, their influence quickly leads to truth-conditional changes in the semantic description of the predicates at issue. It is obvious that such aspects of lexical meaning cry out for further research.

9

Vagaries of lexical meaning

9.1 Cognition as the main disturber of the formal peace

It is probably correct to say that there is no area of linguistic studies so filled with baffling phenomena that are not or badly understood as the lexicon, with its delicately tuned and infinitely flexible lexical items lending themselves to all sorts of creative uses. Since it is our purpose to explore the limits of the formalizable and also to treat logic as a branch of lexical semantics, it is only right that we try to give some idea of the complexities and complications that one is faced with when studying the lexicon.

The cognitive dimension is a basic point. The satisfaction conditions of predicates cannot be specified merely by an appeal to properties of the objects denoted. Very often an appeal is made to criteria whose satisfaction is to be derived from all kinds of mental factors. Thus, a cognitive VIEWPOINT is sometimes constitutive of a lexical meaning (see also Sections 9.7 and 7.4). Also, socialized general knowledge, often in the form of psychological PROTOTYPES, is frequently required for the specification of lexical meanings. Likewise for socially shared and context-bound EVALUATIONS, or for EMOTIONS. A further factor disturbing a purely 'objective', model-theoretic view of lexical meaning is the fact that a socially recognized FUNCTION is often essential for the definition of lexical predicate meanings, especially when they apply to artefacts.

Then, lexical meanings are often VAGUE, in that, when applied to certain objects, the resulting L-proposition is neither convincingly true nor convincingly false. A further disturbing factor of overriding importance is the fact that lexical meanings are open to all kinds of extension, specialization, transfer by association or comparison, first creatively and ad hoc, at the level of token interpretation, but then often conventionalized to type level, either for specific contexts or for general use. Such phenomena are both truth-conditional, in that they co-determine truth and falsity, and infuriatingly opaque, not only because they follow paths of human cognition that have so far not been explored to any depth, but also because we still lack an adequate insight into the inborn principle, introduced in Section 3.4.2, that makes humans (and perhaps also higher vertebrates) try all the rules in the book to get the most out of incoming signals. This principle, which we have called the *optimization of sense, truth, and actuality*, or OSTA, spurs listeners to search for cues in such a way that optimal sense is made of the incoming message, given the circumstances of utterance, and to do so in such a way that the incoming message can, if at all possible, be taken to be true with regard to actual reality.

Sometimes there is some light in the overall darkness. For example, there appear to be forms of mutual interaction between, on the one hand, the meaning of the main predicate of an L-proposition and, on the other, the way in which its argument terms select reference objects. Phenomena of this nature are discussed in Section 9.6.3 under the label 'dynamic filtering'. Another area where greater clarity has been achieved over recent years is that of lexical preconditions, discussed in Section 9.4. Lexical meanings are, in most cases, reserved for use in specific contexts or for specific classes of objects: they have, as part of their satisfaction conditions so-called preconditions which generate presuppositions of the sentences in which they are used. This mechanism is now better understood than was the case until recently.

Yet the prevailing experience in lexical studies is one of perplexity. One never stops being amazed. Consider, for example, the English noun *threshold* and its German translation equivalent *Schwelle*. In their normal uses they denote the ridge or sill usually found between doorposts at floor level. Yet, for reasons that have so far remained mysterious, these two words differ in their capacity for semantic extension. Everyone knows the elevations in roads and streets that are now widely called *speedbumps*. The German word for these things is *Schwelle*. Yet although the nearest English translation equivalent of *Schwelle* is *threshold*, it is unthinkable that speedbumps should be called *thresholds* in English. Why does the German word lend itself to this extended use, while the English equivalent does not? One is inclined to think that, at some ill-understood level of interpretation, the word *threshold* implies containment within a space or a transition from one kind of space to another, perhaps as a result of its etymology. *Schwelle*, by contrast, is a swelling in the ground that forms an obstacle to be got over—which is also its etymology, although, on the whole, German speakers do not realize that. But do such etymological facts, which, one would think, must have disappeared from the linguistic awareness of the speakers of the languages in question, really play a role in processes of semantic extension? Can ghosts of the past go on quivering for so long in people's minds? If so, that would shed a remarkable light on both cognition and language.

In any case, no matter how this difference is lodged in cognition, it is not a question of the ontological properties of the objects concerned, but of different ways in which they are conceived of or looked at. While such phenomena should be accounted for in any theory of lexical meaning, they fall by definition outside the scope of truth-conditional, model-theoretic semantics, which analyses the merely stipulated but empirically nonexistent relation between linguistic forms and things in worlds (Ogden and Richards 1923: 11). Cognition is an indispensable element in any sound form of semantics, whether lexical or sentential. And it is lacking in current forms of formal semantics.

Whether the facts of lexical meaning will eventually force us to give up the format of satisfaction conditions as a means of specifying lexical meanings in some recognised metalanguage is an open question. There is probably no valid reason why conditions derived from viewpoint, socialized knowledge, psychological prototypes, emotions, socialized, and context-bound evaluations, or from criteria of functionality, could not be incorporated as satisfaction conditions. These phenomena, therefore, do not seem to pose a basic threat to the satisfaction-condition format for the description of lexical meanings. Nor do presuppositions resulting from lexical preconditions. But one is bound to feel less secure with regard to phenomena of semantic extension or transfer, whose underlying mechanism is still largely obscure, or, worse, with regard to phenomena such as those illustrated above in connection with the words *threshold* and *Schwelle*. Even more problematic, if not outright counterevidential, for the descriptive format of satisfaction conditions are predicates expressing epistemic modalities or varieties of socially binding commitment or appeal, such as the verbs *appear*, *seem*, *will*, *may*, *must*. The meanings of such predicates are only partly describable in terms of conditions to be fulfilled so that truth results. Yet pending these uncertainties, we will proceed as we have done so far and maintain the format of satisfaction conditions as a means for specifying lexical meanings, taking the risk that this format may turn out to have only limited applicability. Until the ideas about the non-truth-conditional aspects of lexical meaning have crystallized into some form of manageable theory, one will try to get the most out of the format of satisfaction conditions, hoping that the storm can be braved when it comes.

9.2 Derivation and compounding

An obvious complication requiring comment is the fact that many lexemes are morphologically complex, consisting of a stem morpheme plus one or more affixes, as in *unreadable* (a form of composition known as *derivation*) or of two or more stem morphemes, as in *front-door-window* or *afterthought* or *ill-tempered* (a form of composition known as *compounding*). Both derivation and compounding result in new, complex word stems, often belonging to a different word class from the original stem or stems. In many languages, stems can be augmented or modified by means of processes of *inflection.* Inflection for nominal stems is usually called *declension.* For verbal stems it is usually called *conjugation.*[1] Both declension and conjugation are sometimes semantically based, such as plural formation for nouns, or tense or mood morphology for verbs. But sometimes they are an automatic consequence of word class distinctions (gender) or of syntactic function (case) or of agreement (person, number, gender for verbs). The morphology of declensions and conjugations is usually presented in the form of the well-known *paradigms* in grammar books. The result of inflection is a syntagmatic word. In some languages, full inflected words can be used again in processes of compounding and derivation. German and English, for example, incorporate singular or plural genitives into compounds, as in German *Landesregierung* (state government) versus *Länderkunde* (geography), or German *Volkskunde* (study of folklore) versus *Völkerkunde* (ethnology), with the plural form *Völker* (peoples). In English one finds *women's hospital* or *trades union.* Dutch has examples of derivational incorporation of plurals, as in *jongensachtig* (boyish), derived from the plural noun *jongens* (boys) and the derivational suffix *-achtig* (-like).

Derivation is usually not fully productive—that is, automatically applicable—but restricted to limited classes of items, sometimes according to some identifiable criterion but more often than not just arbitrarily. For example, it seems totally arbitrary that English happens to have the word *affreightment*, meaning 'the hiring of a ship to carry cargo', just as arbitrary as the fact that most speakers of English do not know of the word's existence. Compounding, by contrast, appears to be potentially fully productive, but in practice dependent on the emergence of a conventionalized specific meaning for the compound, a meaning that reflects a recognizable phenomenon in a given society (compare (8.24)–(8.26) in Section 8.6.2).

In most cases, compounding is not *semantically compositional* since the meaning of the compound is not an automatic result of the component parts. Yet most compounds remain *semantically transparent* in that the component parts somehow show the way towards the specialized meaning of the compound in question. For example, the relation between the component elements in *footman* is very different from that in *footwear*: one has to know

[1] The modern systematic distinction between compounding, derivation, and inflection (declension and conjugation) goes back to the German polymath and grammarian Jacob Grimm (1785–1863), who introduced it in his four-volume *Deutsche Grammatik*, published between 1819 and 1837.

both the language and the culture to know the precise relation between the component parts. Yet the recognizable elements in these compounds guide the listener to a quick identification of the correct meaning. In this respect, compounding is like prelexical syntax, discussed in Section 8.6.1, where the component prelexical elements provide no more than a pointer to the final meaning. Derivation, for that matter, is likewise not, or not fully, compositional, as one sees, for example, from the semantic difference between the two adjectives *legible* and *readable*: when applied to an inscription both mean 'of such a nature that it can be read or deciphered', but when applied to a text only the latter can mean 'pleasant to read'. This difference cannot be reduced to the different component parts. It seems that in morphology only inflectional processes come close to full compositionality. These are also the processes that are most predictable in a language and thus most productive.

Sometimes one feels tempted to assume a zero affix, as in the case of many English verbs or adjectives that occur both as intransitives and as causatives, such as the verbs *drop*, *sink*, *grow*, and so on. As regards adjectives, one thinks of *sad*, *happy*, *cheerful*, and so on, which occur intransitively in phrases like *a sad man* and causatively in phrases like *a sad story*. One ground for assuming a zero affix could be the fact that many languages other than English, such as Gothic or Turkish, have an overt causative affix that may turn an intransitive predicate into its causative counterpart. But more parsimoniously inclined lexicologists will try to make do without such an assumption for English, and relegate phenomena of that nature to the semantics of the verbs and adjectives in question. We can stay neutral on this issue: whether one does or does not wish to assume a zero affix, the process is like derivation in that it is restricted to certain words and excluded for others, without it being clear on what grounds the division is made. Thus, for example, the adjective *nervous* does not occur causatively in English (**a nervous noise*), but its Modern Greek equivalent *nefrikó* does, as in *nefrikó thórivo* (a noise that makes one feel or become nervous).

Derivation often changes the grammatical category of a word. Thus, some English adjectives can be turned into an inchoative or causative verb by the addition of the suffix *-en*, as in *dark-en*, *red-(d)en*, *wid-en*, *stiff-en*, *sweet-en*, *short-en*, *quick-en*, but not **slow-en*, **grey-en*, **long-en* (for which English has *length-en*). This kind of derivation is found without an overt affix in, for example, Sranan, the Creole language of Surinam, where adjectives like *nati* (wet), *weri* (tired), *stefi* (stiff) can be used as corresponding inchoative or causative verbs meaning either 'to begin to be wet/tired/stiff' or 'to make wet/tired/stiff' (Seuren 2001: 459). Neither lexicologists nor other wizards like myself have any idea why this is possible with some adjectives and not with others.

English shows many cases of a regular correspondence between nouns and corresponding verbs, as in *hammer*, *picture*, *hand*, *back*, and so on. Lexicologists treat most such cases as derivations from nouns to verbs. Such derivations are often characterized by a voicing of a final fricative sound, as in *bathe*, *house*, *thieve*, *calve*, depending on the period in which the derived form was lexicalized. (Occasionally, the original noun has been lost, as with the verb *parse*, derived from the Latin noun *pars*, used in medieval schools for 'part of speech'.) Sometimes, however, lexicologists feel that the noun should be considered derived from the verb, as in *close*, whose noun version still has the voiced final fricative, as opposed to the adjective *close*, which has a voiceless final *-s* (and which may have been the derivational source of the verb *close*). But we will leave the phenomena of derivation and compounding to the specialists and pass on to lexical ambiguity.

9.3 Vagueness and nonintersectivity in adjectival predicates

Many predicates are vague, in the sense that they are neither clearly true nor clearly false in borderline cases. The predicate *bald*, for example, implies that whoever is bald has no hair on the top of their head. But, as was observed by the Greek philosopher Eubulides in the fourth century BCE, the predicate *bald* still applies when a man has one hair on his head, and also when he has two hairs, and three hairs. How many hairs must a man have for the predicate *bald* to be no longer applicable? Or how many grains of sand must there be for them to form a heap? This quandary is known as the SORITES PARADOX (from Greek *sōrós* 'heap').

Some vague predicates, moreover, depend on more or less subjective evaluation criteria in given contexts. Thus, a model of the Queen Mary may be a huge object in my study, but a small object in a spacious hall. Or, to take an example from the Greek sophists, what is a salubrious breeze to you or me may be a fatal breeze to a nonagenarian. In other words, the satisfaction conditions of predicates are not always expressible exclusively in terms of 'objective' properties adhering to the objects involved but are often subject to 'subjective' criteria as well. This means that adjectival predicates of this category do not, to many linguists' despair, specify a class of (huge, small, salubrious) objects, even though they manifest themselves grammatically as one-place predicates. Therefore, a noun phrase like *the huge model* cannot be analysed intersectively as the x such that x is both huge and a model.

The problem of nonintersectivity presents itself with regard to a large class of adjectives, not all of which involve subjective evaluations. A good teacher, for example, is not a person who is both good and a teacher, but rather someone who teaches well, and a high climber is not someone who is both high and a climber, but someone who climbs high. In similar manner one would like to be able to say that an old friend is not someone who is both old and a friend, but someone who . . .—but here the analogy comes to a halt, since English grammar does not allow for an expression like **someone who friends oldly.* This is because the nouns *teacher* and *climber* are agentive nominal derivations from the corresponding verbs *teach* and *climb*, respectively, whereas there is no corresponding verb for the noun *friend.* This, however, is but a surface difference, since all lexical content words are predicates, including the word *friend.* It makes sense, therefore, to see if the parallelism between *an old friend* and *a good teacher* can be upheld in the semantic analysis of these expressions.

It looks as if it can be. We can even go further and suggest that adjectives like *good* or *old* are, in fact, adjectival predicates requiring a transitive semantic structure. Sentences like (9.1a) or (9.1b) will then correspond to a semantic structure of the form 'John goods his being a teacher' and 'John olds his being a friend', as in (9.2a) and (9.3a), respectively:

(9.1) a. John is a good teacher.

b. John is an old friend.

(9.2)

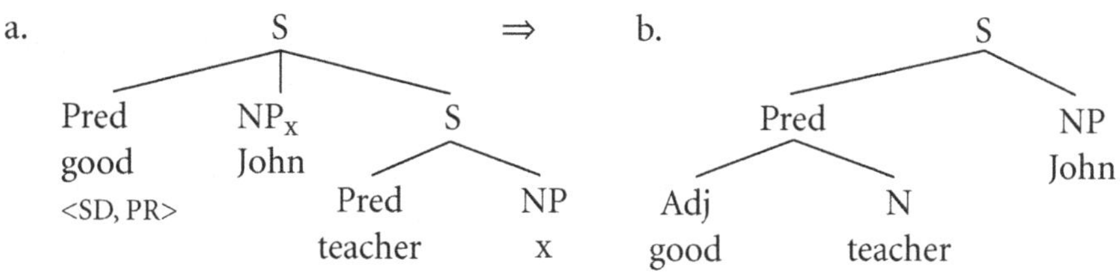

(9.3)

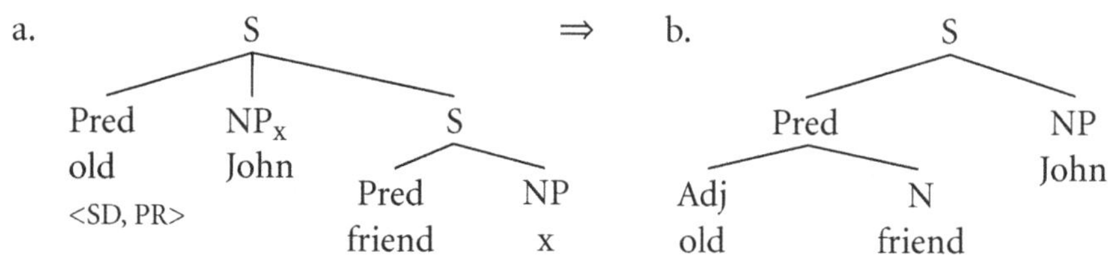

In this analysis, the highest predicate in the L-propositional structures (9.2a) and (9.3a), respectively, is the adjectival predicate (*good*, *old*), not the noun (*teacher*, *friend*). The subject term of the adjectival predicate is *John* and the object term is an embedded S-structure whose subject term is coreferential with the higher subject term (expressed by means of the variable x). The grammar will then delete the anaphoric $_{NP}[x]$ through the well-known rule of SUBJECT DELETION (SD) and subsequently raise the lower predicate through the well-known rule of PREDICATE RAISING (PR) (see Section 8.6.1, where PR is discussed in relation to prelexical syntax). The relabelling of *good* and *old* as 'Adj', and of *teacher* and *friend* as 'N' is taken to be an automatic result of the transformational process, driven by the lexicon where the surface categories of the words in question are specified (Seuren 1996: 56). This gives (9.2b) and (9.3b), respectively, where the now complex predicates *good teacher* and *old friend* have become intransitive owing to the application of the two rules mentioned.

The first reaction of many readers will doubtless be to boggle at such extravagance. Yet it is perhaps worth their while to have a closer look. They will then see that this analysis has a great deal going for it. The first and most obvious advantage is that it provides a direct and principled answer to the question of what the quality of goodness, or oldness, actually applies to. In this analysis, it applies to the pair of objects consisting of (a) the person called *John* and (b) the fact of his being a teacher, or a friend. And it does so in the sense that John 'goods' his being a teacher, or 'olds' his being a friend, meaning that John, in being or doing what he is or does, satisfies the criteria expressed by the predicate (*good, old*) for the property expressed by the object term (being a teacher, or a friend). Such paraphrases may sound odd, but so do many paraphrases in formal analytical language. There is no principle requiring that paraphrases in semantic theory should correspond to colloquially current surface sentences of the same language (though it helps when the paraphrase is current in at least some natural language). The adjectives in question thus take the form of two-place predicates. This avoids ad hoc and otherwise unsatisfactory analyses of the kind 'John is good as a teacher' or 'John is old as a friend', but at the cost of a certain unnaturalness in the paraphrase.

Then, the analysis of these adjectives as two-place predicates makes it possible to formulate satisfaction conditions for them in the usual way, without any appeal to pragmatic or other complicating factors. *Good*, for example, can now be described as applying to pairs of entities and properties in such a way that the entity, in being or doing what it is or does, must satisfy certain quality criteria defined for the predicate *good* with regard to the property (being a teacher or a friend). More formally (o is an object, P is a property):

(9.4) $[\![\text{Good}]\!] = \{ \langle o,P \rangle \mid$ o, in being what it is or doing what it does, satisfies pre-established quality criteria with regard to P$\}$

One notes that the condition 'pre-established quality criteria' refers to general knowledge, just like the predicates discussed in (9.17)–(9.20) in Section 9.6.3 (see Section 7.4 in Volume II for further discussion).[2]

One should note, furthermore, that the analysis given above requires no ad hoc rules of grammar. SUBJECT DELETION and PREDICATE RAISING are well-known rules in the open syntax of large numbers of languages. In this respect, the present analysis contrasts starkly with all other attempts at an analysis of the phenomena concerned, which all need to invoke ad hoc rules or processes.

Finally, there is the highly relevant fact that, in this analysis, normal propositional negation functions precisely the way it is supposed to function. As we have seen in previous chapters, the propositional negation NOT takes a whole L-proposition *S* as its scope, denying the truth of *S*. (*S* may be a propositional function, in which case NOT denies the truth of each proposition that comes about as a result of replacing the variable with a referring term.) Yet even though it is correct to say this, there is the additional fact, mentioned earlier in Section 3.2 in connection with example (3.12), that when *S* contains, at the L-propositional level of semantic analysis, one or more embedded S-terms then the negation applies specifically to the predicate of the highest S-term under the negation operator, as it denies the assignment of the property expressed by that predicate to its argument term(s). The more deeply embedded S-terms thus remain unaffected. For example, when I utter (9.5), with the (simplified and partial) L-propositional form (9.6), then what I deny is that it is at five in the morning that John gets up, not that John gets up:

(9.5) John does not get up at five in the morning.

(9.6)

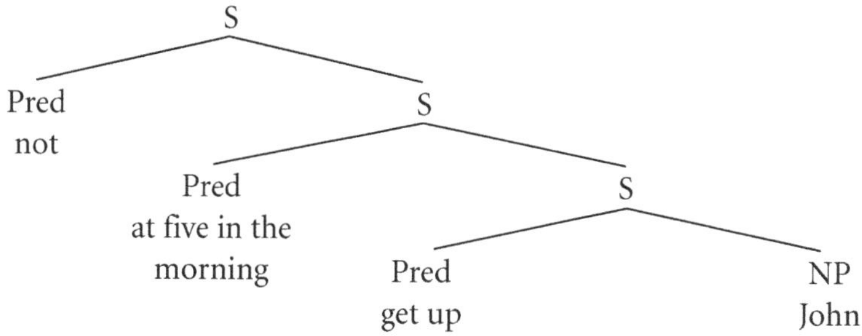

[2] The critical reader will wonder how a sentence like *John is good* is to be analysed. Unless *good* is taken in a philosophical sense, meaning 'morally good', such sentences are usually elliptical in that the object term referring to the property whose criteria for the predicate (*good*, *old*, *nice*, or what not) are said to be satisfied by the referent of the subject term is tacitly understood. Thus, when listening to someone playing the piano one can say *She is good!* meaning that she is a good pianist, or, in analytical terms, that she satisfies the criteria of quality required by the predicate *good* for piano-playing.

This observation confirms old observations made in the nineteenth and early twentieth centuries by scholars like Steinthal, Meyer-Lübke, Von der Gabelentz, Wegener, Lipps, Stout, Wundt, Mathesius, and others (see Seuren 1998: 120–33 for quotations, references, and extensive discussion). The behaviour of the negation in sentences like (9.5) bears out the observations made by these eminent scholars, in that, in (9.5), the negation specifically cancels the assignment of the property of being at five in the morning to John's getting up, which remains given.[3]

The point here is that if the analysis proposed in (9.2) and (9.3) is accepted, the negation as used in sentences like (9.7a, 9.7b) behaves precisely according to this principle:

(9.7) a. John is not a good teacher.
b. John is not an old friend.

(9.8)

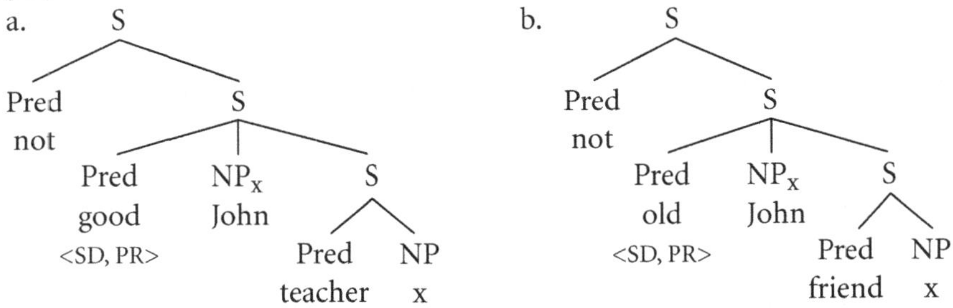

In L-propositional structures like (9.8a) or (9.8b), the negation specifically cancels the property of 'being good' or 'being old' to John's being a teacher or being a friend, respectively, which remain given. This amounts to saying that (9.7a) does not deny that John is a teacher but rather that John is a *good* teacher, and analogously for (9.7b), which is now seen to deny not so much that John is a friend as that he is a friend of long standing.

Looked at in the light of these distinctive advantages, the analysis given of the adjectives in question appears to deserve some consideration. Moreover, as it is good methodology in science to generalize from the most difficult case,

[3] Compare also sentences with the aspectual verbs *stop* or *start*, as in the near-synonymous (i) and (ii):

(i) It has stopped raining softly.

(ii) It has started raining hard.

Sentence (i) does not entail that it has stopped raining. On the contrary, it roughly entails sentence (ii). And likewise for (ii), which does not entail that it has started raining, but rather that it has stopped raining softly. The manner adverbs *softly* and *hard* must thus be taken to be the highest predicate in the respective semantic analyses of these sentences.

one might even think of treating all adjectives that can be used predicatively in this way, including those that appear to be fully intersective. A sentence like *This is a blue flag* would then be analysed as 'this blues its being a flag'. The negation of this sentence, *This is not a blue flag*, would seem to support this hypothesis, as it does not deny flaghood to the object referred to but only blueness; the flaghood remains intact.

9.4 Presuppositional aspects of lexical meaning

One specific aspect of lexical meanings is the fact that they often contain satisfaction conditions that are taken as PRECONDITIONS, to be satisfied before the others can be taken into consideration. We speak of the PRESUPPOSITIONAL ASPECTS of lexical meanings, because it is in the preconditions of lexical meanings that we locate the source of presuppositions.

Presuppositions of a carrier sentence *S* are conditions to be fulfilled by preceding discourse for *S* to be interpretable (a more extensive analysis of the notion of presupposition is given in Chapter 10 of Volume II; see also Seuren 1985, 1988, 1994*b*, 2000). Consider the following examples ('≫' stands for 'presupposes'):

(9.9) a. Dan is divorced. ≫ Dan was married. (categorial)

b. Dan knows that he is broke. ≫ Dan is broke. (factive)

c. Dan is bald. ≫ there is a person "Dan." (existential)

d. Only Dan laughed. ≫ Dan laughed. (focus)

The presuppositions of these sentences embody conditions that must be fulfilled by preceding discourse, in the sense that the discourse must either contain, or allow for the post hoc suppletion (accommodation) of, the information contained in the presuppositions. This makes for coherent bits of discourse when the presupposition is presented first, followed by the carrier sentence (properly provided with anaphors and deictic expressions):

(9.10) a. Dan was once married but he is divorced now.

b. Dan is broke and he knows it.

c. There is a person called 'Dan', and he is bald.

d. Dan laughed, and only he did.

In order to explain the essence of this phenomenon we must have a quick shot at presupposition theory, anticipating the full discussion in Chapter 10 of Volume II.

The source of this phenomenon appears to lie in a particular distinction within the satisfaction conditions of the predicates in question. One remembers that the satisfaction conditions of a predicate F are the conditions that must be fulfilled by the object or objects **o** denoted by the argument terms of F for the proposition resulting from applying F to **o** to be true. Thus, it is part of the satisfaction conditions of the predicate *bald* that the object said to be bald must, if human, lack hair on his head. The satisfaction conditions embody the information or cognitive content made available to the listener on interpretation of the utterance in question. That is, given an L-proposition $F(a_1, \ldots a_n)$, the discourse-bound incrementation of $F(a_1, \ldots a_n)$ to the discourse domain **D** involves the proper distribution of the information conveyed by the satisfaction conditions of the predicate F to the discourse representations (addresses) of the terms $a_1, \ldots a_n$.

Now assume a distinction between two kinds of satisfaction condition of a predicate F, the PRECONDITIONS and the UPDATE CONDITIONS jointly forming the satisfaction conditions of F. The distinction is to do with the process of discourse incrementation, elaborated in Chapters 7 and 8 of Volume II. It consists in the fact that the preconditions must be incremented first, before the update conditions can be incremented. Thus, in the case of the predicate *be divorced*, if I say *Dan is divorced*, then, if I am to be properly understood, the information that there is a person called 'Dan' and that this person has been married must have been incremented before the information that he is divorced can be added. This means that there must first be a representation of—that is, a discourse address for—a married person called 'Dan', in both the discourse domain **D** of me, the speaker, and that of the listener, containing all the information necessary for an update with the information that this person is divorced. As already adumbrated in Section 7.3, it is assumed that such an address has the form (9.11), where '**d–1**' is the address label, '**a**' functions as the (Russellian) existential quantifier, and 'P(a)' (for any predicate P) is a propositional function over the bound variable a:

(9.11) **d–1** [**a** | Human(a), Called 'Dan'(a), Have been married(a)]

Example (9.11) is thus to be read as: 'there is an object *a* such that *a* is human and *a* is called "Dan" and *a* has been married.' Sentence (9.11) is now ready for the addition of the information that Dan is divorced.

The new update information that Dan is divorced can be added without (9.11) already containing the information that Dan has been married. That missing information is then retrieved from the preconditions in the meaning specification of the predicate *be divorced* and slipped into (9.11) by way of what is known as ACCOMMODATION or POST HOC SUPPLETION.

Address representations must have open access to relevant knowledge stored in cognition. This makes it possible to go further than the semantic specification of preconditions and insert information post hoc without that information even having to be expressed in the relevant discourse address. For example, if I say that Bob's face is disfigured by scars, the fact that Bob has a face need not be either mentioned or incremented, because it is a matter of general knowledge that every human has exactly one face (physically speaking, of course).

This account functions well for cases like (9.9a–9.9c), with their main predicates *divorced*, *know*, and *bald*, respectively. But how about (9.9d), with its focus presupposition induced by *only*? The answer is not provided in a few words. It requires a grammatical analysis of focus sentences in which the focus particle, in this case *only*, is seen to be the main predicate. For (9.9d) this would amount to an analysis of the sentence in terms of 'Dan was the only one who laughed', with *Be the only one* as the main predicate. We shall not go into this question here, as it would take us too far from the issue at hand, but see Seuren (1996: 315–23) for some comment.

A discourse domain **D** may 'refuse' a precondition. It does so when either **D** or the knowledge base with which it is connected contains information that is incompatible with the precondition in question. When a precondition is refused, the utterance in question does not fit into **D** and will be uninterpretable: it 'makes no sense'—unless there is the possibility of a metaphorical or otherwise nonliteral interpretation, about which there will be more discussion in Section 9.5. When, on the other hand, **D** accepts all preconditions, the update conditions will be added without a hitch by the process of accommodation (post hoc suppletion).

It will be clear that the automatic post hoc suppletion of presuppositions into the discourse domain is a direct consequence of OSTA, the overall principle of optimization of sense, truth, and actuality introduced in Section 3.4.2, which makes listeners go for the best and most useful integration of the incoming message into the cognitive model constituted by episodic knowledge of the speech situation, general encyclopedic knowledge, and the semantically determined discourse domain of the moment.

The functionality of presuppositions in language is obvious. Since the preconditions are built into the satisfaction conditions of predicates as fixed for the language in question, presuppositions are recoverable from the carrier sentences for any competent speaker of the language in question. This makes it unnecessary to spell out the presuppositions in full to make the discourse comprehensible, since, owing to OSTA, presuppositions are automatically inserted *post hoc* (accommodated), as long as that is at all feasible. In this way, much of what has to be said can be left unsaid because it is silently understood from the little that is said. Sentence sequences such as (9.10a–9.10d) above are somehow redundant, since the first member of each pair of sentences is recoverable from the second. This redundant overlap in a sentence *S* not only makes it unnecessary to actually pronounce its presuppositions, it also provides a test to decide whether *S* is suitable for any given **D**. Presuppositions, in other words, are implicit bits of text.

9.5 Metaphor

From presuppositions one passes naturally to metaphor. One element in the vast issue of metaphor is the study of the presuppositional aspects of lexical meaning illustrated in the preceding section. The presuppositional machinery has proved to be functional not only because it makes for an enormous amount of energy saving in processes of linguistic interaction, but also for a different reason, to do with metaphorical use. When **D** rejects a precondition of a predicate on account of world or situational (episodic) knowledge, rather than on account of inconsistency with the pre-existing **D** itself, one speaks of a CATEGORY MISTAKE. Thus a sentence like (9.12), if taken literally, is unincrementable and hence uninterpretable in any **D** because it contains a gross category mistake, engines not being the kind of object of which it makes sense to say that they laugh:

(9.12) The engine had a good laugh.

Does this mean that (9.12) is irredeemably lost to linguistic interaction? The answer is no, because there is the possibility of a nonliteral interpretation, in particular an interpretation in the form of METAPHOR. As has been shown several times earlier on, humans do not give up their effort at interpreting utterances that quickly: the OSTA principle makes humans seek interpretations that make sense and, if at all possible, can even be taken to be true. In the case of (9.12), sense can be provided by the OSTA principle allowing listeners to imagine a fictitious setup in which the actual engine is endowed with enough human qualities to be able to laugh—a technique widely used in comic drawings. Such an interpretation thus contains an implicit comparison of the engine in question with a human being. This may seem far-fetched for (9.12), but it is not impossible. When such an interpretation is in any way 'evocative'—a notion still far removed from formal treatment—one speaks of METAPHOR. Metaphor can thus be defined as a form of interpretative adaptation whereby term referents that fail to satisfy certain categorial preconditions of the main predicate are treated, for the purpose of the current interaction, as if they did satisfy the preconditions at hand, so that an implicit comparison arises.

Such an interpretation is subject to limits. Outside these limits there is no metaphor, or an intended metaphor does not 'work'. A general definition of these limits clearly falls outside the realm of the study of language alone: it can only be given as a result of a cooperative effort by linguists and cognitivist psychologists, and perhaps also sociologists and anthropologists. In any case, no such general definition is aimed at in the present study. All we can do here is give a tentative and partial *definition* of what constitutes metaphor, not of the conditions under which it may occur. We take the constitutive element of metaphor to be the fact that it is an instance of category mistake, which becomes effective as a metaphor only if the implicit comparison arising from it has sufficient evocative power—a condition we are in no position to elaborate. But the fact that we can treat category mistakes, and thus metaphors, as violations of preconditions, and hence as part of presupposition theory, gives the study of metaphor a well-defined place in semantics, which may be a step forward in the direction of a fuller understanding of the phenomena concerned.

The examples discussed in the literature on metaphors are not all of the same kind. Suppose a man says to his sweetheart:

(9.13) You are my sunshine.

Whether or not this is a case of metaphor is largely a question of terminology. It would seem that if one uses the term *metaphor* for such cases, it had better be specified that what is at issue here is, say, *declarative metaphor*. But one may feel that it is preferable not to speak of metaphor at all in such cases, but only of comparison in a wider sense. The term *metaphor* is more properly restricted to cases like (9.14a–9.14c), which one may call *implicit metaphors* if one wishes, since they involve a specific form of implicit comparison, as has been shown. Examples (9.14a–9.14c) are all taken from high quality, truly creative literature, not from ordinary, conventionalized usage. But that is how metaphors start, as creative language use:

(9.14) a. And the train which had picked them at sunrise out of a waste of glaciers and hotels was waltzing at sunset round the walls of Verona. (E. M. Forster, *Where Angels Fear to Tread*)

b. At least let my delusions be allowed to fool me,
so that I will forget how my soul is empty.
(K. Kaváfis, 'September 1903')

c. Sweet love, renew thy force; be it not said
Thy edge should blunter be than appetite, . . .
(W. Shakespeare, Sonnet 56)

Metaphors such as (9.14a–9.14c) have in common that to a banal mind they are no more than instances of category mistake. The predicate *waltz* in (9.14a) requires, as a precondition, that the object said to be waltzing be an animate being capable of dance. Clearly, trains do not satisfy that condition, but Forster, by saying that the train was waltzing, appeals to OSTA, which forces his readers to consider this train, for the purpose of the story, as an animate being capable of dance. A reader who follows this lead will then picture the rhythmic rumbling of the train as it makes its way round the city walls of Verona as a happy dance ending the journey that had started so grimly in the icy Alps, which is precisely the force of the metaphor. Had Forster written about a train that was cycling or galloping or crawling round the walls of Verona, either no metaphor or a different one would have arisen.

In like manner, the predicate *empty* in (9.14b) preconditionally requires a container-like object as the reference object of the subject term. But in this sentence, the subject term refers to the writer's soul, which is not literally a container. To picture it as one, however, has a certain evocative power, which makes the metaphor effective. The same mechanism is at work in (9.14c), where Shakespeare applies the predicate *blunt*, which is preconditionally restricted to cutting instruments, to feelings of love, as well as to the sensation of appetite, thereby comparing both implicitly to a knife or sword.

At least some implicit metaphors behave peculiarly under negation or under certain speech-act operators. Sentences like the following, for example, are distinctly odd:

(9.15) a. The train was not waltzing at sunset round the walls of Verona.
b. Why was the train waltzing at sunset round the walls of Verona?

There does not seem to be much discussion in the existing literature of such matters. To my knowledge, questions of why the negation or a different speech-act operator should have the power of modifying or even destroying the original metaphor are not broached. Further investigation into such matters will no doubt reveal interesting new insights.[4]

Sometimes metaphoric usage spreads and becomes conventionalized, in which case the usage is codified in the lexicon of the language. The conventionalization may even lead to the total demise of the original metaphor. The word *obvious*, for example, was originally used for anything that is 'in the way' so

[4] One is immediately struck by the fact that the negation *not* in (9.15a) appears to be the metalinguistic *not* of Horn (1985), causing an echo-effect. This may give rise to the idea that, perhaps, metaphors assume the character of a POSITIVE POLARITY ITEM, allowing only for metalinguistic—that is, echo-inducing—negation. But this is no more than a first thought, to be tested out by further research.

that it can't be missed along a road, either as an obstacle or as a sign or meeting point. When it began to be applied to insights, ideas, answers, problems, and so on, listeners were forced to regard processes of thought as journeys, with a beginning and an end and clearly visible signs or obstacles in the way. The comparison of thought processes to a journey then became an easy source for further, less original, metaphors, which became conventionalized even more quickly.

It is generally so that most if not all of the nonconcrete vocabulary of human languages, especially the languages of culture, is based on conventionalized extensions, specializations, and cognitive transfer of pristine meanings, which have sometimes been lost completely and have to be reconstructed by methods of comparative philology. Let us consider a few examples from the vast multitudes that are available, especially in the Indo-European languages, whose history is best known. German *lesen*, like Latin *legere*, now means 'read', but the original meaning was 'pick out, select', as is still apparent in the German *Auslese* and its English counterpart *selection* (which has the Latin prefix *se-* meaning 'apart'). The meaning 'read' was conventionalized and new words had to be sought for the expression of the concept of 'picking out'. In Modern Greek, *read* is translated as *diavázo*, which originally meant 'go through'. The Classical Greek word for 'truth' is *alētheia*, meaning 'the property of not being hidden', containing the root *lath-*, also known from *lethargy* and *latent*, whose meaning is associated with the notion of hiding or being hidden. The magic-laden Latin adjective *fēlix* ('happy') originally meant 'fertile', with the root *fē*— of the same meaning, also found in Latin words like *fēmina* ('woman', literally 'she who suckles') or *fēnus* ('interest on capital'). The word *capital* is derived from Latin *caput* ('head'). The word *town* originally meant 'fence' (like its German cognate *Zaun*), then, by metonymy, 'enclosure', then, by specialization, 'town'. We *express* our thoughts, not using physical pressure in acts of squeezing, but using grammar in acts of speech. The word *metaphor* itself is a case in point. The original meaning of the Greek *metaphorá* is 'transport', but it then also started to mean 'conveyance' (of property), 'change' (but not in the sense of 'small cash') and finally 'metaphor'. With the exception of 'change', all these meanings have survived in Modern Greek, although the word is now also used for 'removal' in the again specialized sense of moving furniture from one house to another.

Are all such examples instances of metaphor, artistic in origin but worn and conventionalized with time? No, but the point is that the examples that may reasonably be categorized as live metaphors form an integral and inseparable part of the entire vast dynamism of lexical meaning and change of meaning as described in, for example, Stern (1931), where careful and comprehensive analyses and categorizations are presented of lexical semantic change, following a tradition that started in ancient studies on rhetoric. In rough outline, Stern distinguishes conventionalized semantic extensions, specializations (narrowings), transfers (from part to whole or vice versa, or by association of adjacency or similarity, or by comparison and so on), giving large numbers of examples of all of these categories of change, mostly taken from the lexicon of English. The picture that emerges is one that demands an explanation in terms of both cognitive and cultural categories—a task yet to be accomplished.

In their 1980 study, Lakoff and Johnson present a wealth of examples of half-worn metaphors, conventionalized but not quite dead yet, as a modicum of reflection shows their original metaphorical character. Typical examples are: *the [foundation, buttressing, construction, collapse] of a theory*, where theories are seen as buildings (1980: 46); *make an [impression, mark]*, where sensations are seen as instances of physical impact (1980: 50); *a change [from/out of] X into Y* (1980: 74), where the prepositional predicates *from*, *out of* and *into* see their preconditions of place violated by conditions or states, as one changes from one state to another. Time is seen as an object that 'flies' towards us from the unknown, so that one *looks forward* to the future, which is *ahead of* us, while an event *x* that is later than an event *y* is said to *follow* the event *y* (1980: 41–2).[5] Examples of this nature can be multiplied ad nauseam.

Along with other forms of creative language use, metaphor is thus seen to be a powerful source of new lexemes where they are needed. It is hardly speculative to surmise that metaphor is an important factor in the complex machinery that allows 'primitive' vocabularies of simple societies, where linguistic needs are largely limited to the mundane necessities of everyday business around the house, the land and the clan, and perhaps also to the recounting of memories, to be expanded so as to cater for the needs of more advanced societies that begin to organize states and to cultivate the arts and sciences. It would be a worthwhile exercise to study the lexicons of the languages of present-day highly developed societies from this anthropological point of view. The results of such a study may yield important insights into the workings of language and of the mind.

[5] In preclassical ancient Greek, man is seen as walking backward into the future. The future is behind and the past in front of us, because we can 'see' the past but the future can only be guessed. See Liddell and Scott (1951) under *opísô*: 'I. of Place, *backwards*. ... II. of Time, *hereafter*, since the future is unseen and was therefore regarded as *behind* us, whereas the past is known and therefore before our eyes.' The same under *ópisthen*: 'I. of Place, *behind, at the back*. ... II. of Time, *after, in future, hereafter*.' Liddel and Scott then give a large number of loci in Homer and other preclassical authors. Núñez and Sweetser (2006) show the same phenomenon for the west-Amazonian language Aymara.

9.6 Polysemy as 'semantic relatedness'

9.6.1 *What is 'semantic relatedness'?*

Other than metaphor, which is the topic of a vast body of literature, polysemy is something writers on lexical semantics appear to approach with a certain apprehension—and rightly so. The literature dealing specifically with polysemy is scarce compared with that on metaphor, no doubt because, unlike metaphor, polysemy has no place in literary studies. The oldest discussion of polysemy phenomena I have come across is by the medieval French philosopher Peter Abelard (1079–1142), who speaks of 'figurative or improper use' in his earlier, and of 'translatio' in the sense of 'transfer' in his later writings. As is shown in Rosier-Catach's admirable study (Rosier-Catach 1999), Abelard's discussion has a decidedly modern touch, not only because of his well-chosen examples (such as the use of *Homer* for Homer's works, as in *This word occurs only once in Homer*; Rosier-Catach 1999: 149) but also because he places polysemy phenomena in the context of the question of the compositional computation of meanings and truth-values from composite structures. Abelard's conclusion is that meanings and truth-values cannot be compositionally computed from composite structures because the correct interpretation of polysemous words in sentences depends on discourse factors and world knowledge.[6] But that is where Abelard stops: he offers no further analysis—which, incidentally, is very much like the position modern formal semantics finds itself in, in that some semanticists at least realize that polysemy phenomena prove that truth-value assignments are noncompositional, but no formal semanticist has much to say on the question of the underlying mechanism.

[6] Rosier-Catach (1999: 130, 148) quotes, for example, the following texts from Abelard (translations mine). In Abelard's *Theologia 'Summi Boni'* one reads:

Et hic quoque pluribus confirmavimus exemplis quod nomen ex adiuncto significationem commutat.

[And here we have also shown by means of a large number of examples that a word may change its interpretation according to surrounding context.] (Buytaert and Mews 1987: III § 63)

In his *Dialectica* Abelard writes:

At vero cum totius constructionis sententia pensatur ac simul verba in sensu alterius enuntiationis confunduntur, non iam singularium dictionum significatio attendenda est, sed tota magis orationis sententia intelligenda; atque in eo impropria dicitur orationis constructio quod eius sententia ex significatione partium non venit.

[But when the meaning of the whole construction is considered and at the same time words are wrongly taken in the sense of some other utterance, one must no longer concentrate on the interpretation of the individual words but rather try to understand the total intended meaning of the sentence. That is why a linguistic construction may be called 'improper', namely because its intended meaning does not follow from the meanings of its parts.] (De Rijk 1956: 136)

In modern times, one often sees polysemy treated as a special form of lexical ambiguity—the property of one lexical form having more than one meaning—where the distinct meanings are somehow 'related' (a position rejected by Abelard; Rosier-Catach 1999: 163–4). But what this 'relatedness' consists in, remains unclear.[7]

John Lyons, having rejected etymology as a criterion for 'relatedness', writes:

> The second major criterion that is traditionally invoked by linguists and lexicographers in drawing the distinction between homonymy and polysemy is unrelatedness *vs.* relatedness of meaning; and it is clear that this is a relevant and important consideration. (Lyons 1977: 551)

But he then proceeds to show that no reliable, generally applicable and precise criteria are available to decide whether two senses are or are not related. And this has not changed since Lyons's book was published.

It is presumably for that reason that many authors prefer to do without the notion of 'relatedness' and simply equate polysemy with lexical ambiguity, even though they often admit that this is not satisfactory. Pustejovsky defines polysemy simply as lexical ambiguity:

> Many words in a language have more than one meaning, a property usually called polysemy. (Pustejovsky 1995: 27)

The same is found in Geeraerts:

> In lexical semantics, *polysemy* refers to the situation in which a lexical item has more than one meaning, for example, *late* 'after the usual, expected, or agreed time', 'advanced in day or night', 'no longer alive', etc. (Geeraerts 1994*b*: 3227)

Cruse (1986: 52) speaks of *modulation* for what is commonly called polysemy, but he offers no criteria that go beyond (otherwise quite sensible) folk psychology. He prefers to avoid the terms *polysemous* and *polysemy* altogether, as they 'are not entirely ideal' (1986: 80) in his method of analysing lexical meanings.

Hudson, who rejects the view of the lexicon as a list of items, finds it unnecessary to distinguish lexical entries, 'the notorious distinction between polysemy and homonymy being the main culprit' (Hudson 1984: 3). The next occurrence of the word *polysemy* is on page 40 of the book, where one reads the following curious passage, in the context of a discussion of what unites the things that go under the name of 'furniture':

[7] For some strange reason, reams of paper have been filled with philosophical or quasi-philosophical discussions on the notion of 'family resemblance' introduced by Wittgenstein in his *Philosophical Investigations* (1968: 32; remark 67) to account for the 'relatedness' of different senses of one word. But 'family resemblance' is, of course, nothing but an alternative label for the equally unclear 'polysemy', not an explanation.

Our analysis says that there is indeed one property which is shared by all instances of furniture, namely the fact that you can apply the word *furniture* to them. This analysis puts the relations between, say, tables and cupboards on all fours with the relations between the horn of a cow and the hooter of a car: if it were not for the shared word, there would be no reason for linking them. But the word exists, and they are linked. In other words, 'family resemblances' are just a special case of what is traditionally called polysemy. (Hudson 1984: 40)

So Hudson's answer to the question of what common property makes the single word *furniture* 'applicable to a range of instances which have no properties in common at all' (ibid.) is the property that 'you can apply the word *furniture* to them'. And that is what is meant by polysemy. Any common properties that might make for a Wittgensteinian 'family resemblance' are just a bonus.

A more serious author is Dowty, who speaks of 'the thorny problems of polysemy versus homophony' (1979: 62) and admits:

But I have made no attempt to capture the linguists' distinction between *polysemy* (identical forms with different but 'related' senses) and *homophony* (identical forms with different but 'unrelated' senses), and for purposes of model-theoretic semantics, I believe this distinction is not (at present) a useful one anyway (though it may be for purposes of a psychological theory of semantics). (Dowty 1979: 363–364)

This admission is refreshingly frank when set off against the whole of the formal-semantic and philosophical-semantic literature, where polysemy is simply never mentioned (though Wittgenstein's magical but vacuous 'family resemblances' are all over the place in the philosophical literature). The distinction Dowty makes between 'related senses' for polysemy and 'unrelated senses' for homophony or ambiguity reflects the traditional but all too intuitive view mentioned above. Not surprisingly, Dowty finds this distinction 'not useful' in a formal-semantic framework, but he sensibly leaves room for alternative approaches. And his implicit hint that psychological theory has so far not provided a satisfactory answer to the question of what makes different word senses 'related' is certainly well taken.

An immediate conclusion to be drawn from this state of affairs is that a direct appeal to the study of cognition is a necessary—though probably not a sufficient—requirement for an adequate understanding of lexical meaning. And one has to admit that cognitive science has so far not come up with the goods.

In the following two subsections questions of ambiguity and of dynamic meaning application are considered in greater detail, so as to make the notion of polysemy a little more precise from certain points of view.

9.6.2 *Lexical ambiguity, homonymy, and specialization*

Let us first look at the distinction taken to hold between true lexical ambiguity, often called *homonymy*, on the one hand, and polysemy in the sense of different but 'related' senses on the other. Lexemes are often ambiguous in that they denote classes of objects that are clearly distinct not only on account of their 'objective' properties but also because their cognitive representations are in no way related. In the formal-semantic and the analytical philosophical literature, the cognitive factor is usually not recognized. The American philosopher Quine defined lexical ambiguity as follows:

> Vague terms are only dubiously applicable to marginal objects, but an ambiguous term such as *light* may be at once clearly true of various objects (such as dark feathers) and clearly false of them. (Quine 1960: 129)

Quine thus used the intuitive criterion of 'clearly true' or 'clearly false' as a yardstick for lexical ambiguity. This has the appearance of being a cut-and-dried criterion, yet it really is too terse. For besides clear cases of lexical ambiguity where two or more clearly distinct meanings are associated with the same word form—as with the English nouns *fence*, which, as a noun, stands either for a separation between two pieces of land or for a person dealing in stolen goods, or *scale* in its various meanings including the outer covering of fish—there are other, much more numerous, cases where one feels that, on the one hand, different meanings are involved while, on the other, there clearly is an element binding the various senses together.

A case in point is the English noun *primer*, which is either an introductory school book, or the powder used to explode a charge, or a first protective layer of paint. Although one has here three distinct meanings, there is also a common element that stands out with relative clarity, the semantic element 'first' or 'initial'. Or consider the English verb *nurse*, which has quite distinct satisfaction conditions in the phrases *nurse a patient*, *nurse a cold*, and *nurse a baby*. Yet although, in Quine's terms, these differences would make the verb *nurse* ambiguous, such an analysis would miss out on the fact that the three different 'senses' are clearly related in that they revolve around the notion of 'caring so as to ensure physical welfare'. One notes, moreover, that the verb *cure* has a partly analogous sense distinction, as is apparent in the phrases *cure a patient* and *cure a cold*. According to Quine's criterion (and given his unrefined, strictly bivalent notions of truth and falsity), *cure* should be considered ambiguous. Yet this goes against the sound intuitive grain of any linguist or lexicographer who has not had the benefit of a formal training in logic. We linguists want a more refined analytical apparatus, which will allow us, on the one hand, to show the semantic diversity of *cure a patient* and *cure a cold*, while preserving the semantic unity of the two on the other. Small wonder many linguists have felt attracted to prelexical syntax. Prelexical syntax would allow the phrases *nurse a patient* and *nurse a cold* to be caught under the umbrella prelexical analysis [TRY X [CAUSE X [NOT-HAVE Y,Z]]] (especially for illnesses), incorporating *cure*, which is analysed as [CAUSE X [NOT-HAVE Y,Z]] (especially for illnesses).

The main problem, however, seems to be conventionalized meaning shift, by metaphor, metonymy or other forms of semantic specialization. Do we want to say that a verb like *frame* is homonymous because, when applied to a person, it may mean either that the person in question has been put in a position where they can be falsely accused of a crime or that their portrait has been given a frame? Most lexicographers and lexicologists feel that it would be inappropriate to speak of homonymy or ambiguity in such a case, because English speakers are somehow aware of the fact that the former meaning is derived from the latter by way of conventionalized metaphor. Only when native speakers no longer see the connection, will lexicographers opt for homonymy, as, presumably, in the case of the noun *spirit*, which denotes either some form of concentrated alcohol, or what is regarded by some as the immaterial part of a human being, or what is regarded as the underlying idea of a text or a person's ideology. Likewise for the plural noun *premises*, which stands either for those elements in an argument that lead to a conclusion, or for a house or a building. Or indeed for the noun *argument*, which applies either to a form of reasoning, or to a more or less violent dispute, or, as mostly in the present book, to terms of a predicate. And what to think of the many uses of the noun *draft* (*draught*)? On the whole, one has to admit that the borderline between genuine homonymy and different applications, creative or conventionalized, of the same meaning is fluid, even when one feels that, under Quine's criterion, there is truth in one reading but falsity in the other.

Semantic specialization may lead to a split in the form of words or morphemes. A clear case is the Englah adjective pair *urban* versus *urbane*, both derived from the Latin adjective *urbanus* 'belonging to or associated with a city or town'. Another example is the pair *salon* versus *saloon*. Originally, *salon* meant 'large and somewhat luxuriously appointed room, possibly for representative or public use'. In Britain, this word has specialized for hairdressers' or beauticians' salons, next to its phonological variant *saloon*, which has come to mean, roughly, a four-door family motor car or a large public room on a passenger ship. Across the Atlantic, however, the primary meaning of *saloon* has become 'public drinking place', called *pub* in Britain, whereas *salon* is mostly used in the meaning of 'gathering of fashionable artists or intellectuals'. English *equation* is pronounced [iˈkweizhˈn] in mathematical contexts, but [iˈkweishˈn] in the meaning 'act of equating'. In Oxford, one speaks of one's [rŭm], with short *u*, when referring to one's *room* in college, whereas in other uses the word may be pronounced with long *ū* as well. A picture may *represent*—that is, [reprɪˈzent]—someone's grandfather, but a bank cheque (US check) that bounced the first time may be *represented*—that is, [ri:prɪˈzentɪd]. Cases of this nature naturally merge with cases discussed in Section 9.6.4, which deals with semantic specialization of co-existing variant forms. This process is widely observable in the development of languages, as one sees in English words like *disk*, *record*, *tape*, *plane*, *car* (or its Italian equivalent *macchina*, which has specialized for 'car' from the general meaning 'machine'). Scholars and scientists have a knack for such specialization.

A hypothetical case is the English noun *slab*, whose standard meaning is something like 'thick flat rectangular piece, preferably of hard material'. Slabs of different materials are used in gardens, kitchens, and other places, but suppose the first association of many speakers, in a large enough area, is with garden slabs, or with the typical slab used as a dissection table in mortuaries. Then it would not be surprising if the word *slab* were to specialize, for all speakers in the area in question, for the meaning 'garden slab', or 'mortuary slab', as the case might be, resulting either in an ill-defined form of polysemy or, possibly, in the introduction of a different lexeme for the other uses of the word.

The problem of homonymy versus polysemy is aggravating. Take, for example, the English noun *uncle*, which may denote a parent's brother or a parent's sister's husband. Is this word ambiguous? Intuition tells us that it is not, and this intuition is probably based on the fact that the relation of an 'ego' to a parent's brother is felt to be of the same nature as the relation to a parent's sister's husband. But there is no theory, to my knowledge, that accounts for that sort of intuition. It is nevertheless real, as one can see from the Italian word *nipote*, which may denote either a grandchild or a niece or nephew, and which, unlike the word *uncle* for speakers of English, is considered ambiguous by Italian speakers. When one asks an Italian speaker of the right age how many *nipoti* he or she has, that person will retaliate with a question to the effect that it should be clarified first what kind of *nipote* is intended. Likewise with the Dutch words *neef* or *nicht*, which stand for either one's aunt or uncle's son or daughter (*cousin*) or for the son or daughter of one's sibling or one's spouse's sibling (*nephew/niece*). Here again, speakers of Dutch will be unable to answer the question how many *neven* (*nichten*) one has as long as it is not specified what kind of *neef* or *nicht* is intended, cousin or nephew/niece.

In general, whether there is full ambiguity or a milder form of semantic differentiation, listeners disambiguate, or select the proper reading, on the basis of the OSTA principle mentioned earlier. This principle makes listeners seek optimal sense of the incoming message, and preferably in such a way that the message, if in the form of an assertion, can be taken to be true. The problem is that OSTA, though obviously present, active, and relevant, is extremely hard to catch in precise terms.

9.6.3 *Dynamic filtering*

Another form of polysemy is what is called here 'dynamic filtering'. By this we mean the phenomenon that the lexically fixed predicate meaning may influence the way in which ρ-values of argument terms are selected and, vice versa, the phenomenon that the nature of the ρ-values of argument terms may influence the interpretation of the predicate. In both cases one recognizes the OSTA principle: listeners try every rule in the book to make optimal sense of the incoming message.

When predicate meaning influences the selection of ρ-values of argument terms, the conditions imposed by the satisfaction conditions of the predicate on the ρ-values of its terms make the interpreting subject select the appropriate ρ-value within the semantic constraints of the argument term in question—a process called 'type coercion' by Pustejovsky (1995: 59, 111). For example, since the category of objects required as ρ-value by the predicate *be on fire* is different from the category of objects required as ρ-value by the predicate *have a day off*, the term *the office* will take different ρ-values in (9.16a) and (9.16b). In (9.16a) the NP *the office* is naturally taken to refer to a building, whereas in (9.16b) it is naturally taken to refer to a group of people:

(9.16) a. The office is on fire.
b. The office has a day off.

The following are examples of the second type of filtering, where the nature of the ρ-values of the argument terms influences the interpretation of the main predicate:

(9.17) a. This hotel room has a bathroom.
b. This student has a supervisor.

(9.18) a. John is fond of his dog.
b. John is fond of cherries.
c. John is fond of mice.

(9.19) a. John was baking bread.

b. John was baking bananas. (Pustejovsky 1995: 122)

(9.20) a. The road we were on was rather flat.

b. Your left front tyre is flat.

Sentence (9.17)—repeated from Section 1.3.2.2—shows the dependency of possession predicates on general cognition. In (9.17a) the possession predicate *have* expresses the well-known relation between hotel rooms and bathrooms. A many-to-one or many-to-many mapping is excluded: it has to be a one-to-one, or, exceptionally, a one-to-many mapping (when one room has several bathrooms). In (9.17b), however, the same possession predicate *have* expresses a relation whose nature is determined by the generally known function of supervisors with regard to their students. This relation allows for a many-to-many mapping between students and supervisors. One notes that (9.17a, 9.17b) are unambiguous: (9.17a) is false, not 'misleading' or whatever, when the room in question has to share a bathroom with other rooms. Anyone suing a hotel that advertises 'rooms with a bathroom' while the occupants of the rooms in question are referred to a bathroom down the corridor, to be shared with others, will win their case in court because the advertisement contains a lie. But a student suing his faculty over the fact that he has to share his supervisor with other students will be laughed out of court.

Even so, one does not feel that, therefore, the lexicon should contain different, homonymous items for *have* and similar possession predicates. A lexicographer is better advised to have a single item *have*, described, more or less, as expressing a *well-known relation of appurtenance* (Janssen 1976) between the kind of object referred to in subject position and the kind of object referred to in object position. The semantic description (satisfaction condition) of *have* and other possessive predicates is thus taken to contain a parameter for 'what is well known', making the interpretation of this predicate in each token occurrence truth-conditionally dependent on world knowledge. (For further comment see Section 7.4, Volume II.)

The conditions are not identical for all possession predicates. Possessive pronouns, for example, may express a relation of 'being responsible for' or 'taking care of', which other possession predicates cannot express. An example is sentence (9.21) uttered by a gardener with regard to the flower beds he is tending:

(9.21) Please don't mess up my flower beds.

This sentence can be uttered appropriately without the speaker implying that the flower beds are owned by him.

In (9.18a), obviously, John's fondness is of a rather different nature from what is found in (9.18b): the fondness expressed in the one is clearly incompatible with the fondness expressed in the other. Are (9.18a) and (9.18b) ambiguous, the different readings being defined truth-conditionally? I am inclined to think that (9.18a) is unambiguous, possibly because its object term has a definite determiner. But (9.18b), with a generic object term, seems to be ambiguous, although with one preferred reading. Sentence (9.18c) appears to be genuinely ambiguous, no doubt because mice may, though not necessarily with equal plausibility, be the object of one's affection and of one's appetite. Do we, therefore, need two distinct lexical entries for the predicate *be fond of*? Probably not, provided it is possible to insert a knowledge-related parameter into the semantic description of the predicate in question, perhaps in combination with the grammatical distinction between definite and generic object terms.

The occurrences of the predicate *bake* in (9.19a, 9.19b) differ according to whether it is interpreted as involving 'creation' (in such cases it is 'effizierend' in the terminology of certain schools of lexicology), as in (9.19a), or as involving a treatment leading to a change of state, as in (9.19b). Here, too, one single lexical entry for *bake* will probably suffice. But it will require an appeal to the generally available but selectively used possibility of deriving an 'effizierend' predicate meaning from a merely relational meaning by the optional insertion of an element CAUSE-TO-EXIST into the satisfaction conditions of the predicate in question. Such an appeal will have to be part of a general theory of internal semantic derivations within given lexical meanings, such as a derivation from intransitive to causative (by the insertion of an element CAUSE), or from stative to inchoative (by the insertion of BEGIN), or, as in the case at hand, from relational to 'effizierend' (by the insertion of CAUSE-TO-EXIST).

A further example is provided by (9.20a, 9.20b) with regard to the predicate *flat*, whose meaning moulds itself according to what is said to be flat. One may say of a road, a tyre, a mountain, a face, the world, and quite a few more categories of objects, that they are flat, but in each case the satisfaction conditions differ. There is an overall element 'spread out, preferably horizontally, without too much in the way of dips, protrusions, or elevations'. But that in itself is insufficient to determine what 'being flat' amounts to in the cases at hand. Sentence (9.20a), for example, is ambiguous between a reading in which the road was more or less horizontal and one in which there were not too many bumps and holes in the road's surface. The correct interpretation comes about only if it is known (a) what roads, tyres, mountains, faces, the world, and all the other things that may be said to be flat are normally thought to be like, and (b) what makes most sense in the context at hand (the OSTA principle).

Dictionaries, even the best ones, limit themselves to giving examples, hoping that the user will get the hint.

Colour names provide standard examples in the literature on lexical meaning. It has been known to lexicographers and lexicologists for a very long time that colour names have whimsical extensions, often associated with specific classes of objects and depending on specific contexts. Prototype theory, created during the 1970s (Berlin and Kay 1969; Rosch 1973, 1977; Rosch and Mervis 1975; Geeraerts 1986: 189–94), added significantly to the existing insights of lexicologists, as it provided an experimental basis for a linguistically fixed 'kernel' meaning of colour names and other terms with similar behaviour.[8]

Related cases are predicates like *enter*, *build*, *paint*, *write*, *sit*, *carry on*, *stop*, *continue*, *enough*, *ready* (the latter three being stock examples in recent pragmatic–semantic publications)—in general, all predicates that allow for overt direct or prepositional, nominal or clausal, object terms but often occur without them, including, of course, gradable predicates like *big*, *strong*, *low*. When these predicates occur in their 'incomplete' form—that is, without an object term—it is left to the listener to infer, on grounds of situational and/or encyclopedic knowledge, the values of the parameters that are not made explicit. It would seem that the *syntactic* specification of these predicates should merely specify that they can occur without an object term, unlike, for example, transitive *grow* or *cultivate*, which cannot.

Propositions containing predicates whose satisfaction co-depends on contextual and encyclopedic knowledge are only partially encoded in both the (syntactically underlying) L-propositions and the resulting utterances. It would be nonsense, of course, to propose that the dictionary meaning of predicates that can legitimately occur in an 'incomplete' form should specify, for each class of occurrences, what exactly is meant, since that kind of knowledge is not linguistic knowledge. The most one can require of a *linguistic* description of such 'incompletely used' predicates is that they contain, besides a specification of the prototype whenever applicable, open parameters referring the listener to his or her contextual and encyclopedic knowledge. Equally nonsensical would be a proposal to the effect that the syntax of sentences containing such predicates should be taken to contain one or more constituents specifying the values of all parameters relevant to their interpretation but left unexpressed—even though, on the whole, it remains possible to express such further specifications linguistically.

[8] Curiously, pragmatist philosophers, such as Bezuidenhout (2002) or Cappelen and Lepore (2005), fail to acknowledge this literature. They freely use hackneyed examples from the prototype-theory literature of colour names applied to wines, faces or apples, which are presented as original findings and amply discussed, though without any substantial answer.

Robin Carston may well be right when she writes:

> Underdeterminacy is an essential feature of the relation between linguistic expression and the propositions (thoughts) they are used to express; generally, for any given proposition/thought, there is no sentence which fully encodes it. (Carston 2002: 29)

Perhaps it is sensible to be a little more cautious and leave open the possibility of Quinean 'eternal sentences', which need no appeal at all to any kind of knowledge beyond what is encoded in the lexically fixed linguistic meaning of the words used in the sentences in question. One may think, for example, of statements like *All humans are mortal.* But Carston is no doubt right to the extent that she wants the dependence on extralinguistic knowledge to be a central feature in utterance comprehension. I also think she is right in positing that the extralinguistic knowledge needed for proper comprehension is part of the proposition (thought) expressed, even if it is not encoded in the corresponding L-proposition and sentence. It follows that, as a matter of principle, the listener has a great deal of cognitive reconstruction to carry out in order to achieve an adequate interpretation of incoming utterances. (See Sections 7.2.3 and 7.4 in Volume II for further discussion.)

9.6.4 *Semantic specialization of doublets and synonyms*

Often pairs of doublets, sometimes original synonyms, have the extension of one variant narrowed to specific uses. For example, the English adjective *etesian*, meaning 'annually occurring' and derived from Greek *étos* 'year', is used exclusively for annual summer winds that blow over the Mediterranean—despite the ready availability of the perfectly good and much more common synonym *annual.* Less recondite examples are easily found. The colour called *blond* for humans is called *pale* for ale and *fawn* for canines, cervines, or coats. The untanned skin of a Caucasian human is called *white* or, especially for children and women, *fair*, not *pale*, although the colour is the same as that of a Caucasian's frightened face, which is called *pale.*[9] We *trim* the hedge but *cut* the grass and our hair, and we *mow* the lawn. We *raise* children and cattle but *grow* plants, although we *rear* all three. We *brew* beer and also tea, when we want to sound manly, although normally we *make* tea, just as we *make* coffee, soup, or wine. Music is *composed* but books are *written.* A hundred years ago, the adjectives *broad* and *wide* were closer synonyms than they are in present-day English, where *broad* signifies an open expanse, whereas *wide* evokes a passing through or across. And so on ad infinitum.

[9] A Caribbean lampoon derides Caucasians, who, it says, call themselves 'white' although they turn red in the sun, blue with cold, purple with anger, yellow with fear, and green with envy.

The same is found for morphological variants. English, for example, has recently selected the weak past and past-participle form of the verb *hang* (*hanged*) for hanging people, and kept the corresponding strong form (*hung*) for hanging other things, such as paintings on a wall. A similar specialization has occurred in the doublet *struck* and *stricken* as past participles of the verb *strike*. And as was pointed out in Section 9.2, the adjectives *legible* and *readable*, both derived from a stem meaning 'read', have come to mean 'decipherable' and 'pleasant to read', respectively. *Bereaved* has become a specialization of *bereft*, and [ˈlərnɪd] of [ˈlərnt], both written *learned* (the latter also *learnt*). By contrast, *dreamed* and *dreamt* have not specialized.

Or take the pair of English adjectival suffixes *-ic* and *-ical*. Often, the choice between the two is lexically fixed, as in the adjectives *nonsensical, theatrical, tactical, clinical, umbilical, logical, technical, grammatical, physical, numerical, topical, biblical, practical, hypothetical*, with the suffix *-ical*, as opposed to, for example, *aquatic, diplomatic, eclectic, dogmatic, forensic, lunatic, terrific, generic, specific, automatic, artistic, basic, symphonic, operatic, metallic, domestic, synthetic, algebraic, meteoric*, which obligatorily take the suffix *-ic*. Sometimes, however, there is a free choice between the two: *symmetric(al), metric(al), rhythmic(al), metaphoric(al), theoretic(al)*. These cases are interesting in that there often occurs a split between the cases in which the *-ic* and the *-ical* forms can be used. *Symbolic(al)* and *theoretic(al)*, for example, still occur in their two guises, but it clearly is *symbolic logic*, not **symbolical logic*. Likewise for the adjective *comic(al)*, whose two forms are freely interchangeable except in certain special uses, as in *comic opera*, which does not at all mean the same as *comical opera*. And we speak of *electrical appliances* and *electrical current*, but of the *electric chair* and of *optical art* or an *optical illusion*, but of an *optic disk* or an *optic nerve*. English has the two adjectival variants *rhetoric* and *rhetorical*, but only the latter is usable in *rhetorical questions*. Sometimes the whole adjective changes meaning, as with *classical* versus *classic*, or *historical* versus *historic*, or *economical* versus *economic*. *Semantical* and *syntactical* used to exist side-by-side with their shorter variants *semantic* and *syntactic*, respectively, but anyone using *semantical* or *syntactical* in linguistics or semantics nowadays shows that he or she is not a professional.

Over time, such semantic differentiation of form pairs has often led to wide divergences between forms that are no longer readily recognizable as alternative variants, thus creating enriched lexicons. For example, the words *hospital, hostel*, and *hotel* used to be the same word with the same meaning 'shelter for the poor and sick', but now the three variant forms have assumed quite different meanings. Likewise, the word *roster* used to be a variant of *roaster*—a grid of parallel irons for roasting food—but the two words are no longer felt to be related. Both *hood* and *hat* derive from Old-High-German *huot* 'head covering', but they now have different meanings, probably likewise through phonological differentiation. Sometimes also, one variant becomes obsolete and is replaced by a totally different word. Thus, Old-English *scrift*, meaning 'writing' (cp. German *Schrift* and so on; derived from Latin *scriptum*) specialized for 'prescribed punishment' and became *shrift* 'confessional hearing' or 'remission of sins'. The original meaning was then rendered by *writing*, of Old-Germanic origin. Subsequently, modern English revived the Latin *scriptum* as *script*.

Semantic specialization of variant forms is a further possible source of irregularity of phonological development and thus creates a further class of exceptions to the Young Grammarians' principle of the 'absolute exceptionlessness of sound laws'.

9.7 Further cognitive elements in lexical meanings

The cases discussed up to this point far from exhaust the entire repertoire of the ways in which lexical meanings depend on cognition. For example, context-bound or situational knowledge plays a role in the interpretation of predicates that involve a 'viewpoint' or 'perspective', such as the pair *come* and *go*, or predicates like *to the right (left) of*, *in front of*, *behind*. The two versions of (9.22) are truth-conditionally identical, but they differ semantically in that something like a 'mental camera' is involved. In the *went*-version, the camera has, so to speak, stayed in the corridor but in the *came*-version it has moved into the office along with Dick.

(9.22) Dick and Harry were waiting in the corridor. Then Dick was called into the office. Five minutes later, Harry [went/came] in too.

Often the location of the 'mental camera' depends on the persons involved in the linguistic interaction, as is shown by (9.23a, 9.23b):

(9.23) a. Can I come and help you?
b. Can I go and help him?

In (9.23a), *come* is the normal word to use: the use of *go* is not excluded but would require some further explanation. In (9.23b), *go* and *come* are equally possible. The use of *come* would suggest that the sentence is directed at a listener who is in the same location as the person to be helped. Much of all this still awaits sorting out, especially because languages differ in their criteria for coming and going.

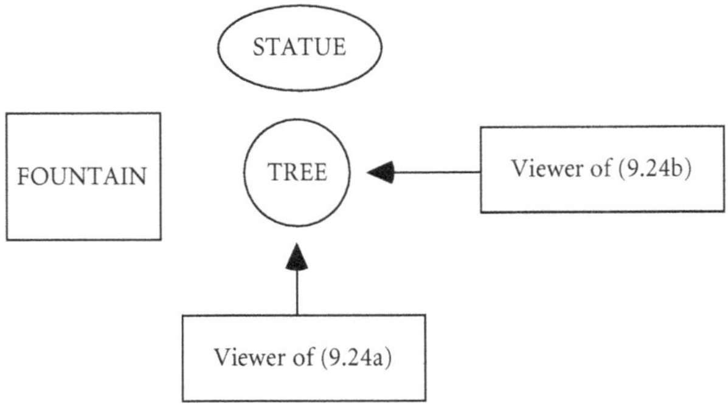

FIGURE 9.1 Viewpoint differences for locative prepositions

In similar manner, the sentences (9.24a) and (9.24b) may describe the same situation, but from different points of view, as shown in Figure 9.1. In (9.24a), the viewer, the tree, and the statue are in a straight line; in (9.24b) it is the viewer, the tree, and the fountain that are in a straight line:

(9.24) a. There was a statue behind the tree, and a fountain to the left of the tree.
b. There was a fountain behind the tree, and a statue to the right of the tree.

A further cognitive criterion for the lexical meaning of predicates, especially those denoting artefacts, seems to be the *function* of the objects denoted. What makes a table a table is hard to define on the basis of mere physical criteria. Anyone familiar with furniture design will quickly realize how different the physical shapes of tables and the materials they are made of can be. The answer is that what makes a table a table is its function: tables are used for certain typical purposes, and anything fulfilling one or more of these purposes is a table. The same goes for the concept of 'furniture', which (*pace* Hudson (1984: 3), discussed in Section 9.6.1) is clearly defined to a large extent on grounds of function. Function also seems to be a defining factor for the concept of luxury. Laws imposing special taxation on luxury goods or luxury activities often enumerate the goods and activities in question, making exceptions for special cases (like frock coats for undertakers). Yet what defines luxury is not a list of goods or activities, but socially recognized function—roughly, anything relatively expensive and exceeding the necessities of life, often replacing cheaper or simpler alternatives.

9.8 Higher-order predicates

Given the special focus on logic in Volume II, it is only fair that a special section should be devoted to higher-order predicates, since they are not only a feature of logic but also abound in the nonlogical vocabularies of natural languages. To see what the notion of higher-order predicate amounts to, consider first what is meant by a first-order predicate.

A first-order predicate is a predicate whose argument terms are free to take single objects as their ρ-values. For example, the predicate *be asleep* is a first-order predicate because it is primarily said of single persons that they are asleep. Of course, one may say of ten persons that they are asleep. In that case the predicate *be asleep* is interpreted as being applied to each person individually: $person_1$ is asleep, $person_2$ is asleep, ..., $person_{10}$ is asleep. Higher-order predicates, by contrast, are predicates that cannot be applied to single objects, but only to sets or classes or groups of objects. There are many such predicates in natural languages. In English, for example, one has predicates like *disperse, congregate, come together, come trickling in, be numerous, be unanimous, be(come) extinct.* To say of a person that he or she 'dispersed' or 'came trickling in' is eerie, to say the least. But to say of a group of people that they 'dispersed' or 'came trickling in' is perfectly normal.

Some predicates are higher-order only with respect to one term. For example, the predicate *collect* is first-order with regard to its subject term but higher-order with regard to its object term, since one can say *John collects English loanwords in Russian*, but not *John collects the English loanword on page 12*.

There appears to be a typical difference between, on the one hand, verbal and adjectival predicates and, on the other, nominal predicates (nouns). Since this difference is, to my knowledge, never mentioned in the literature, it is worth mentioning here. It consists in the fact that it is typical of *verbal* and *adjectival* predicates that they allow for TYPE-RAISING, from being first-order to being higher-order predicates. The predicate *bark*, for example, is normally applied to individual dogs. But it may also be applied to a whole pack of dogs, as when one says *The whole pack was barking.* Or one may say *John left*, but also *The congregation left.* Likewise for most adjectives. One may say *John was scared* but also *The whole nation was scared*, or *John was asleep* but also *The whole bus was asleep* (where the NP *the whole bus* refers to a group of people as a result of dynamic filtering forced by the preconditions of the predicate *be asleep*). This possibility of type-raising may make for ambiguities. For example, a sentence like *Five men were carrying a piano* is interpretable as saying that five men were each carrying a piano, but also as saying that a group of five men was carrying a single piano (see Section 8.1.2 of Volume II). The vast majority of *nominal* predicates, however, which appear as nouns at surface structure level, do not allow for such type-raising. A predicate like *dog* is an obligatorily first-order predicate as one can say of an individual animal that it is a dog, or of a number of individual animals that they are dogs, but not of a pack of dogs that it is a dog. Likewise, one can say of an individual room that it is a room, but not of a whole apartment that it is a room.

Some nouns denote second-order entities, such as *pack* or *team*. There are also nouns, though not many, that denote third-order entities, such as *league*, in the sense of a set of teams. There are even nouns, but they are extremely rare, that achieve sixth-order status, or even higher. Such nouns are to be sought in strictly hierarchical organizations such as the armed forces. A platoon is a group of men, which makes *platoon* a second-order predicate. *Company* is a third-order predicate, as a company consists of platoons. *Battalion* is a fourth-order predicate, as a battalion consists of companies. *Regiment* and *brigade* are fifth-order predicates, since, I understand, regiments and brigades consist of battalions. And *division* is a sixth-order predicate, since a division is built up from brigades or regiments—if I am well informed on these matters. Yet none of these nouns can undergo type-raising: it will never be possible to say of a battalion that it is a company.

Fortunately, the linguist need not worry about a noun being of a fifth- or sixth-order type. The only relevant lexical and grammatical differences found in natural language relate to the distinction between first-order and higher-order predicates. When it is said that *the battalion dispersed*, the semantics of that sentence does not differentiate between readings in which the dispersing took place companywise or platoonwise or individually.

Bibliography

Adam, C. and P. Tannery (eds) (1909) *Œuvres de Descartes*. Vol. XI. Léopold Cerf, Paris.

Aikhenvald, A. (2003) 'Evidentiality in typological perspective', in Aikhenvald and Dixon (eds): 1–31.

—— and R. M. W. Dixon (eds) (2003) *Studies in Evidentiality*. Benjamins, Amsterdam/Philadelphia.

Akmajian, A. (1970) 'On deriving cleft sentences from pseudo-cleft sentences', *Linguistic Inquiry* 1/2: 149–68.

Aksu-Koç, A. and D. I. Slobin (1985) 'Acquisition of Turkish', in D. I. Slobin (ed.), *The Crosslinguistic Study of Language Acquisition*. Vol. I: *The Data*. Lawrence Erlbaum, Hillsdale, NJ: 839–78.

Antal, L. (1964) *Content, Meaning, and Understanding*. (Ianua Linguarum, Series Minor, 31) Mouton, the Hague.

Antunes de Araujo, G. (2004) 'A Grammar of Sabanê, a Namikwaran Language'. Ph.D. thesis, Free University, Amsterdam. LOT, Utrecht.

Arbib, M. A. (2002) 'The mirror system, imitation, and the evolution of language', in K. Dauten-Hahn and C. Nehaniv (eds), *Imitation in Animals and Artifacts*. The MIT Press, Cambridge, MA: 229–80.

Asher, R. E. and J. M. Y. Simpson (eds) (1994) *The Encyclopedia of Language and Linguistics*. Pergamon Press, Oxford.

Austin, J. L. (1962) *How to Do Things with Words*. The William James Lectures delivered in Harvard University in 1955. Edited by J. O. Urmson. Clarendon Press, Oxford.

—— (1970) *Philosophical Papers*. 2nd edn. Ed. J. O. Urmson and G. J. Warnock. Clarendon Press, Oxford.

Bach, K. and R. M. Harnish (1979) *Linguistic Communication and Speech Acts*. MIT Press, Cambridge, MA.

Baker, C. L. (1970) 'Double negatives', *Linguistic Inquiry* 1/2: 169–86.

Bauer, B. M. (2004) 'Vigesimal numerals in Romance: An Indo-European perspective', in B. Drinka (ed.), *Indo-European Language and Culture in Historical Perspective: Essays in Memory of Edgar C. Polomé. General Linguistics* 41: 21–46.

Bäuerle, R., U. Egli, and A. von Stechow (eds) (1979) *Semantics from Different Points of View*. Springer, Berlin-Heidelberg-New York.

Beauzée, N. (1767) *Grammaire générale, ou exposition raisonnée des éléments nécessaires du langage, pour servir de fondement à l'étude de toutes les langues*. (2 vols.) J. Barbou, Paris. References in this volume to modern edition: B. E. Bartlett (ed.), (1974), Stuttgart: Frommann-Holzboog, 2 vols.

Bendix, E. H. (1966) *Componential Analysis of General Vocabulary. The Semantic Structure of a Set of Verbs in English, Hindi, and Japanese.* Mouton, The Hague.

Berlin, B. and P. Kay (1969) *Basic Color Terms: Their Universality and Evolution.* University of California Press, Berkeley.

Bertolet, R. (1994) 'Are there indirect speech acts?', in Tsohatzidis (ed.): 335–49.

Bezuidenhout, A. L. (2002) 'Truth-conditional pragmatics', *Philosophical Perspectives* 16: 105–34.

Bierwisch, M. and R. Schreuder (1992) 'From concepts to lexical items', *Cognition* 42/1: 23–60.

Bloch, A. A. (1986) *Studies in Arabic Syntax and Semantics.* Harrassowitz, Wiesbaden.

Block, N. (2002) 'The harder problem of consciousness', *Journal of Philosophy* 99/8: 1–35.

Borsley, R. D. (2006) 'Review of Postal (2004)', *Language* 82/2: 442–6.

Botha, R. P. (1989) *Challenging Chomsky. The Generative Garden Game.* Blackwell, Oxford.

—— (1997*a*) 'Neo-Darwinian accounts of the evolution of language: 1. Questions about their explanatory focus', *Language and Communication* 17/3: 249–67.

—— (1997*b*) 'Neo-Darwinian accounts of the evolution of language: 2. Questions about complex design', *Language and Communication* 17/4: 319–40.

—— (1998*a*) 'Neo-Darwinian accounts of the evolution of language: 3. Questions about their evidential basis, logic and rhetoric', *Language and Communication* 18/1: 17–46.

—— (1998*b*) 'Neo-Darwinian accounts of the evolution of language: 4. Questions about their comparative merit', *Language and Communication* 18/3: 227–49.

—— (1999) 'On Chomsky's "fable" of instantaneous language evolution', *Language and Communication* 19/3: 243–57.

—— (2003) *Unravelling the Evolution of Language.* Elsevier, Oxford.

Bottani, A. and R. Davies (eds) (2006) *Modes of Existence: Papers in Ontology and Philosophical Logic.* (Philosophical Research 5). Ontos Verlag, Frankfurt.

Brown, C. M. and P. Hagoort (eds) (1999) *The Neurocognition of Language.* Oxford University Press, Oxford.

Bühler, K. (1934) *Sprachtheorie. Die Darstellungsfunktion der Sprache.* Gustav Fischer, Jena.

Bülow, E. (1972) *Versuch einer semantischen Faktorenanalyse der verbalen Ausdrücke des Sehens.* Schwann, Düsseldorf.

Butterworth, B. (1999) *The Mathematical Brain.* Macmillan, London.

Buytaert, Éloi M. and C. J. Mews (1987) 'Theologia "Summi Boni"', *Petri Abaelardi Opera Theologica* Vol. III. Corpus Christianorum; 13. Continuatio Mediaevalis. Brepols, Turnhout.

Cappelen, H. W. and E. Lepore (2005) *Insensitive Semantics. A Defense of Semantic Minimalism and Speech Act Pluralism.* Blackwell, Oxford.

Carnap, R. (1956) *Meaning and Necessity. A Study in Semantics and Modal Logic.* Enlarged edn. The University of Chicago Press, Chicago.

Carston, R. (1998) 'Negation, "presupposition" and the semantics/pragmatics distinction', *Journal of Linguistics* 34: 309–50.

—— (2002) *Thoughts and Utterances: The Pragmatics of Explicit Communication.* Blackwell, Oxford.

Castañeda, H.-N. (1985/86) 'Objects, existence, and reference. A prolegomenon to Guise Theory', in Haller (ed.): 3–59.

Chafe, W. L. (1970) *Meaning and the Structure of Language.* The University of Chicago Press, Chicago-London.

Chomsky, N. (1959) 'Review of Skinner (1957)', *Language* 35/1: 26–58.

—— (1972) *Studies on Semantics in Generative Grammar.* (Janua Linguarum, series minor 107). Mouton, The Hague.

—— (1995) *The Minimalist Program.* MIT Press, Cambridge, MA.

Clark, B. (2005) 'On stochastic grammar', *Language* 81/1: 207–17.

Clark, H. H. (1992) *Arenas of Language Use.* The University of Chicago Press, Chicago.

—— (1996) *Using Language.* Cambridge University Press, Cambridge.

—— and D. Wilkes-Gibbs (1990) 'Referring as a collaborative process', in P. R. Cohen, J. Morgan, and M. E. Pollack (eds), *Intentions in Communication.* MIT Press, Cambridge, MA: 463–93.

Collins, B. and I. M. Mees (1999) *The Real Professor Higgins. The Life and Career of Daniel Jones.* Mouton de Gruyter, Berlin-New York.

Croft, W. (1991) *Syntactic Categories and Grammatical Relations. The Cognitive Organization of Information.* Chicago University Press, Chicago.

—— (2001) *Radical Construction Grammar.* Oxford University Press, Oxford.

Cruse, D. A. (1986) *Lexical Semantics.* Cambridge University Press, Cambridge.

Cutler, A. and C. Clifton (1999) 'Comprehending spoken language: A blueprint of the listener', in Brown and Hagoort (eds): 123–66.

Darnell, R., J. T. Irvine, and R. Handler (1999) *The Collected Works of Edward Sapir III, Culture.* De Gruyter, Berlin.

De Rijk, L. M. (1956) *Petrus Abaelardus, Dialectica. First Complete Edition of the Parisian Manuscript.* Van Gorcum/Hak & Prakke, Assen.

De Rijk, R. P. G. (1974) 'A note on prelexical Predicate Raising', in Seuren (ed.): 43–74.

De Saussure, F. (1916) *Cours de linguistique générale.* Edited by C. Bally and A. Sechehaye, with the collaboration of A. Riedlinger. Payot, Paris-Lausanne.

De Waal, F. (2006) *Primates and Philosophers. How Morality Evolved.* Princeton University Press, Princeton.

Dehaene, S. (1997) *The Number Sense.* Oxford University Press, New York.

—— (2005) 'Evolution of human cortical circuits for reading and arithmetic: The "neuronal recycling" hypothesis', in S. Dehaene, J.-R. Duhamel, M. D. Hauser, and G. Rizzolatti (eds), *From Monkey Brain to Human Brain.* MIT Press, Cambridge, MA: 133–57.

—— V. Izard, P. Pica, and E. Spelke (2006) 'Core knowledge of geometry in an Amazonian indigene group', *Science* 311 (January): 381–4.

Donnellan, K. (1966) 'Reference and definite descriptions', *Philosophical Review* 75: 281–304.

Dowty, D. (1979) *Word Meaning and Montague Grammar*. Reidel, Dordrecht.

—— R. E. Wall, and S. R. Peters (1981) *Introduction to Montague Semantics*. Reidel, Dordrecht.

Dummett, M. E. (1964) 'Truth', in G. Pitcher (ed.), *Truth*. Prentice-Hall, Englewood Cliffs, 93–111.

—— (1973) Frege. *Philosophy of Language*. Duckworth, London.

—— (1981) *The Interpretation of Frege's Philosophy*. Duckworth, London.

—— (2000) *Elements of Intuitionism*. 2nd edn. Clarendon Press, Oxford.

Evans, J. St. B. T. (1989) *Bias in Human Reasoning*. Erlbaum, Hove-London (UK)-Hillsdale (USA).

—— (2004) 'Biases in deductive reasoning', in R. F. Pohl (ed.), *Cognitive Illusions. A Handbook on Fallacies and Biases in Thinking, Judgement and Memory*. Psychology Press, Hove-New York: 127–44.

Evers, A. (1975) 'The Transformational Cycle in Dutch and German'. Ph.D. thesis, University of Utrecht.

Fadiga, L., L. Craighero, G. Buccino, and G. Rizzolatti (2002) 'Speech listening specifically modulates the excitability of tongue muscles: A TMS study', *European Journal of Neuroscience* 15/2: 399–402.

Fauconnier, G. (1985) *Mental Spaces: Aspects of Meaning Construction in Natural Language*. MIT Press, Cambridge, MA.

Federmeier, K. D. (2007) 'Thinking ahead: The role and roots of prediction in language comprehension', *Psychophysiology* 44: 491–505.

Ferreira, F. and N. D. Patson (2007) 'The "good enough" approach to language comprehension', *Language and Linguistics Compass* 1/1–2: 71–83.

Findlay, J. N. (1963) *Meinong's Theory of Objects and Values*. Clarendon Press, Oxford.

Firth, J. R. (1957) *Papers in Linguistics 1934–1951*. Oxford University Press, Oxford.

Fisher, S. E. and G. F. Marcus (2006) 'The eloquent ape: Genes, brains and the evolution of language', *Nature Reviews/Genetics* 7: 9–20.

Fodor, Janet D., Jerry A. Fodor, and M. F. Garrett (1975) 'The psychological unreality of semantic representations', *Linguistic Inquiry* 6/4: 515–31.

Fodor Jerry A. (1970) 'Three reasons for not deriving "kill" from "cause to die"', *Linguistic Inquiry* 1/4: 429–38.

—— (1975) *The Language of Thought*. Harvester Press, Hassocks, Sussex.

—— (1983) *The Modularity of Mind. An Essay on Faculty Psychology*. A Bradford Book. MIT Press, Cambridge, MA.

—— (1998) *Concepts: Where Cognitive Science Went Wrong*. Oxford University Press, New York.

—— (1999) 'Against definitions', in E. Margolis and S. Laurence (eds), *Concepts. Core Readings*. MIT Press, Cambridge, MA: 491–512. (Excerpt of previous)

—— and J. J. Katz (eds) (1964) *The Structure of Language. Readings in the Philosophy of Language*. Prentice Hall, Englewood Cliffs, New Jersey.

—— M. F. Garrett, E. C. T. Walker, and C. H. Parkes (1980) 'Against definitions', *Cognition* 8: 263–367.

Frege, G. (1884) *Die Grundlagen der Arithmetik: eine logisch-mathematische Untersuchung über den Begriff der Zahl.* W. Koebner, Breslau.

—— (1892) 'Ueber Sinn und Bedeutung', *Zeitschrift für Philosophie und philosophische Kritik* 100: 25–50. (Also in Patzig 1969: 40–65)

—— (1974^2) *The Foundations of Arithmetic: A logico-mathematical enquiry into the concept of number.* 2nd edn. Translation of Frege 1884. Blackwell, Oxford.

Gardiner, A. H. (1932, 1951^2) *The Theory of Speech and Language.* Clarendon Press, Oxford.

Gardner, H. (1985) *The Mind's New Science. A History of the Cognitive Revolution.* Basic Books, New York.

Geeraerts, D. (1986) *Woordbetekenis. Een overzicht van de lexicale semantiek.* Uitgeverij Acco, Leuven/Amersfoort.

—— (1994*a*) 'Lexical field', in Asher and Simpson (eds): 2144–6.

—— (1994*b*) 'Polysemy', in Asher and Simpson (eds): 3227–8.

Giannakidou, A. (2004) 'How much can we weaken downward entailment?' University of Chicago. Draft, March.

Goldberg, A. E. (1995) *Constructions. A Construction Grammar Approach to Argument Structure.* Chicago University Press, Chicago.

—— (2006) *Constructions at Work. The Nature of Generalization in Language.* Oxford University Press, Oxford.

Graffi, G. (2004) 'Per la storia di alcuni termini e concetti grammaticali: il declino di *oratio* e l'ascesa di *propositio* come termini per "frase"', in C. Milani (ed.), *Per una storia della grammatica in Europa.* Pubblicazioni ISU dell'Università Cattolica, Milano.

Grice, H. P. (1975) 'Logic and conversation', in P. Cole and J. L. Morgan (eds), *Speech Acts.* (Syntax and Semantics 3) Academic Press, New York-San Francisco-London: 41–58.

Grigorjewa, O. (1973) 'Semantisch-syntaktische Charakteristik der Verben der sinnlichen Wahrnehmung in der deutschen Sprache'. Ph.D. thesis, Leipzig University.

Groenendijk, J. A. G. and M. B. J. Stokhof (1984) 'On the semantics of questions and the pragmatics of answers', in F. Landman and F. Veltman (eds), *Varieties of Formal Semantics.* (Groningen-Amsterdam Studies in Semantics 3) Foris, Dordrecht: 143–70.

Gruber, Jeffrey (1967) 'Look and see', *Language* 43/4: 937–48.

—— (1976) *Lexical Structures in Syntax and Semantics.* North-Holland, Amsterdam.

Hagoort, P. and J. van Berkum (2007) 'Beyond the sentence given', *Philosophical Transactions of the Royal Society* B 362: 801–11.

—— M. Wassenaar, and C. Brown (2003) 'Real-time semantic compensation in patients with agrammatic comprehension: Electrophysiological evidence for multiple-route plasticity', *Proceedings of the National Academy of Sciences of the United States of America (PNAS)* 100/7: 4340–5.

—— L. Hald, M. Bastiaansen, and K.-M. Petersson (2004) 'Integration of word meaning and world knowledge in language comprehension', *Science* 304 (5669): 438–41.

Hall-Partee, B. (1979) 'Semantics—mathematics or psychology?' in Bäuerle et al. (eds): 1–14.

Haller, R. (ed.) (1985/86) 'Non-Existence and Predication', *Grazer Philosophische Studien* 25/26. Rodopi, Amsterdam.

Hamblin, C. L. (1976) 'Questions in Montague English', in B. H. Partee (ed.), *Montague Grammar*. Academic Press, New York-San Francisco-London: 247–59.

—— (1987) *Imperatives*. Blackwell, Oxford.

Harris, R. A. (1993) *The Linguistics Wars*. Oxford University Press, New York.

Harris, Z. S. (1951) *Methods in Structural Linguistics*. The University of Chicago Press, Chicago.

Hawkins, J. A. (1983) *Word Order Universals*. Academic Press, New York-London.

—— (1988) 'Explaining language universals', in J. A. Hawkins (ed.), *Explaining Language Universals*. Blackwell, Oxford: 3–28.

—— (1994) *A Performance Theory of Order and Constituency*. Cambridge University Press, Cambridge.

Helenius, P., R. Salmelin, E. Service, and J. F. Connolly (1998) 'Distinct time courses of word and sentence comprehension in the left temporal cortex', *Brain* 121: 1133–42.

Hintikka, J. (1974) 'Questions about questions', in M. K. Munitz and P. K. Unger (eds), *Semantics and Philosophy*. New York University Press, New York: 103–58.

Hjelmslev, L. (1953) *Prolegomena to a Theory of Language*. (Translated from the Danish *Omkring sprogteoriens grundlaeggelse*, 1943, by Francis J. Whitfield) Indiana University Publications in Anthropology and Linguistics. Memoir 7 of the International Journal of American Linguistics. Waverly Press, Baltimore, MD.

Hoepelman, J. (1981) 'On questions', in J. A. G. Groenendijk, T. M. V. Janssen, and M. B. J. Stokhof (eds), *Formal Methods in the Study of Language*. 2 vols. Mathematisch Centrum, Amsterdam, Vol. I: 207–35.

Horn, L. R. (1972) 'On the Semantic Properties of Logical Operators in English'. Ph.D. thesis, University of Southern Califiornia, Los Angeles, CA.

—— (1978) 'Remarks on Neg-Raising', in Peter Cole (ed.), *Syntax and Semantics*. Volume 9. *Pragmatics*. Academic Press, New York-San Francisco-London: 129–220.

—— (1985) 'Metalinguistic negation and pragmatic ambiguity', *Language* 61/1: 121–74.

—— (1989) *A Natural History of Negation*. The University of Chicago Press, Chicago-London.

Householder, F. W. (1981) *The Syntax of Apollonius Dyscolus*. Translated, and with commentary by Fred W. Householder. (Studies in the History of Linguistics 23). Benjamins, Amsterdam/Philadelphia.

Huck, G. J. and J. A. Goldsmith (1995) *Ideology and Linguistic Theory. Chomsky and the Deep Structure Debate*. Routledge, London and New York.

Hudson, R. (1984) *Word Grammar*. Blackwell, Oxford.

Isard, S. (1975) 'Changing the context', in E. L. Keenan (ed.), *Formal Semantics of Natural Language*. Cambridge University Press, Cambridge: 287–96.

Jackendoff, R. (1993) *Patterns in the Mind*. Harvester Wheatsheaf, New York.

—— (2002) *Foundations of Language. Brain, Meaning, Grammar, Evolution*. Oxford University Press, New York.

Jackson, F. (1982) 'Epiphenomenal qualia', *Philosophical Quarterly* 32: 127–36.

Janssen, T. A. J. M. (1976) 'Hebben-konstrukties en indirekt-objektskonstrukties'. Ph.D. thesis, Radboud University Nijmegen. HES Publishers, Utrecht.

—— (2007) 'A speaker/hearer-based grammar: The case of possessives and compounds', in M. Hannay and G. Steen (eds.), *Structural-Functional Studies in English Grammar*. Benjamins, Amsterdam/Philadelphia: 353–87.

Johnson-Laird, P. N. (1977) 'Reasoning with quantifiers', in P. N. Johnson-Laird and P. C. Wason (eds), *Thinking. Readings in Cognitive Science*. Cambridge University Press, Cambridge: 129–42.

—— (1989) 'Mental models', in M. I. Posner (ed.), *Foundations of Cognitive Science*. MIT Press, Cambridge, MA: 469–500.

Joly, A. (1967) *Negation and the Comparative Particle in English*. Cahiers du psychomécanique du langage, 9. Les Presses de l'Université Laval, Quebec.

Joseph, B. D. (2003) 'Evidentials. Summation, questions, prospects', in Aikhenvald and Dixon (eds): 307–27.

Joüon, P. (1947) *Grammaire de l'hébreu biblique*. Institut Biblique Pontifical, Rome.

Kamp, H. (1981) 'A theory of truth and semantic interpretation', in J. A. G. Groenendijk, T. M. V. Janssen, and M. B. J. Stokhof (eds), *Formal Methods in the Study of Language*. 2 vols. Mathematisch Centrum, Amsterdam, Vol. I: 277–322.

—— and U. Reyle (1993) *From Discourse to Logic. Introduction to Model-Theoretic Semantics of Natural Language, Formal Logic and Discourse Representation Theory*. Kluwer, Dordrecht.

Karttunen, L. (1977) 'Syntax and semantics of questions', *Linguistics and Philosophy* 1/1: 3–44.

Katz, J. J. and J. A. Fodor (1963) 'The structure of a semantic theory', *Language* 39/2: 170–210. (Also in Fodor and Katz (eds) 1964: 479–518.)

Kempen, G. and E. Hoenkamp (1987) 'An Incremental Procedural Grammar for sentence formulation', *Cognitive Science* 11: 201–58.

Kindinger, R. (1965) *Philosophenbriefe; Aus der Wissenschaftlichen Korrespondenz von Alexius Meinong*. Akademische Druck- u. Verlagsanstalt, Graz.

Kiprian, A. (2002) *Vospominanija o Mitropolite Antonii i Episkope Gavriile* (Memories of Metropolitan Antonius and Bishop Gavrili). Institute of Orthodox Theology St. Tikhon, Moscow.

Kneale, W. and M. Kneale (1962) *The Development of Logic*. Clarendon Press, Oxford.

Konieczny, L., B. Hemforth, C. Scheepers, and G. Strube (1997) 'The role of lexical heads in parsing: Evidence from German', *Language and Cognitive Processes* 12/2/3: 307–48.

Kripke, S. (1972) 'Naming and necessity', in D. Davidson and G. Harman (eds), *Semantics of Natural Language*, Reidel, Dordrecht: 253–355.

—— (1980) *Naming and Necessity*. Blackwell, Oxford (=Kripke 1972).

Kühner, R. and C. Stegmann (1955) *Ausführliche Grammatik der lateinischen Sprache. Satzlehre*. 2 volumes. 3rd edn, rev. by Andreas Thierfelder. Gottschalk, Leverkusen.

Labov, W. (1966) *The Social Stratification of English in New York City*. Center for Applied Linguistics, Washington DC.

—— (1972) *Language in the Inner City. Studies in Black English Vernacular*. University of Pennsylvania Press, Philadelphia.

Ladusaw, W. A. (1979) 'Polarity Sensitivity as Inherent Scope Relations'. Ph.D. thesis, University of Texas at Austin.

Lakoff, G. (1971) 'On generative semantics', in D. D. Steinberg and L. A. Jakobovits (eds), *Semantics. An Interdisciplinary Reader in Philosophy, Linguistics and Psychology*. Cambridge University Press, Cambridge, 232–96.

—— and M. Johnson (1980) *Metaphors We Live By*. The University of Chicago Press, Chicago-London.

Lakoff, R. (1969) 'A syntactic argument for Negative Transportation'. Papers from the Fifth Regional Meeting of the Chicago Linguistic Society. April 1969: 140–7.

Langacker, R, W. (1987) *Foundations of Cognitive Grammar*. Vol. 1: *Theoretical Prerequisites*. Stanford CA: Stanford University Press.

—— (1991) *Foundations of Cognitive Grammar*. Vol. 2: *Descriptive Application*. Stanford CA: Stanford University Press.

—— (2003) 'Construction grammars: Cognitive, radical, and less so'. Plenary paper ICLC 8, Logroño (Spain), June 25, 2003. MS.

Lashley, K. S. (1951) 'The problem of serial order in behavior', in L. A. Jeffress (ed.), *Cerebral Mechanisms in Behavior. The Hixon Symposium*. John Wiley & Sons, New York /Chapman & Hall Ltd., London: 112–46.

Laury, R. and T. Ono (2005) 'Data is data and model is model: You don't discard the data that doesn't fit your model!' *Language* 81/1: 218–25.

Lehrer, A. (1974) *Semantic Fields and Lexical Structure*. North-Holland, Amsterdam.

Lejewski, C. (1985/86) 'Logic and non-existence', in Haller (ed.): 209–34.

Levelt, W. J. M. (1989) *Speaking. From Intention to Articulation*. A Bradford Book. MIT Press, Cambridge, MA.

—— (1999) 'Producing spoken language: A blueprint of the speaker', in Brown and Hagoort (eds): 83–122.

Lewis, C. I. (1946) *An Analysis of Knowledge and Valuation*. Open Court, La Salle, Illinois.

Lewis, D. (1970) 'General semantics', *Synthese* 22/1: 18–67.

Liddell, H. G. and R. Scott (1951[9]) *A Greek-English Lexicon*. A New Edition Revised and Augmented by Sir Henry Stuart Jones. Clarendon Press, Oxford.

Linsky, L. (1967) *Referring*. Routledge & Kegan Paul, London-New York.

Lyons, J. (1977) *Semantics*. Cambridge University Press, Cambridge.

McCawley, J. D. (1967) 'Meaning and the description of languages', *Kotoba no Uchu* 2/9: 10–18; 2/10: 38–48; 2/11: 51–7. (Also in McCawley 1973: 99–120.)

—— (1968) 'Lexical insertion in a transformational grammar without deep structure'. *Papers from the Fourth Regional Meeting Chicago Linguistic Society*. Department of Linguistics, University of Chicago: 71–80. (Also in McCawley 1973: 155–66.)

—— (1970) 'English as a VSO language', *Language* 46/2: 286–99. (Also in McCawley 1973: 211–28, and in Seuren (ed.) 1974: 75–95.)

—— (1972) 'A program for logic', in D. Davidson and G. Harman (eds), *Semantics of Natural Language*, Reidel, Dordrecht: 498–544.

—— (1973) *Grammar and Meaning. Papers on Syntactic and Semantic Topics*. Taishukan Publishing Company, Tokyo.

MacCorquodale, K. (1970) 'On Chomsky's review of Skinner's Verbal Behavior', *Journal of the Experimental Analysis of Behavior* 13: 83–99.

Marcus, G. F. (2004) *The Birth of the Mind. How a Tiny Number of Genes Creates the Complexities of Human Thought*. Basic Books, New York.

—— (2006) 'Cognitive architecture and descent with modification', *Cognition* 101: 443–65.

—— and S. E. Fisher (2003) 'FOXP2 in focus: What can genes tell us about speech and language?' *Trends in Cognitive Sciences* 7/6: 257–62.

Marmo, C. (1995) 'A pragmatic approach to language in Modism', in S. Ebbesen (ed.), *Sprachtheorien in Spätantike und Mittelalter*. Narr, Tübingen.

Mathesius, V. (1928) 'On linguistic characterology with illustrations from modern English', in *Actes du Premier Congrès International de Linguistes à La Haye*. Reprinted in J. Vachek (ed.), *A Prague School Reader in Linguistics*. Indiana University Press, Bloomington, Indiana, 1964: 59–67.

May, R. (1977) 'The grammar of quantification'. Ph.D. dissertation, MIT.

Meijering, T. C. (1981) 'Naturalistic Epistemology: Helmholtz and the Rise of a Cognitive Theory of Perception'. Ph.D. thesis, The University of California at Berkeley.

Meinong, Alexius (1988) *Über Gegenstandstheorie. Selbstdarstellung*. Felix Meiner, Hamburg.

—— (ed.) (1904) *Untersuchungen zur Gegenstandstheorie und Psychologie*. J. A. Barth, Leipzig.

Meyer, C. F. and H. Tao (2005) 'Response to Newmeyer's "Grammar is grammar and usage is usage"', *Language* 81/1: 226–8.

Montague, R. (1973) 'The proper treatment of quantification in ordinary English', in K. J. J. Hintikka, J. M. E. Moravcsik, and P. Suppes (eds) *Approaches to Natural Language. Proceedings of the 1970 Stanford Workshop on Grammar and Semantics*. Reidel, Dordrecht: 221–42.

Morgan, J. L. (1969) 'On arguing about semantics', *Papers in Linguistics* 1/1: 49–70.

Musacchio, J. M. (2005) 'The ineffability of qualia and the word-anchoring problem', *Language Sciences* 27/4: 403–35.

Newmeyer, F. J. (1999) 'Some remarks on the functionalist-formalist controversy in linguistics', in M. Darnell, E. Moravcsik, F. Newmeyer, M. Noonan, and K. Wheatley (eds), *Functionalism and Formalism in Linguistics.* Volume I: *General Papers.* Benjamins, Amsterdam/Philadelphia: 469–86.

—— (2003) 'Grammar is grammar and usage is usage', *Language* 79/4: 682–707.

—— (2005) 'A reply to the critiques of "Grammar is grammar and usage is usage"', *Language* 81/1: 229–36.

Nieuwland, M. S., K. M. Petersson, and J. van Berkum (2007) 'On sense and reference: Examining the functional neuroanatomy of referential processing', *NeuroImage* 37: 993–1004.

Nuchelmans, G. (1973) *Theories of the Proposition. Ancient and Medieval Conceptions of the Bearers of Truth and Falsity.* North-Holland, Amsterdam.

—— (1980) *Late-Scholastic and Humanist Theories of the Proposition.* North-Holland, Amsterdam.

—— (1983) *Judgment and Proposition. From Descartes to Kant.* North-Holland, Amsterdam.

Núñez, R. E. and E. Sweetser (2006) 'With the future behind them: Convergent evidence from Aymara language and gesture in the crosslinguistic comparison of spatial construals of time', *Cognitive Science* 30: 401–50.

Ogden, C. K. and I. A. Richards (1923) *The Meaning of Meaning. A Study of the Influence of Language upon Thought and of the Science of Symbolism.* Routledge & Kegan Paul, London.

Parsons, T. (1980) *Nonexistent Objects.* Yale University Press, New Haven.

Patzig, G. (ed.) (1969) *Gottlob Frege. Funktion, Begriff, Bedeutung. Fünf logische Studien.* Vandenhoeck & Ruprecht, Göttingen.

Payne, T. (1997) *Describing Morphosyntax.* Cambridge University Press, Cambridge.

Pembeci, I., C. Bozşahin, and D. Zeyrek (1998) 'Computer-aided learning of Turkish morphology', *Proceedings of the Joint Conference of the Association for Computing in Humanities (ACH) and the Association for Literary & Linguistic Computing (ALLC),* Debrecen, Hungary.

Penrose, R. (2000) *The Large, the Small, and the Human Mind.* With Abner Shimony, Nancy Cartwright, and Stephen Hawking. Edited by Malcolm Longair. Canto, Cambridge University Press, New York.

Pica, P., C. Lemer, V. Izard, and S. Dehaene (2004) 'Exact and approximate arithmetic in an Amazonian indigene group', *Science* 306 (October): 499–503.

Pickering, M. J. and S. C. Garrod (2007) 'Do people use language production to make predictions during comprehension?' *Trends in Cognitive Sciences* 11/3: 105–10.

Postal, P. M. (2004) *Skeptical Linguistic Essays.* Oxford University Press, New York.

Pullum, G. K. and B. C. Scholz (2002) 'Empirical assessment of stimulus poverty arguments', *The Linguistic Review* 19/1: 9–50.

PUSTEJOVSKY, J. (1995) *The Generative Lexicon*. MIT Press, Cambridge, MA.

QUINE, W. V. O. (1953) *From a Logical Point of View*. Harvard University Press, Cambridge, MA.

—— (1960) *Word and Object*. MIT Press, Cambridge, MA.

RITCHIE, G. (1983) 'Semantics in parsing', in M. King (ed.), *Parsing Natural Language*. Academic Press, New York: 199–217.

RIZZOLATTI, G., L. FOGASSI, and V. GALLESE (2001) 'Neurophysiological mechanisms underlying the understanding and imitation of action', *Nature Reviews Neuroscience* 2: 661–70.

ROGERS, A. (1973) 'Physical Perception Verbs in English'. Ph.D. thesis, The University of California at Los Angeles.

ROSCH, E. (1973) 'On the internal structure of perceptual and semantic categories', in T. E. Moore (ed.), *Cognitive Development and the Acquisition of Language*. Academic Press, New York: 111–44.

—— (1977) 'Classification of real world objects: Origins and representations in cognition', in P. N. Johnson-Laird and P. Wason (eds), *Thinking*. Cambridge University Press, Cambridge: 212–22.

—— and C. B. MERVIS (1975) 'Family resemblances: Studies in the internal structure of categories', *Cognitive Psychology* 7/4: 573–605.

ROSIER-CATACH, I. (1999) 'La notion de translatio, le principe de compositionalité, et l'analyse de la prédication accidentelle chez Abélard', in J. Biard, (ed.), *Langage, sciences, philosophie au XIIe siècle*. Paris, Vrin: 125–64.

ROSS, J. R. (1970) 'On declarative sentences', in R. Jacobs and P. Rosenbaum (eds), *Readings in English Transformational Grammar*. Ginn & Co, Boston: 222–72.

RUSSELL, B. (1903) *The Principles of Mathematics*. Cambridge University Press, Cambridge. (2nd edition by Allen & Unwin, London 1937).

—— (1904) 'Meinong's theory of complexes and assumptions', *Mind* 13: 204–19; 336–54; 509–24.

—— (1905*a*) 'On denoting', *Mind* 14: 479–93.

—— (1905*b*) 'Review of *Untersuchungen zur Gegenstandstheorie und Psychologie*. Edited by A. Meinong. Johann Ambrosius Barth, Leipzig, 1904', *Mind* 14: 530–8.

—— (1920) *Introduction to Mathematical Philosophy*. Allen and Unwin, London.

—— (1940) *An Inquiry into Meaning and Truth*. Allen & Unwin, London.

—— (1944) 'My mental development', in P. A. Schilpp (ed.), *The Philosophy of Bertrand Russell*. Open Court, La Salle, Ill.

RUWET, N. (1972) *Théorie syntaxique et syntaxe du français*. Du Seuil, Paris.

SADOCK, J. M. (1975) *Toward a Linguistic Theory of Speech Acts*. Academic Press, New York-San Francisco-London.

—— (1991) *Autolexical Syntax: A Theory of Parallel Grammatical Representations*. The University of Chicago Press, Chicago.

SADOCK, J. M. (1994) 'Toward a grammatically realistic theory of speech acts', in Tsohatzidis (ed.): 393–406.

SAMPSON, G. (1997) *Educating Eve. The 'Language Instinct' Debate*. Cassel, London.

Sanford, A. J. and S. C. Garrod (1998) 'The role of scenario mapping in text comprehension', *Discourse Processes* 26/2–3: 159–90.

—— and P. Sturt (2002) 'Depth of processing in language comprehension: Not noticing the evidence', *Trends in Cognitive Sciences* 6/9: 382–6.

—— E. J. Dawydiak, and L. M. Moxey (2007) 'A unified account of quantifier perspective effects in discourse', *Discourse Processes* 44/1: 1–32.

Scharten, R. (1997) 'Exhaustive Interpretation: A Discourse-Semantic Account'. Ph.D. thesis, Radboud University, Nijmegen.

Schepping, M.-T. (1982) *Kontrastive semantische Analyse von Verben des Visuellen im Französischen und Deutschen*. Narr, Tübingen.

Scholz, B. C. and G. K. Pullum (2002) 'Searching for arguments to support linguistic nativism', *Linguistic Review* 19/2: 185–224.

Schreuder, R. (1978) 'Studies in Psycholexicology, with Special Reference to Verbs of Motion'. Ph.D. thesis, Radboud University, Nijmegen.

Searle, J. R. (1969) *Speech Acts. An Essay in the Philosophy of Language*. Cambridge University Press, Cambridge.

Seuren, P. A. M. (1969) *Operators and Nucleus. A Contribution to the Theory of Grammar*. Cambridge University Press, Cambridge.

—— (1972*a*) 'Predicate Raising and Dative in French and sundry languages'. Unpublished paper, Magdalen College, Oxford. (Published as Chapter 7 in Seuren 2001.)

—— (1972*b*) 'Taaluniversalia in de transformationele grammatika', *Leuvense Bijdragen* 61/4: 311–70. (Partially republished in English as Chapter 12 in Seuren 2001.)

—— (1972*c*) 'Autonomous versus semantic syntax', *Foundations of Language* 8/2: 237–65. (Also in Seuren (ed.) 1974: 96–122 and as Chapter 3 in Seuren 2001.)

—— (1973) 'The comparative', in F. Kiefer and N. Ruwet (eds), *Generative Grammar in Europe*. Reidel, Dordrecht: 528–64.

—— (1974*a*) 'Negative's travels', In Seuren (ed.): 183–208. (Reprinted as Chapter 8 in Seuren 2001.)

—— (ed.) (1974*b*) *Semantic Syntax. Oxford Readings in Philosophy*. Oxford University Press, Oxford.

—— (1975) *Tussen Taal en Denken. Een bijdrage tot de empirische funderingen van de semantiek*. Oosthoek, Scheltema & Holkema, Utrecht.

—— (1980) 'Dreiwertige Logik und die Semantik natürlicher Sprache', in *Sprache der Gegenwart. Schriften des Instituts für deutsche Sprache, Band L. Grammatik und Logik*. Jahrbuch 1979 des Instituts für deutsche Sprache. Schwann Verlag, Düsseldorf: 72–103.

—— (1981) 'Tense and aspect in Sranan', *Linguistics* 19/11–12: 1043–76. (Reprinted as Chapter 21 in Seuren 2001.)

—— (1982) 'Internal variability in competence', *Linguistische Berichte* 77: 1–31. (Reprinted as Chapter 5 in Seuren 2001.)

—— (1983) 'The auxiliary system in Sranan', in F. Heny and B. Richards (eds), *Linguistic Categories: Auxiliaries and Related Puzzles*, Vol. 2. Reidel, Dordrecht: 219–51.

—— (1984*a*) 'The comparative revisited', *Journal of Semantics* 3/1–2: 109–141.

—— (1984*b*) 'Operator Lowering', *Linguistics* 22/5: 573–627. (Reprinted as Chapter 9 in Seuren 2001.)
—— (1985) *Discourse Semantics*. Blackwell, Oxford.
—— (1987) 'The self-styling of Relevance Theory', *Journal of Semantics* 5/2: 123–43.
—— (1988) 'Presupposition and negation', *Journal of Semantics* 6/3–4: 175–226. (Reprinted as Chapter 15 in Seuren 2001.)
—— (1989) 'Notes on reflexivity', in F. J. Heyvaert and F. Steurs (eds), *Worlds behind Words. Essays in honour of Prof. Dr. F.G. Droste on the Occasion of his Sixtieth Birthday*. Leuven University Press, Louvain, 85–95.
—— (1993) 'Why does 2 mean "2"? Grist to the anti-Grice mill', in E. Hajičová (ed.), *Proceedings of the Conference on Functional Description of Language*, Prague 24–27 November, 1992. Faculty of Mathematics and Physics, Charles University, Prague, 225–35. (Reprinted as Chapter 16 in Seuren 2001.)
—— (1994*a*) 'Sign', in Asher and Simpson (eds): 3885–8. (Also in P.V. Lamarque (ed.), *Concise Encyclopedia of Philosophy of Language*. Pergamon Press, Oxford, 1997: 244–8.)
—— (1994*b*) 'Presupposition', in Asher and Simpson: 3311–20.
—— (1996) *Semantic Syntax*. Blackwell, Oxford.
—— (1998) *Western Linguistics. An Historical Introduction*. Blackwell, Oxford.
—— (2000) 'Presupposition, negation and trivalence', *Journal of Linguistics* 36/2: 261–97.
—— (2001) *A View of Language*. Oxford University Press, Oxford.
—— (2003) 'Verb clusters and branching directionality in German and Dutch', in Seuren and Kempen (eds): 247–96.
—— (2004*a*) *Chomsky's Minimalism*. Oxford University Press, New York.
—— (2004*b*) 'The importance of being modular', *Journal of Linguistics* 40: 593–635.
—— (2005) 'Eubulides as a 20th-century semanticist', *Language Sciences* 27/1: 75–95.
—— (2006) 'McCawley's legacy', *Language Sciences* 28/5: 521–6.
—— (2007) 'The theory that dare not speak its name: A rejoinder to Mufwene and Francis', *Language Sciences* 29/4: 571–3.
—— and H. C. Wekker (1986) 'Semantic transparency as a factor in Creole genesis', in P. Muysken and N. Smith (eds), *Substrata versus Universals in Creole Genesis*. Papers from the Amsterdam Creole Workshop, April 1985. Benjamins, Amsterdam/Philadelphia: 57–70.
—— and C. Hamans (2008) 'Antifunctionality in language development and language contact'. Paper read at the 41st Annual Meeting of the Societas Linguistica Europaea, Forli (Italy), 17–20 September 2008.
—— and G. Kempen (eds) (2003) *Verb Constructions in German and Dutch*. (CILT 242) Benjamins, Amsterdam/Philadelphia.
Seyfarth, R. M., D. L. Cheney, and P. R. Marler (1980) 'Monkey responses to three different alarm calls: Evidence of predator classification and semantic communication', *Science* 210 Nr. 4471: 801–3.
Sidgwick, A. (1895) 'Context and meaning', *Mind* 4: 281–306.

SKINNER, B. F. (1957) *Verbal Behavior*. Appleton-Century-Crofts, New York.

SPERBER, D. and D. WILSON (1986) *Relevance. Communication and Cognition*. Blackwell, Oxford (2nd edn, 1995).

STALNAKER, R. C. (1978) 'Assertion', in P. Cole (ed.), *Pragmatics*. (Syntax and Semantics 9) Academic Press, New York-San Francisco-London.

STEGMÜLLER, W. (1957) *Das Wahrheitsproblem und die Idee der Semantik. Eine Einleitung in die Theorien von A. Tarski und R. Carnap*. Springer, Vienna.

—— (1986^7) *Hauptströmungen der Gegenwartsphilosophie. Eine kritische Einführung*. Band II. Kröner, Stuttgart.

STEINTHAL, H. (1860) *Charakteristik der hauptsächlichsten Typen des Sprachbaues*. (Neubearbeitung von Dr. Franz Misteli). Dümmler, Berlin.

STERN, G. (1931) *Meaning and Change of Meaning. With Special Reference to the English Language*. Göteborgs Högskolas Årsskrift XXXVIII, Gothenburg. (Reprint: Greenwood Press, Westport, CN, 1975.)

STOUT, G. F. (1896, 1909^3) *Analytical Psychology*. 2 vols. Sonnenschein, London; Macmillan, New York.

STRAWSON, P. F. (1950) 'On referring', *Mind* 59: 320–44.

—— (1952) *Introduction to Logical Theory*. Methuen, London.

—— (1954) 'A reply to Mr Sellars', *Philosophical Review* 63/2: 216–31.

—— (1964) 'Identifying reference and truth-values', *Theoria* 30/2: 96–118.

—— (1974) *Subject and Predicate in Logic and Grammar*. Methuen, London.

TOMASELLO, M. (2003) *Constructing a Language. A Usage-Based Theory of Language Acquisition*. Harvard University Press, Cambridge, MA.

TOPBAŞ, S., I. MAVIŞ, and M. BAŞAL (1997) 'Acquisition of bound morphemes: Nominal case morphology in Turkish', in K. Imer and N. E. Uzun (eds), *Proceedings of the VIIIth International Conference on Turkish Linguistics*, 7–9 August, Ankara University: 127–37.

TOWNSEND, D. J. and T. G. BEVER (2001) *Sentence Comprehension. The Integration of Habits and Rules*. MIT Press, Cambridge, MA.

TRIER, J. (1931) *Der deutsche Wortschatz im Sinnbezirk des Verstandes. Die Geschichte eines sprachlichen Feldes*. Band. I: *Von den Anfängen bis zum Beginn des 13. Jahrhunderts*. Carl Winter, Heidelberg.

—— (1934) 'Das Sprachliche Feld. Eine Auseinandersetzung', *Neue Jarhbücher für Wissenschaft und Jugendbildung* 10: 428–49.

—— (1973) *Aufsätze und Vorträge zur Wortfeldtheorie*. (Janua Linguarum, series minor 174). Edited by A. van der Lee and O. Reichmann. Mouton, The Hague.

TSOHATZIDIS, S. L. (ed.) (1994) *Foundations of Speech Act Theory. Philosophical and Linguistic Perspectives*. Routledge, London-New York.

VAN BERKUM, J., P. ZWITSERLOOD, C. M. BROWN, V. KOOIJMAN, and P. HAGOORT (2005) 'Anticipating upcoming words in discourse: Evidence from ERPs and reading times', *Journal of Experimental Psychology: Learning, Memory, and Cognition* 31/3: 443–67.

——A. W. Koornneef, M. Otten, and M. S. Nieuwland (2007) 'Establishing reference in language comprehension: An electrophysiological perspective', *Brain Research* 1146: 158–71.

Van Dalen, D. (1999) *Mystic, Geometer, and Intuitionist: The Life of L. E. J. Brouwer. Vol. 1, The Dawning Revolution.* Oxford University Press, Oxford.

Van Fraassen, B. (1971) *Formal Semantics and Logic.* Macmillan, New York-London.

Van Heijenoort, J. (ed.) (1967) *From Frege to Gödel. A Source Book in Mathematical Logic, 1879–1931.* Harvard University Press, Cambridge, MA.

Vliegen, M. (1988) *Verben der auditiven Wahrnehmung im Deutschen.* Narr, Tübingen.

Watkins, K. E., A. P. Straffella, and T. Paus (2003) 'Seeing and hearing speech excites the motor system involved in speech production', *Neuropsychologia* 1558: 1–6.

Wegener, P. (1885) *Untersuchungen über die Grundfragen des Sprachlebens.* Niemeyer, Halle. (Reprinted 1991, with an introduction by Clemens Knobloch. Benjamins, Amsterdam/Philadelphia).

Weinreich, U. (1972) *Explorations in Semantic Theory.* (Janua Linguarum, series minor 89). Mouton, The Hague.

Whitney, W. D. (1870) *A Compendious German Grammar.* (5th edition, revised) Holt & Williams, New York.

Wicha, N. Y. Y., E. M. Moreno, and M. Kutas (2004) 'Anticipating words and their gender. An event-related brain potential study of semantic integration, gender expectancy, and gender agreement in Spanish sentence reading', *Journal of Cognitive Neuroscience* 16/7: 1272–88.

Wierzbicka, A. (1972) *Semantic Primitives.* (Linguistische Forschungen 22) Athenäum, Frankfurt.

——(1997) *Understanding Cultures through their Key Words. English, Russian, Polish, German, and Japanese.* Oxford University Press, New York.

——(1996) *Semantics. Primes and Universals.* Oxford University Press, Oxford.

——(1998) 'The semantics of English causative constructions in a universal-typological perspective', in M. Tomasello (ed.), *The New Psychology of Language. Cognitive and Functional Approaches to Language Structure.* Erlbaum, Mahwah, NJ: 113–53.

Wittgenstein, L. (1968) *Philosophical Investigations.* Translated by G. E. M. Anscombe. Blackwell, Oxford.

Zalta, E. N. (1983) *Abstract Objects: An Introduction to Axiomatic Metaphysics.* Reidel, Dordrecht.

——(1988) *Intensional Logic and the Metaphysics of Intentionality.* A Bradford Book. MIT Press, Cambridge, MA.

Index

The manufacturer's authorised representative in the EU for product
safety is Oxford University Press España S.A. of El Parque Empresarial
San Fernando de Henares, Avenida de Castilla, 2 - 28830 Madrid
(www.oup.es/en or product.safety@oup.com). OUP España S.A. also acts
as importer into Spain of products made by the manufacturer.
Printed and bound by CPI Group (UK) Ltd, Croydon, CR0 4YY
14/09/2026
02216286-0009